Springer Theses

Recognizing Outstanding Ph.D. Research

For further volumes:
http://www.springer.com/series/8790

Aims and Scope

The series "Springer Theses" brings together a selection of the very best Ph.D. theses from around the world and across the physical sciences. Nominated and endorsed by two recognized specialists, each published volume has been selected for its scientific excellence and the high impact of its contents for the pertinent field of research. For greater accessibility to non-specialists, the published versions include an extended introduction, as well as a foreword by the student's supervisor explaining the special relevance of the work for the field. As a whole, the series will provide a valuable resource both for newcomers to the research fields described, and for other scientists seeking detailed background information on special questions. Finally, it provides an accredited documentation of the valuable contributions made by today's younger generation of scientists.

Theses are accepted into the series by invited nomination only and must fulfill all of the following criteria

- They must be written in good English.
- The topic should fall within the confines of Chemistry, Physics, Earth Sciences, Engineering and related interdisciplinary fields such as Materials, Nanoscience, Chemical Engineering, Complex Systems and Biophysics.
- The work reported in the thesis must represent a significant scientific advance.
- If the thesis includes previously published material, permission to reproduce this must be gained from the respective copyright holder.
- They must have been examined and passed during the 12 months prior to nomination.
- Each thesis should include a foreword by the supervisor outlining the significance of its content.
- The theses should have a clearly defined structure including an introduction accessible to scientists not expert in that particular field.

Bing-Jie Ni

Formation, Characterization and Mathematical Modeling of the Aerobic Granular Sludge

Doctoral Thesis accepted by
School of Earth and Space Sciences
University of Science and Technology of China

Springer

Author
Dr. Bing-Jie Ni
Advanced Water Management Centre
The University of Queensland
St. Lucia, Brisbane, QLD
Australia

Supervisor
Prof. Han-Qing Yu
Department of Chemistry
University of Science
and Technology of China
Hefei, Anhui
China

ISSN 2190-5053 ISSN 2190-5061 (electronic)
ISBN 978-3-642-31280-9 ISBN 978-3-642-31281-6 (eBook)
DOI 10.1007/978-3-642-31281-6
Springer Heidelberg New York Dordrecht London

Library of Congress Control Number: 2012942256

Printed on acid-free paper

Springer is part of Springer Science+Business Media (www.springer.com)

Parts of this thesis have been published in the following journal articles

1. Ni, B.J., Fang, F., Xie, W.M., Xu, J., Yu, H.Q. (2012) Formation of distinct soluble microbial products by activated sludge: kinetic analysis and quantitative determination. Environmental Science and Technology. 46, 1667–1674. (Reproduced with Permission)

2. Ni, B.J., Xie, W.M., Chen, Y.P., Fang, F., Liu, S.Y., Ren, T.T., Sheng, G.P., Yu, H.Q., Liu, G., Tian, Y.C. (2011) Heterotrophs grown on the soluble microbial products (SMP) released by autotrophs are responsible for the nitrogen loss in nitrifying granular sludge. Biotechnology and Bioengineering. 108, 2844–2852. (Reproduced with Permission)

3. Ni, B.J., Hu, B.L., Fang, F., Xie, W.M., Kartal, B., Liu, X.W., Sheng, G.P., Jetten, M.S.M., Zheng, P., Yu, H.Q. (2010) Microbial and physicochemical characteristics of compact anaerobic ammonium oxidation (anammox) granules in an UASB reactor. Applied and Environmental Microbiology. 76, 2652–2656. (Reproduced with Permission)

4. Ni, B.J., Zeng, R.J., Fang, F., Xie, W.M., Sheng, G.P., Yu, H.Q. (2010) Fractionating soluble microbial products in the activated sludge process. Water Research. 44, 2292–2302. (Reproduced with Permission)

5. Ni, B.J., Yu, H.Q. (2010) Modeling and simulation of the formation and utilization of microbial products in aerobic granular sludge. AIChE Journal. 56, 546–559. (Reproduced with Permission)

6. Ni, B.J., Sheng, G.P., Li, X.Y., Yu, H.Q. (2010) Quantitative simulation of the granulation process of activated sludge for wastewater treatment. Industrial and Engineering Chemistry Research. 49, 2864–2873. (Reproduced with Permission)

7. Ni, B.J., Zeng, R.J., Fang, F., Xu, J., Sheng, G.P., Yu, H.Q. (2009) A novel approach to evaluate the production kinetics of extracellular polymeric substances (EPS) by activated sludge using weighted nonlinear least-squares analysis. Environmental Science and Technology. 43, 3743–3750. (Reproduced with Permission)

8. Ni, B.J., Fang, F., Rittmann, B.E., Yu, H.Q. (2009) Modeling microbial products in activated sludge under feast-famine conditions. Environmental Science and Technology. 43, 2489–2497. (Reproduced with Permission)

9. Ni, B.J., Fang, F., Xie, W.M., Sun, M., Yu, H.Q. (2009) Characterization of extracellular polymeric substances excreted by mixed microorganisms in activated sludge with gel-permeating chromatography, excitation emission

matrix fluorescence spectroscopy measurement and kinetic modeling. Water Research. 43, 1350–1358. (Reproduced with Permission)

10. Ni, B.J., Xie, W.M., Yu, H.Q., Wang, Y.Z., Wang, G., Dai, X.L. (2009) Granulation of activated sludge in a pilot-scale sequencing batch reactor for the treatment of low-strength municipal wastewater. Water Research. 43, 751–761. (Reproduced with Permission)

11. Ni, B.J., Chen, Y.P., Liu, S.Y., Fang, F., Xie, W.M., Yu, H.Q. (2009) Modeling a granule-based anaerobic ammonium oxidizing (ANAMMOX) process. Biotechnology and Bioengineering. 103, 490–499. (Reproduced with Permission)

12. Ni, B.J., Yu, H.Q. (2009) A new approach to analyze the activated sludge process: Application to the soybean wastewater treatment in a sequencing batch reactor. AIChE Journal. 55, 2737–2745. (Reproduced with Permission)

13. Ni, B.J., Fang, F., Xie, W.M., Yu, H.Q. (2008) Growth, maintenance and product formation of autotrophs in activated sludge: Taking the nitrite-oxidizing bacteria as an example. Water Research. 42, 4261–4270. (Reproduced with Permission)

14. Ni, B.J., Yu, H.Q., Sun, Y.J. (2008) Modeling simultaneous autotrophic and heterotrophic growth in aerobic granules. Water Research. 42, 1583–1594. (Reproduced with Permission)

15. Ni, B.J., Yu, H.Q. (2008) Growth and storage processes in aerobic granules grown on soybean wastewater. Biotechnology and Bioengineering. 100, 664–672. (Reproduced with Permission)

16. Ni, B.J., Yu, H.Q., Xie, W.M. (2008) Storage and growth of denitrifiers in aerobic granules: Part II. Model calibration and verification. Biotechnology and Bioengineering. 99, 324–332. (Reproduced with Permission)

17. Ni, B.J., Yu, H.Q. (2008) Storage and growth of denitrifiers in aerobic granules: Part I. Model development. Biotechnology and Bioengineering. 99, 314–323. (Reproduced with Permission)

Supervisor's Foreword

Aerobic granular sludge technology has been recognized as next technological generation for biological wastewater treatment. It will play an important role as an innovative technology alternative to the present activated sludge process in industrial and municipal wastewater treatment in the near future due to its high substrate removal efficiency, low excess sludge disposal, and low cost for capital construction.

In September of 2006, Bing-Jie joined the Laboratory of Environmental Engineering, USTC, as a PhD candidate, and started his research work on the formation, characterization, and mathematical modeling of the aerobic granular sludge through integrating the process engineering tools and advanced molecular microbiology. Under my supervision, he designed the project plans, constructed bioreactors, developed mathematical modeling approaches, and carried out experiments. Also, he has completed data analysis and published lots of high quality papers in world-class journals in the field of biotechnology and environmental engineering, which were included in this dissertation chapters.

Dr. Ni's PhD thesis formulated mathematic structure models to describe the complex aerobic granulation process quantitatively based on experimental observations and revealed the detailed granule formation mechanisms. The simultaneous autotrophic and heterotrophic dynamic growth in aerobic granules was elucidated in-depth, which shed light on the microbial community competitions and distributions in aerobic granules. Comprehensive models for microbial products were also constructed and validated first in this thesis to provide insights into the dynamics of all the soluble and solid components in aerobic granules. More importantly, the aerobic granulation of sludge in pilot-scale reactor for treatment of low-strength municipal wastewater was achieved successfully for the first time and the key factors responsible for this process were identified, which have great implications for aerobic granular sludge technology. Furthermore, the thesis demonstrated accelerated start-up of the Anammox process by seeding the reactor with aerobic granules first, and provided support to the optimization of the process, which is making a significant impact on control system design for biological nitrogen removal.

I believe that the research results of this thesis will contribute significantly to the advance of understanding and optimizing the bacterial granulation processes, the next generation of technology for cost-effective biological wastewater treatment.

February 2012

Han-Qing Yu
Professor of Environmental Engineering

Acknowledgments

There are many people that I want to thank for their professional and personal support throughout the course of my PhD study. I would like to express my grateful appreciation to my supervisor, Prof. Han-Qing Yu, for all of his time and efforts in guiding me during my PhD study. Prof. Yu routinely provided me with insightful comments about my work and taught me a lot through his logical, well-structured, and organized approaches. He always gave me excellent support and encouragement throughout my Ph.D. study.

I greatly appreciate the opportunity to work with colleagues at the Laboratory of Environmental Engineering, University of Science & Technology of China. During my PhD study, they gave me many valuable comments and suggestions that helped to improve this dissertation.

I would like to express my thanks to Prof. Bruce E. Rittmann, Prof. Raymond J. Zeng, Prof. Mike S. M. Jetten, Dr. Bao-Lan Hu, and Dr. Boran Kartal for their useful advice and discussion throughout my PhD study.

I would also like to thank my family. My parents have provided me with much love and support in these years, as they have always done throughout my life.

I would also like to thank the National Natural Science Foundation of China (NSFC, Grant Nos. 50625825 and 50738006) for the partial support of this work.

Bing-Jie Ni
School of Earth and Space Sciences
University of Science and Technology of China

Contents

Chapter 1
Introduction

1.1 Background

In the last decade, intensive research has demonstrated that aerobic granular sludge technology is a novel and promising development in the field of biological wastewater treatment. This process is usually based on sequencing batch reactors (SBRs), with a cycle configuration chosen such that a strict selection for fast settling aerobic granules and a frequent repetition of distinct feast and famine conditions occur. This leads to the growth of stable and dense granules. Compared with the conventional activated sludge system, an aerobic granular sludge-based system has several advantages. An outstanding feature is the excellent settleability (high settling velocity) which is a prerequisite to handle high liquid flows. Moreover, granular sludge provides a high and stable rate of metabolism, resilience to shocks and toxins due to the protection by a matrix of extracellular polymeric substances (EPS) and internal storage products (X_{STO}), long biomass residence times, biomass immobilization inside the granules, and therefore, the possibility for bioaugmentation. In this sense, aerobic granular sludge technology will play an important role as an innovative technology alternative for activated sludge process in industrial and municipal wastewater treatment in the near future.

To facilitate and promote its practical application in wastewater treatment, researchers worldwide have excessively investigated the fundamentals of aerobic granulation. Analysis of the great body of literature published in the last decade shows that the properties of aerobic granules formed in SBRs are influenced by many factors, including substrate composition, organic loading, hydrodynamic shear force, feast-famine regime, feeding strategy, dissolved oxygen (DO), reactor configuration, solids retention time, cycle time, settling time, and volume exchange ratio. Up to now, studies of aerobic granular sludge have mainly focused on its cultivation, the determination of its treatment performance, and impact factors, but information on the internal interactions among process variables, such as growth, storage, microbial products formation, and endogenous respiration,

B.-J. Ni, *Formation, Characterization and Mathematical Modeling of the Aerobic Granular Sludge*, Springer Theses, DOI: 10.1007/978-3-642-31281-6_1,

as well as the sludge characteristics, including biomass detachment, oxygen transfer, and diffusion. In addition, the storage processes in aerobic granules under dynamic conditions remain unclear, and the precise mathematical modeling of complex aerobic granulation processes has not been established well. Little information is available on the production of EPS and soluble microbial products (SMP) in aerobic granular sludge, which is essential in the optimization of the activity and stability of aerobic granules and reactor performance. Furthermore, most of studies concerning aerobic granules have been focused on well-controlled lab-scale reactors with high- or middle-strength synthetic wastewaters. How to cultivate active and compact aerobic granules with a low-strength municipal wastewater is crucial for its full application. Thus, an in-depth investigation into the mechanism, mathematical modeling, and scaling-up should help us better understand the aerobic granular sludge process, design the novel bioreactors, optimize the reactor operating parameters, and improve system stability efficiency through the manipulation by model-based simulations.

1.2 Literature Review

1.2.1 Definition of Aerobic Granular Sludge

Aerobic granules are dense spherical self-immobilized aggregates of microorganisms with a strong compact structure and excellent settling ability (Fig. 1.1). They have a well-defined appearance and are visible as separate entities larger than 0.1 mm in diameter after settling. However, a general definition of aerobic granular sludge had not been made until the first aerobic granular workshop was held in Munich, Germany in 2004 (de Kreuk et al. 2005). After a long discussion in this workshop aerobic granule was defined as: Granules making up aerobic granular activated sludge are to be understood as aggregates of microbial origin, which do not coagulate under reduced hydrodynamic shear, and which settle significantly faster than activated sludge flocs.

In the second aerobic granular workshop in 2006, a further explanation of the definition was discussed, regarding aggregates of microbial origin, no coagulation under reduced hydrodynamic shear, settling faster than activated sludge flocs, minimum size, method of harvesting (de Kreuk et al. 2007). When an aggregate fulfils all characteristics as described above, it can be called aerobic granular sludge. This simplifies the interpretation of experimental results and clarifies when to speak about aerobic granular sludge, activated sludge, or biofilms.

Fig. 1.1 Image of aerobic granular sludge (Reprinted from Yang et al. (2008) with permission from Elsevier)

1.2.2 Formation of Aerobic Granular Sludge

Granulation is affected by a number of operational parameters, such as substrate composition, organic loading rate, feeding strategy, reactor design, settling time, exchange ratio, and aeration intensity (hydrodynamic shear force). It seems that there is a relatively small operational "window" for the successful cultivation of aerobic granules. Under favorable conditions, the much desired "super" granules could be formed.

1.2.2.1 Substrate Composition and Organic Loading Rate

Aerobic granules have been cultivated with a wide variety of wastewater. Generally, aerobic granulation is independent of substrate type. However, the morphology and microstructure of aerobic granules are dependent highly on the composition of the wastewater they are grown on. As an example, the structural instabilities of aerobic granule occurred due to filamentous outgrowth in the treatment of dairy wastewater (Schwarzenbeck et al. 2005). The outgrowth of filamentous organisms was related to the fact that, in the dairy wastewater, easily biodegradable substances were slowly released at a low concentration attributed to slow hydrolysis of the initial polymeric substrates.

It is recognized that carbohydrates in wastewater and a low DO level lead to flocculation, due to the favored growth of filamentous microorganisms (Gaval and Pernell 2003; Martins et al. 2003). The glucose-grown aerobic granules exhibit a filamentous structure, while acetate-grown aerobic granules have a nonfilamentous and very compact bacterial structure in which a rod-like species predominate (Tay et al. 2001). At a high potential growth rate on certain substrates, it is more difficult to cultivate aerobic granules. It is easier to obtain compact granule structure on methanol than on acetate, because of the growth rates of the microorganisms on methanol is lower than on acetate. The only observed exception was the growth of granular sludge on glucose. Usually, microorganisms have a high growth rate on glucose, but in biofilm and granule systems, populations are found to have a low growth rate on glucose and therefore dense and smooth structures are formed (de Kreuk et al. 2005). Aerobic nitrifying granules can be cultivated with an inorganic carbon source (Tsuneda et al. 2003, 2006). The nitrifying granules show excellent nitrification ability. The accumulated evidence suggests that aerobic granules can form across a very wide range of organic loading rate (OLR) from 0.4 to 15 kg COD m^{-3} day^{-1} (Morgenroth et al. 1997; Moy et al. 2002; Liu and Tay 2004). This indicates that the formation of aerobic granule in an SBR is substrate-concentration-independent. However, it has been reported that kinetic behavior and morphology of aerobic granules are also related to the applied substrate loading (Moy et al. 2002; Zheng et al. 2006). The mean size of aerobic granules increases from 1.6 to 1.9 mm with an increasing OLR from 3 to 9 kg COD m^{-3} day^{-1} (Liu and Tay 2004). Furthermore, OLR could affect the stability of aerobic granules. Zheng et al. (2006) reported that bacteria-dominated aerobic granules with a mean diameter of 1 mm could be cultivated in an SBR at a high OLR of 6 kg COD m^{-3} day in 30 d. However, under such high-loading conditions, the bacteria-dominated granules were not stable and readily transited into large-sized filamentous ones. The instability of aerobic granules may be attributed to the mass transfer limitation and possible presence of anaerobes in the large-sized aerobic granules.

1.2.2.2 Reactor Configuration

Reactor configuration influences the flow pattern of liquid and microbial aggregates in reactors (Beun et al. 1999; Liu and Tay 2002). Until now, all aerobic granules are cultivated in pneumatically agitated reactors, including bubble column and airlift reactor with sequencing batch operation. An important operation factor in aerobic granulation and reactor operation is superficial air velocity. It exerts an influence on aerobic granules through oxygen supply and hydrodynamic shear stress. The effect of hydrodynamic shear stress on aerobic granules will be discussed in details in Sect. 3.4. In a column-type upflow reactor a higher ratio of reactor height to diameter (H/D) can ensure a longer circular flow trajectory, which in turn provides a more effective hydraulic attrition to aerobic granules. However, at a higher H/D, the aerobic granules at the top face a low shear stress, and will grow out more readily with filamentous and/or finger-type structure. In this case, no granules would be formed (de Kreuk et al. 2005).

Shear stress provided by aeration rate depends on reactor configuration as well as reactor scale. Beun et al. (2000) reported that much more dense granules with a smaller diameter were obtained in an airlift reactor at the same substrate loading rate compared to a bubble column. To reduce aeration rate to a reasonable value,

Liu et al. (2007) designed a novel airlift loop reactor with divided draft tubes for aerobic granulation.

1.2.2.3 Aeration Intensity

The positive role of a high hydrodynamic shear force provided a great aeration rate in the stable operation of biofilm and aerobic granule systems has been widely recognized (Liu and Tay 2002). Chen et al. (2007) found that at shear forces of 2.4 and 3.2 cm s^{-1}, granules could maintain a robust and stable structure. Granules developed in low shear forces of 0.8 and 1.6 cm s^{-1} deteriorated to large-sized filamentous ones with irregular shape and loose structure, and resulted in poor performance and operation instability. Granules cultivated under high shear forces of 2.4 and 3.2 cm s^{-1} stabilized with clear morphology, dense and compact structure, and good performance in 120-day operation. In most of studies on aerobic granulation, the upflow air superficial velocity in reactors is much higher than 1.2 cm s^{-1}. However, a high aeration rate means a high energy consumption.

Adav et al. (2008) reported that the production of extracellular polysaccharides was closely associated with the shear force and the stability of aerobic granules. The extracellular polysaccharides content increased with the increasing shear force estimated in terms of superficial upflow air velocity. Thus, a high shear force stimulates bacteria to secrete more extracellular polysaccharides.

1.2.2.4 Cycle Time

The cyclic operation of an SBR consists of influent filling, aeration, settling, and effluent removal. The settling time and exchange ratio of liquid volumes at the end of each cycle presents the main screening step to remove non-granular biomass from the reactor. A shorter cycle time results in a shorter hydraulic retention time (HRT), which provides a stronger selective pressure. Sludge loss is observed through hydraulic washout at a short cycle time because bacterial growth is unable to compensate (Pan et al. 2004). Liu et al. (2007) reported the influence of cycle time on the kinetic behavior of aerobic granules. The observed specific biomass growth rate of aerobic granules decreased from 0.266 to 0.031 d^{-1}, while the observed biomass growth yield of granular sludge decreased from 0.316 to 0.063 g VSS g^{-1} COD when the cycle time was increased from 1.5 to 8 h.

The settling time acts as a major hydraulic selection pressure on microbial community in SBRs. A short settling time preferentially selects for the growth of rapid settling bacteria and the sludge with a poor settleability is washed out. It is recognized that the selection pressure imposed by short settling time should be more important in fully aerobic granule systems, but in anaerobic–aerobic alternative systems with phosphate accumulating organism (PAOs), the settling time seemed to be less important because of the inherent tendency of PAOs to aggregates (de Kreuk et al. 2005), meanwhile Meyer et al. (2003) cultivated aerobic granules containing glycogen accumulating organisms at a settling time of 25 min.

1.2.2.5 Feeding Strategy

The unique feature of an SBR over a continuous-flow activated sludge reactor is its cycle operation, which in turn results in a periodical starvation phase during the operation. It is proposed that such a periodical starvation would be somehow important to the aerobic granulation (Tay et al. 2001). Although starvation is proposed not to be a prerequisite for aerobic granulation (Liu et al. 2007), the increase in hydrophobicity on carbon-starvation has been reported (Sanin et al. 2003).

McSwain et al. (2004) enhanced aerobic granulation by intermittent feeding. In fact, pulse feeding to the SBR contributes to compact aerobic granules. Li et al. (2006) observed that the aerobic granulation process was initiated by starvation and cooperated by shear force and anaerobic metabolism. A shorter starvation time resulted in a faster granulation (Yang et al. 2005).

1.2.2.6 Dissolved Oxygen, Temperature, and pH

Aerobic granules are successfully cultivated at a DO concentration above 2 mg L^{-1} (Yang et al. 2005; Adav et al. 2008). However, Peng et al. (1999) observed that small granules (diameter of 0.3–0.5 mm) were agglomerated into big flocs during settling in an SBR at a DO of 1 mg L^{-1}. Mosquera-Corral et al. (2005) reported that reducing the oxygen saturation to 40 % caused deterioration, a decreased density and finally breaking of the granules. Based on the literature available, DO concentration is not a dominating factor for aerobic granulation.

The biological process rates depend on temperature. Most of the studies on aerobic granular sludge were carried out at room temperatures (20–25 °C). In an investigation into the effect of temperature changes on the conversion processes and the stability of aerobic granular sludge, de Kreuk et al. (2005) found that temperature change could affect the performance of an aerobic granular sludge reactor to a large extent. The start-up of a reactor at low temperatures led to the presence of organic COD in the aeration phase, deterioration of granule stability, and even biomass washout. Once a reactor was started up at a higher temperature it was possible to operate a stable aerobic granular sludge system at a lower temperature. Thus, they concluded that start-up should take place preferentially during warm summer periods, and that decreased temperatures during winter periods should not be a problem for granule stability and pollutant removal in a granular sludge system (de Kreuk et al. 2005).

In microbial growth pH is an important environmental factor. However, information regarding the effect of pH on species selection and aerobic granulation is still limited. Yang et al. (2007) evaluated the effect of feeding alkalinity and pH on the formation of aerobic sludge granules. In an SBR with a low alkalinity of 28.7 mg $CaCO_3$ L^{-1} in the influent and a reactor pH of 3.0, rapid formation of fungi-dominating granules was achieved in 1 week. In another SBR with a high alkalinity of 301 mg $CaCO_3$ L^{-1} and a reactor pH of 8.1, formation of bacteria-dominating granules was achieved after 4 weeks of operation. These results suggest that microbial communities and structural features of aerobic granules could be formed through controlling the feeding alkalinity and reactor pH.

The reviews above show that aerobic granulation is a very complex phenomenon with numerous internal interactions among process variables. All of them have significant effects on the overall reactor performance. Thus, cultivating aerobic granules with wastewaters with a complex composition, e.g., soybean-processing and fatty-acids-rich wastewaters, need to be further explored. More work should be performed to answer how can the dynamic aerobic granulation process to be quantitatively characterized regarding to these complex internal interactions. In addition, the review results indicate that many factors are involved in the granulation of activated sludge, but most of studies concerning aerobic granule have been focused on well-controlled lab-scale reactors with high- or middle-strength synthetic wastewaters. Is it feasible to cultivate aerobic granules for the treatment of low-strength municipal wastewater in a pilot-scale SBR? What should be the key factors in the granulation of activated sludge grown on such a low-concentration wastewater in an SBR? These questions need to be answered.

1.2.3 Microbial Products in Microbial Aggregates

1.2.3.1 Microbial Products Formation

The EPS formation of microbial aggregates has various origins (Nielsen and Jahn 1999; Liu and Fang 2003). EPS contain high-molecular-weight secretions from microorganisms, and the products of cellular lysis and hydrolysis of macromolecules. In addition, the organic matters from wastewater can also be adsorbed to the EPS matrix. Lysis and hydrolysis products may be integrated into the matrix, and contribute significantly to the content of EPS.

As shown in Fig. 1.2, carbohydrates and proteins are usually found to be the major components of EPS with a protein/carbohydrate ratio in the range of 0.2–5 (Frolund et al. 1996). Extracellular DNA is found to be an important component in EPS and acts as a cell-to-cell interconnecting matrix component in biofilm (Allesen-Holm et al. 2006). Humic substances may also be a key component of the EPS in sludge in biological wastewater treatment reactors, accounting for approximately 20 % of the total amount (Frolund et al. 1995; 1996). In addition, lipids, nucleic acids, uronic acids, and some inorganic components (Fig. 1.2) have also been found in EPS from various matrixes (Frolund et al. 1996; Dignac et al. 1998). The origins and composition of EPS are very complex, and thus the

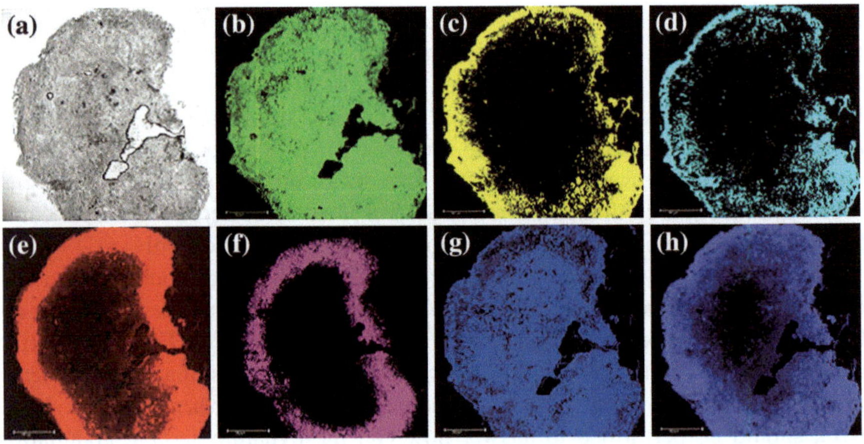

Fig. 1.2 The images of acetate-fed granule. **a** Phase contrast image, **b** *green* (FITC) proteins; **c** *533 yellow* (*Nile red*) lipids, **d** *cyan blue* (ConA)-D-glucopyranose polysaccharides; **e** *534 red* (*SYTO 63*) nucleic acids; **f** *pink* (*SYTOX Blue*) dead cells, **g** *blue* (*calcofluor 535 white*)-D-glucopyranose polysaccharides; **h** *purple blue* (*Calcium Green*) calcium (From Chen et al. (2007), with kind permission from Springer Science+Business Media)

production of EPS may be affected by a number of factors, such as the operational parameters and nutrient levels. EPS production is also closely related to microbial growth and substrate consumption (Laspidou and Rittmann 2002a), and is thus influenced by the factors that govern bacterial metabolism (Veiga et al. 1997). EPS can be degraded by bacteria as sources of carbon and energy when there is a substrate shortage (Kommedal et al. 2001), and thus in certain conditions the production of EPS is the result of the excretion and consumption of microbial cells.

The X_{STO} storage in microbial aggregates occurs in systems where electron donor and acceptor availability are separated (e.g., anaerobic/aerobic dynamics) or because the substrate is not continuously available for the microorganisms. In the first case, the X_{STO} plays a specific role in the ecophysiology of a certain group of bacteria, the most well-known being the PAOs and glycogen accumulating organisms (GAOs). The second case concerns a more general aspect of microbial growth under non-steady-state conditions. It has been shown that microbial aggregates organisms respond to feast/famine regimes by the production of storage polymers, like glycogen and poly-b-hydroxyalkanoates (Reis et al. 2003). In these cases, the storage polymer acts as a buffer for the substrate that is taken up but not directly used for growth. The more the microorganisms are able to store during the feast period and subsequently use it for growth, the more they have a competitive advantage (Reis et al. 2003). The stored polymers are usually polyhydroxyalkanoates (PHAs) (mainly poly-3-hydroxybutyric acid, PHB) when the substrate is rich in fatty acids (Majone et al. 1999; Reis et al. 2003). PHAs, as the main intracellular storage products, are thermoplastics synthesized by bacteria which are biodegradable and are produced from renewable resources (sugars and fatty acids).

Apart from polyhydrobutyrate (PHB), other PHAs synthesised include PHV, polyhydroxymethylvalerate (PHMV), and polyhydroxymethylbutyrate (PHMB).

Boero et al. (1991) stated that SMP resulted from intermediates or end products of substrate degradation and endogenous cell decomposition, whereas Noguera et al. (1994) defined SMP as the pool of organic compounds that resulted from substrate metabolism (usually with biomass growth) and biomass decay during the complete mineralization of simple substrates. The most definitive list on the origin of SMP is provided by Kuo (1993). Microorganisms might excrete soluble organic materials to establish the concentration equilibrium across the cell membrane (Harold 1972; Payne 1976). Bacteria also excrete organic materials during starvation because they must obtain energy for maintenance by endogenous respiration or metabolism of intracellular components when the substrate is essentially absent (Burleigh and Dawes 1967; Boylen and Ensign 1970; Barker and Stuckey 1999). The presence of an increased concentration of exogenous energy source can stimulate the excretion of SMP (Saier et al. 1975; Neijssel and Tempest 1976; Barker and Stuckey 1999). Sudden addition of a carbon and energy source to bacteria starved for carbon and energy may accelerate the death of some bacteria. SMP may be produced as a result of this process (Pirt 1975; Barker and Stuckey 1999). If essential nutrients are present in very low concentrations, SMP may be produced to scavenge the required nutrient (Pirt 1975; Emery 1982; Morel 1983; Barker and Stuckey 1999). Siderophores are a typical example. In addition, SMP are produced in response to environmental stress, such as extreme temperature changes and osmotic shocks (Barker and Stuckey 1999). Kuo (1993) also speculated that SMP were produced in response to toxic substances. Furthermore, SMP, such as exocellular enzymes, are not only produced during stressed conditions but also during normal growth and metabolism (Barker and Stuckey 1999).

1.2.3.2 Metabolism and Mechanisms

In 2002, Laspidou and Rittmann (2002a, b) proposed a unified theory coupling the production and degradation of SMP with the formation of EPS. The unified theory proposes that: (i) SMP and soluble EPS are identical in systems in which hydrolysis is not important, (ii) bound EPS are hydrolyzed to form BAP, and the authors consider this as the only significant source of BAP; (iii) the formation of bound EPS is growth-associated, and is produced in direct proportion to substrate utilization. Figure 1.3 shows a schematic of electron flow that reflects the unified theory of how active biomass, EPS, UAP, BAP, and active and inert biomass relate to each other. All mechanisms for EPS and SMP described in the unified theory are reflected in the figure. There are four possible ways. Part of the external substrate is used for biomass synthesis. Other substrate electrons are diverted to the formation of UAP and EPS. UAP are released to the aqueous solution, while EPS are released as a solid to form the aggregate matrix. The hydrolysis of EPS produces BAP, which are soluble. Substrate oxidation and respiration of the electrons to

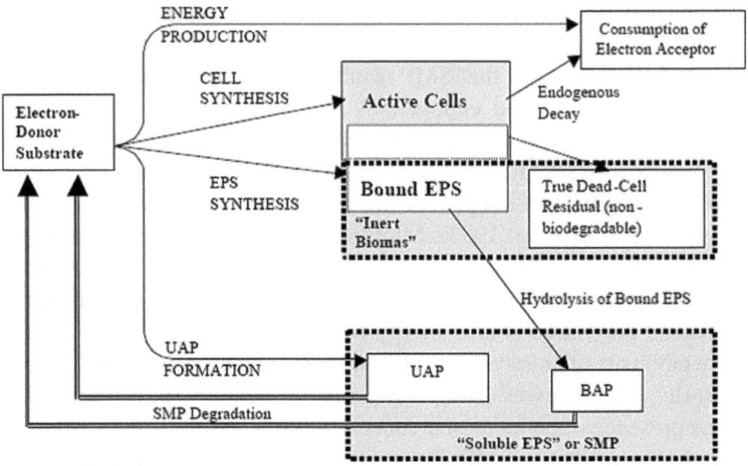

Fig. 1.3 Schematic representation of the unified model for active biomass, EPS, SMP, and inert biomass (Reprinted from Laspidou and Rittmann (2002a) with permission from Elsevier)

reduce O_2 and generate the energy needed to fuel for formation of active biomass, EPS, and UAP.

Recently many studies refer the production of internal storage products by microbial aggregates when exposed to transient carbon supply, such as PHA. The sludge submitted to consecutive periods of external substrate accessibility (feast) and unavailability (famine), generates a so-called unbalanced growth. Under these dynamic conditions, during excess of external carbon substrate, the uptake is driven to simultaneous growth of biomass and polymer storage, and after substrate exhaustion, stored polymer can be used as energy and as a carbon source (see Fig. 1.4, Reis et al. 2003). In these cases, storage polymers are formed under conditions that are not limiting for growth. The storage phenomenon is usually dominant (70 %) over growth, but under conditions in which substrate is present for a long time, physiological adaptation occurs and growth becomes more important (Reis et al. 2003). The ability to store internal reserves gives to these microorganisms a competitive advantage over those without this possibility, when facing transient substrate supply. Although no experimental evidence for mixed cultures exists, it can be anticipated that the metabolic pathways involved in carbon consumption and polymer storage in this case is probably the same as reported for pure culture studies. Considering acetate as the carbon substrate, the produced acetyl-CoA is partially channeled to the TCA for growth and NAD(P)H production, and partially deviated to PHA production. The accumulated intracellular PHA is metabolized when no external substrate is available (Reis et al. 2003).

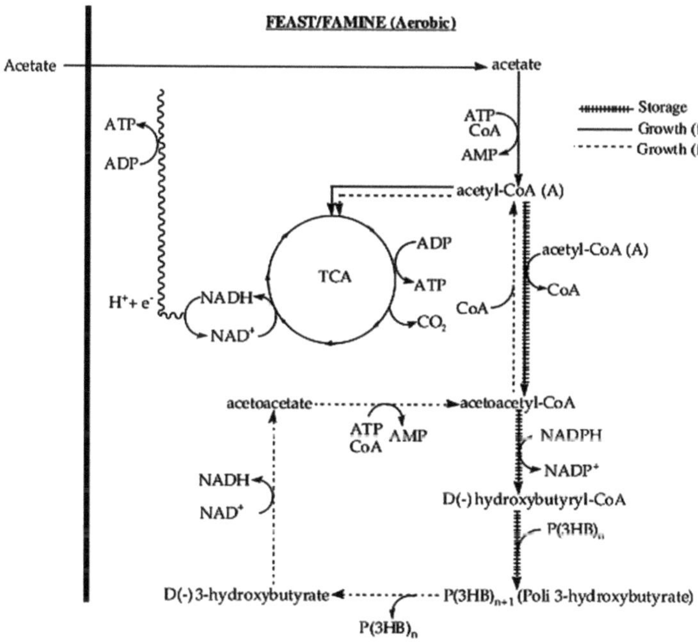

Fig. 1.4 Possible metabolic pathway for acetate consumption under feast/famine conditions (From Reis et al. (2003), with kind permission from Springer Science+Business Media)

1.2.3.3 The Roles of EPS, SMP, and Internal Storage Products in Microbial Aggregates

The flocculation ability of microbial aggregates is a key to the achievement of a low turbidity and a high quality of effluent. The interactions between EPS and cells have a significant effect on microbial flocculation ability (Morgan et al. 1990). In a study of the deflocculation of activated sludge under anaerobic conditions, bacteria and EPS are found to make up the main part of the deflocculated matter (Wilen et al. 2000), which indicates that EPS play an important role in flocculation. Many studies have demonstrated EPS to have a negative effect on the settleability of microbial aggregates (Jin et al. 2003). As EPS are negatively charged, a high concentration of EPS increases the surface charge of microbes, which results in an increase in the repulsive forces between cells and a decrease in the settleability of microbial aggregates (Morgan et al. 1990). EPS can also be regarded as a key factor in the thickening and dewatering processes of sludge (Houghton et al. 2001; Mikkelsen and Keiding 2002). Sludge granulation refers to the self-immobilization of microbes in biological wastewater treatment reactors, which results in a compact structure of aerobic and anaerobic granules. There are plenty of EPS in the interior of these aerobic and anaerobic granules (de Beer et al. 1996; Tay et al. 2001; McSwain et al. 2005). The formation of granular sludge is influenced by complex interactions between EPS and microbial cells. de Beer et al. (1996)

reported that anaerobic granules contain more EPS than flocs. Quarmby and Forster (1995) found that the carbohydrate content in EPS and the granule strength both decreased simultaneously, which suggested that EPS played a crucial role in sludge granulation and the maintenance of the structure of granular sludge. EPS can bind cells closely through ion bridging interactions, hydrophobic interactions, and polymer entanglement, which serves to enhance and promote the formation of microbial granules of biofilm.

Microbial accumulation of storage polymers has been linked to the presence of excess carbon source and nutrient deficiency (Reis et al. 2003). Activated sludge processes can be highly dynamic with respect to the feed regime, especially for sequencing batch reactor (SBR) (van Loosdrecht et al. 1997) and enhanced biological phosphorus removal processes (Oehmen et al. 2005). In these processes, the microorganisms are exposed to significant concentrations of the external substrate only for a relatively short period of time. Internally stored products allow them to take advantage of the dynamic feast and famine periods. The storage of intracellular polymers is usually the main mechanism for the removal of readily biodegradable carbon sources in activated sludge systems under dynamic conditions (van Loosdrecht et al. 1997; Pratt et al. 2004; Oehmen et al. 2005). For some microorganisms, intracellular storage products, such as polyhydroxyalkanoates (PHA), lipids, and polysaccharides, are significant sinks for carbon and electrons removed from substrate (van Loosdrecht et al. 1997; Pratt et al. 2004). Microorganisms capable of quickly storing substrate as internal storage products under feast and then consuming the stored substrate under famine have a strong competitive advantage over microorganisms without such capacity.

SMP have been found to comprise the majority of soluble organic materials in the effluents from biological treatment systems (Barker and Stuckey 1999; Aqino 2004; Rosenberger et al. 2006). They are of crucial importance to microbial aggregates systems because of their significant impacts on both effluent quality and treatment efficiency (Barker and Stuckey 1999; Grunheid et al. 2005; Jarusutthirak and Amy 2006; Labbs et al. 2006). In addition to contributing to the BOD and COD of the effluent, SMP can have further implications on process performance, although the effect of high concentrations of these products is not yet fully known. Washington et al. (1970) observed that accumulation of SMP in a cultivation medium was accompanied by a decrease in specific respiration rates and Chudoba (1985) concluded from his studies that microbial waste products in high concentrations adversely affected the kinetic activity and the flocculating and settling properties of activated sludge microorganisms. Of these microbial waste products, EPS, although strictly not soluble, are believed to have the major influence on the settling and flocculating properties of activated sludge (Barker and Stuckey 1999), as mentioned above. Indeed, they are now manufactured artificially to be used as alternatives to other synthetic or natural water-soluble polymers or as novel polymers in thickening, suspending, and gelling applications (Barker and Stuckey 1999).

1.2.3.4 Process Inter-Relationships

Solid retention time (SRT) has a considerable effect on the production of EPS, but the results reported in literatures are somewhat contradictory. Many researchers have found that the EPS in various microbial aggregates increases with an increasing SRT, implying that bacteria produce more EPS in endogenous conditions. Sesay et al. (2006) found that an increase in SRT had a significant and positive correlation with the total quantity of EPS in activated sludge as well as the contents of proteins and carbohydrates in EPS. The ratio of proteins to carbohydrates also increased from 1.5 to 2.5 with an increase in SRT from 4 to 20 d. However, some researchers have suggested that EPS is independent of SRT. Liao et al. (2001), for example, found that the EPS of activated sludge content did not change significantly with a longer SRT. However, the protein/carbohydrate ratio was found to increase as the SRT increased from 4 to 12 d, but remained unchanged as the SRT increased from 12 to 16 d. Li and Yang (2007) reported that the TB-EPS of activated sludge had no relationship with the SRT, but that the LB-EPS decreased with an increasing SRT. When normalized SMP production (SMP/S_O) is plotted against SRT, SMP/S_O appears to decrease to a minimum and then increase again, indicating the existence of an optimal SRT for minimizing the production of SMP. For aerobic systems this optimum appears to exist between 2 and 15 days (Barker and Stuckey 1999), while for anaerobic systems the optimum appears at approximately 25 days (Barker and Stuckey 1999). For X_{STO}, the total quantity of X_{STO} in microbial aggregates is much less dependent on SRT compared to EPS and SMP because X_{STO} is much more biodegradable than EPS and SMP.

However, the reactor operation has significant effect on the production of X_{STO}. In the feast and famine processes, the electron acceptor used for carbon oxidation can be either oxygen or nitrate. Studies carried out under aerobic or anoxic conditions have demonstrated that the yield is practically equal for both situations while the specific X_{STO} production rate is tentatively higher for the aerobic operation. The feast and famine process is commonly carried out in a SBR. SBRs are ideal reactors for a selection of robust populations with high ability of X_{STO} storage, because biomass grows under transient (unsteady) conditions. If a continuous system for X_{STO} production is desired, then a possible configuration could include two reactors in series in which a plug flow reactor (PFR) is followed by a continuous stirred tank reactor CSTR, coupled to a settler or membrane filter (Reis et al. 2003). The concentrated biomass is recirculated to the PFR. The effluent proceeding from the PFR, containing biomass with a maximum PHA content, is divided in two streams: one flow directly to the CSTR and the other one flow to a sludge concentration step (membrane or settler). The concentrated sludge is harvested for X_{STO} extraction. In this case, the PFR corresponds to the feast period and therefore its length (or hydraulic retention time-HRT) should be determined by the time required for the complete substrate exhaustion. The CSTR mimics the famine conditions and its HRT should be defined by the time needed for sludge starvation. Beccari et al. (1998) used a CSTR operated with intermittent feeding. This process selects and produces sludge with high storage capacity. The PHB storage yield ranged from 0.06 to 0.5 g COD g^{-1} COD and the sludge stored up to 40–50 % of the total dry weight. In addition, the nature of the substrate not only determines the X_{STO} content but also its composition, which subsequently affects the final polymer properties.

1.2.3.5 Conversion Kinetics

Robinson et al. (1984) investigated the kinetics of cellular reproduction and the rate of EPS formation for *Pesudomonas aeruginosa* in a continuous culture under the carbon-limiting conditions. The production of EPS could be modeled using the Leudeking–Piret equation:

$$\gamma_{\mathrm{EPS}} = k_1 \mu X + k_2 X \tag{1.1}$$

where μ is the specific rate of microbial growth, X is the biomass concentration, k_1 is the growth-associated EPS formation coefficient, and k_2 is the non-growth-associated EPS formation coefficient. According to the rate-expression of Eq. (1.1), coefficient k_1 specifies the fraction of substrate electrons or carbon shunted to EPS formation, while coefficient k_2 implies a first-order transformation from biomass to the EPS. This kinetics has been successfully used to describe the production of EPS for various cultures (Wang and Yu 2007).

The Leudeking–Piret kinetics, however, may not be a satisfactory representation of the kinetics of EPS formation, as it does not include any mechanisms for EPS loss. Both experimental findings and theoretical analysis suggest that EPS are subjected to hydrolysis/dissolution and can serve as a carbon and energy source, which is not taken into account in the Leudeking–Piret equation. In a later modification (Turakhia and Characklis 1989) for the EPS production of *Pesudomonas aeruginosa* in biofilms, a term about the degradation (hydrolysis and dissolution) of EPS was added:

$$\gamma_{\mathrm{EPS}} = k_1 \mu X + k_2 X - k_{\mathrm{hyd}} \mathrm{EPS} \tag{1.2}$$

where k_{hyd} is the specific EPS degradation rate. This EPS kinetics could fit the experimental data well.

Another alternative kinetic expression for EPS formation was developed by Hsieh et al. (1994). According to their mechanistic model, the rate of EPS production in suspended biomass can be written as:

$$\gamma_{\mathrm{EPS}} = \mu_{\mathrm{EPS}} \left(\frac{S}{K_{\mathrm{EPS}} + S} \right) X + f_{\mathrm{XP}} k_d \left(\frac{K_d}{K_d + S} \right) X - \mu_{\mathrm{EPSdiss}} \left(\frac{\mathrm{EPS}/X}{K_{\mathrm{EPSdiss}} \mathrm{EPS}/X} \right) \mathrm{EPS} \tag{1.3}$$

where μ_{EPS} is the maximum specific EPS production rate and $\mathrm{K_{EPS}}$ is the corresponding saturation constant, S is the substrate concentration, f_{XP} is the fraction of biomass that is converted to EPS, k_d is the biomass maintenance coefficient and K_d

is the saturation constant, $\mu_{EPSdiss}$ is the maximum specific rate of EPS degradation by hydrolysis, dissolution, or shearing off from the cell surface, and $K_{EPSdiss}$ is a saturation constant in relation to the EPS/X ratio. The first term on the right hand of Eq. (1.3) is a growth-associated term for EPS formation. Its underlying assumption is a direct and immediate conversion from the substrate to EPS polymers with biomass as a catalyst. The second term is a non-growth-associated term for EPS formation resulted from the cellular maintenance. The third term is an EPS loss term accounting for the endogenous decay of the microbial EPS. The dissolution of EPS from the cell surface is modeled with the Monod kinetics with respect to the ratio of EPS to X. A higher ratio of EPS to biomass, or a "thickener" EPS coat around the cells, would result in faster EPS hydrolysis or more release of the EPS into the solution.

Under dynamic feeding conditions (feast/famine regime), the microorganisms grow and store X_{STO} at the same time, resulting in the formation of storage compounds. These two processes are stoichiometrically independent, and their relative rates can vary. The X_{STO} formation rate closes the balance between substrate uptake rate and the use of substrate for growth processes. In this way, the stored substrate can be seen as a buffer offering the bacteria the potential of rapid substrate uptake (ecologically a significant requirement for survival), combined with a more balanced growth process (Reis et al. 2003). X_{STO} degradation rates are clearly dependent on the polymer content of the cells (Reis et al. 2003). This means that the conversion process should be described by a substrate consumption rate and not by a microbial growth rate. For growth on storage polymers no intrinsic kinetics has been derived. A first-order rate in the polymer to active biomass ratio can give a good description of the experimental observations (Reis et al. 2003):

$$\frac{df_{STO}}{dt} = -k^{*}f_{STO}, \quad f_{STO} = \frac{X_{STO}}{X_{H}} \tag{1.4}$$

where f_{STO} is the fraction of X_{STO} of the total biomass, X_H is the biomass concentration and k is the rate constant.

Most of prior studies have incorporated SMP into mass-balance models that also include mass balances on active and inert biomass. The first type of SMP is UAP, which are produced as a direct result of substrate utilization. Rittmann and McCarty (2001) present the formation kinetics of UAP as follows:

$$r_{UAP} = k_{UAP}qX_{H} \tag{1.5}$$

where k_{UAP} is the UAP-formation coefficient and q is the specific substrate-utilization rate. The second category is BAP, and they are formed from biomass, presumably as part of decay. The BAP rate expression is (Rittmann and McCarty 2001):

$$r_{BAP} = k_{BAP}X_{H} \tag{1.6}$$

where k_{BAP} is the BAP-formation coefficient. The total SMP-production rate expression is:

$$r_{\mathrm{SMP}} = r_{\mathrm{UAP}} + r_{\mathrm{BAP}} = k_{\mathrm{UAP}} q X_H + k_{\mathrm{BAP}} X_H \tag{1.7}$$

However, the unified theory proposed by Laspidou and Rittmann (2002a, b) established that BAP is produced by hydrolysis of bound EPS. Therefore, the only reaction source of BAP is:

$$r_{\mathrm{BAP}} = k_{\mathrm{hyd}} \mathrm{EPS} \tag{1.8}$$

where the first-order hydrolysis rate coefficient is k_{hyd}.

Rittmann and McCarty (2001) summarized that SMP originally formed are relatively biodegradable, and that UAP and BAP have distinct degradation kinetics. They can be described with separate Monod-degradation expressions.

$$r_{\mathrm{deg-UAP}} = \frac{-q_{\mathrm{UAP}} \mathrm{UAP}}{K_{\mathrm{UAP}} + \mathrm{UAP}} X_H \tag{1.9}$$

$$r_{\mathrm{deg-BAP}} = \frac{-q_{\mathrm{BAP}} \mathrm{BAP}}{K_{\mathrm{BAP}} + \mathrm{BAP}} X_H \tag{1.10}$$

in which q_{UAP} and q_{BAP} are maximum specific rates of UAP and BAP degradation, respectively; and K_{UAP} and K_{BAP} are half-maximum rate concentrations for UAP and BAP, respectively.

Microbial products are very complex, and the knowledge regarding microbial products is far complete and much work is still required to fully understand their precise roles in aerobic granules. More detailed metabolic study on the formation of EPS, SMP, and X_{STO} by aerobic granules is needed. The biomass is able to divert organic matter to the forms of EPS, SMP and X_{STO} simultaneous. However, very little is currently known about the metabolism behind this observation. Further research is needed to clarify the key factors governing the each formation processes and then reveal the mechanisms. We need to characterize and quantitatively measure EPS, SMP, and X_{STO} using more advanced and sophisticated chemical analytical techniques and approaches, then to isolate and measure some of microbial products compounds to precisely quantify the amount of them with their fractions from all the different sources. More convenient approach should be developed to evaluate the microbial products formation kinetics. In addition, the possibility for interactions between heterotrophs and autotrophs when the multiple components are included (such as EPS, SMP, or/and X_{STO}), in both autotrophic and mixed system, need to be clarified since the review strongly indicates aerobic granular sludge is inherently multi-component because it has active biomass, EPS, residual inert material, and SMP produced under usual metabolic conditions.

1.2.4 Mathematical Modeling of Aerobic Granular Sludge

From an extensive laboratory-scale investigation into this system and the scaling up of aerobic granular sludge reactors (Schwarzenbeck et al. 2005; Zheng et al. 2006; de Kreuk and van Loosdrecht 2006), the system holds considerable promise for full-scale implementation. To aid the design, operation, and optimization of, and further research into this system, mathematical simulation model would be invaluable as a process evaluation tool. Moreover, mathematical modeling has proven to be very useful to study complex processes, such as the aerobic granular sludge systems (Beun et al. 2001; Su and Yu 2006a, b; de Kreuk et al. 2007; Xavier et al. 2007). Biological processes in aerobic granules are determined by concentration gradients of oxygen and diverse substrates. The substrate and DO concentration profiles are the result of many factors, e.g., diffusion coefficient, conversions rate, granule size, biomass spatial distribution, and density. All of these factors significantly influence each other; thus the effect of separate factors cannot be studied experimentally. Model simulation and prediction can provide a solid foundation for design and operation of biological treatment systems. Accordingly, it is essential to develop appropriate models to describe the aerobic granular sludge systems.

1.2.4.1 Modeling the Dynamic Aerobic Granulation Processes

To simulate the aerobic granulation process, some empirical models have been developed initially, respectively based on logistic curves (Su and Yu 2005) or a linear phenomenological equation (LPE) (Yang et al. 2004). A modified Logistic model has been employed by Su and Yu (2005) to describe the granulation process by diameter:

$$D(t) = \frac{D_{\max}}{1 + e^{-k(t-t_0)}} \tag{1.11}$$

where t is the operating time (days), $D(t)$ is the mean diameter of aerobic granules on day t (mm), $D_{\max}$ is the asymptote of the curve, i.e., the maximum mean diameter of aerobic granules (mm), k is the specific growth rate by diameter (day^{-1}), t_0 is the lag time (day).

A regression was performed with the experimental data by Su and Yu (2005) and the granulation process fed with soybean processing wastewater could be well simulated by the modified Logistic model with a high correlation coefficient of 0.988. A lag time of around 40 days was calculated for the granulation process, whereas the maximum mean diameter of matured granules and specific growth rate by diameter were estimated to be 1.24 mm and 0.12 day^{-1}, respectively. These values were in good agreement with the observed values.

On the basis of the LPE, Yang et al. (2004) developed a kinetic model that can describe the growth of aerobic granules developed at different conditions by applying the LPE to aerobic granulation process.

$$D - D_0 = (D_{eq} - D_0)[1 - e^{-\mu(t-t_0)}] \quad (1.12)$$

where D is the size of microbial aggregate at time t, D_{eq} is size of microbial aggregate at equilibrium, and μ is the specific growth rate of aggregate by size (day^{-1}), t_0 is the time at end of lag phase, and D_0 is the size of microbial aggregates at time t_0. The size of microbial aggregates at the end of lag phase can be used as the initial value for microbial growth.

The simple kinetic model based on the LPE was successfully derived to describe the growth of aerobic granules by Yang et al. (2004). It was found that the growth of aerobic granules in terms of equilibrium size and size-dependent growth rates were inversely related to shear force imposed to microbial community, while a high organic loading favored the growth of aerobic granules, leading to a large size granule. However, these experiment-oriented models might not be applicable for describing the complex aerobic granulation process and optimizing the operating conditions for granulation in a generalized way.

However, aerobic granulation is a very complex phenomenon of microbial immobilization. There are numerous internal interactions among process variables, such as growth, storage and endogenous respiration, and sludge characteristics, including biomass detachment, oxygen transfer, and diffusion. All of them have significant effects on the overall reactor performance. Therefore, the empirical models are generally descriptive and are not able to simulate the aerobic granulation process in other reactors or cultivated under different conditions. In addition, there still some important aspects of aerobic granulation, such as simultaneous increase in density and bioparticle size, breakage due to starvation of inner biomass, and selection of bioparticle size by sedimentation, have not been addressed in this model, resulting in that total number, mean settling velocity, density, size distribution, and SS in effluent during aerobic granulation process cannot be well predicted using this model. Thus, a more comprehensive mathematical model, which takes all these parameters into consideration, should be established.

1.2.4.2 Modeling the Granule-Based Sequencing Batch Reactor

Beun et al. (2001) have developed a simulation model to evaluate the effects of several operating factors on nitrogen removal in a granule-based SBR. This model is based on balance equations connecting conversion processes and transport processes, and thus it is able to predict the N-conversion processes in the reactor under different conditions. It has been shown that nitrification, denitrification, and removal of COD can occur simultaneously in such a granule-based SBR and that the exact location of the autotrophic biomass influences the net N-removal. The

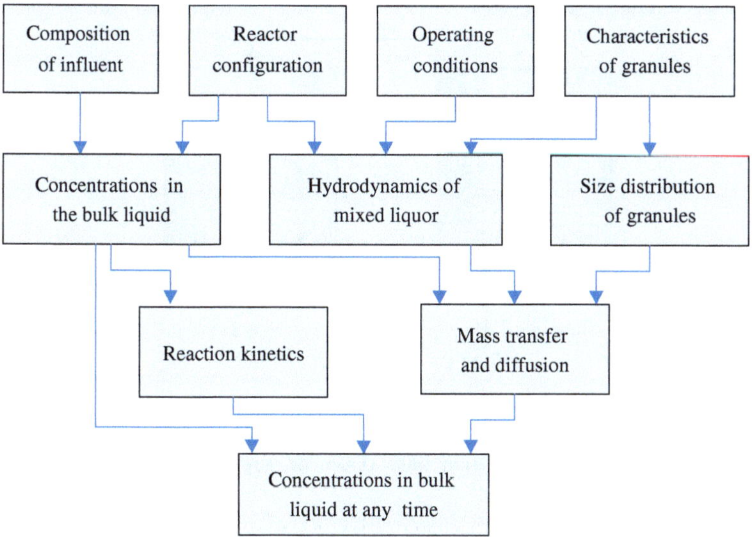

Fig. 1.5 Internal interactions of the generalized model established by Su and Yu (2006a) to simulate an aerobic granule-based SBR with considerations of biological processes, reactor hydrodynamics, mass transfer, and diffusion (Reprinted from Su and Yu (2006a) with permission of American Chemical Society)

distribution of the autotrophs in granules is influenced by the DO in the reactor. It is also found that storage and subsequent degradation of poly-b-hydroxybutyrate PHB benefit for denitrification. In particular, PHB is stored in bacteria situated in deeper layers of granules, below where the autotrophic growth occurs, serves as a C-source for denitrification.

The granule-based SBR is a complex bioreactor with numerous internal interactions among process variables and sludge characteristics. In addition to the biological reactions, mass transfer, hydrodynamic of reactors,s and the characteristics of granules have been proven to be influential to the overall performance of a granule-based SBR. Su and Yu (2006a, b) have established a generalized model to simulate the granule-based SBR with considerations of biological processes, reactor hydrodynamics, mass transfer, and diffusion (Fig. 1.5). A discretization of time, size, and segment of sludge has been employed for the discontinuous expressions of different processes. Granules in the SBR are first classified into various size fractions, and each granule is then sliced up along the radius. The concentrations of model components and biological reactions are calculated based on each slice of granules in each size fraction.

Input values of the model include the influent composition, reactor configuration, operating conditions, and characteristics of granules. With these given values, the mass transfer and diffusion as well as the biological reaction kinetics could be determined. Thus, the concentrations of components (substrate, ammonia, and nitrate) in the bulk liquid at any operating time could be calculated. The SBR is

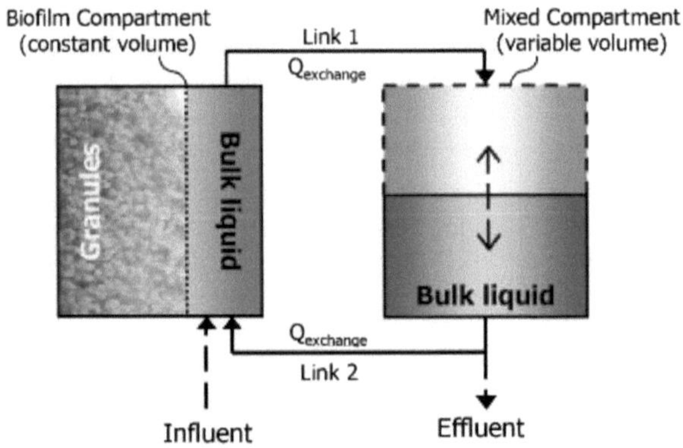

Fig. 1.6 Model set-up in AQUASIM: each compartment contains the soluble components and particulate components involved in the metabolic processes. All soluble and particle components are found in effluent and recirculation links (From de Kreuk et al. (2007), reprinted with permission from Wiley-Blackwell)

considered to be a series of CSTRs in time sequence. In the granule-based SBR, the rates of gas–liquid oxygen transfer are assumed to be proportional to the difference in the oxygen concentration between gas and liquid interfaces. The ASM1 is modified and applied for describing the biological reactions. On the basis of the difference between the calculated and measured results, the model structure is further improved through introducing simultaneous consumption of soluble substrates by storage and heterotrophs growth with a changeable reaction rate. The established model is verified with the experimental results for four granule-based SBRs with various granule sizes and fed with different wastewaters (Su and Yu 2006b). The verification results show that the established model is appropriate to simulate and predict the nitrogen and COD removal in the granule-based SBR.

A mathematical model has also been developed by de Kreuk et al. (2007) to simulate the performance of a granule-based SBR for simultaneous removal of COD, nitrogen, and phosphate. A combination of completely mixed reactor and biofilm reactor compartments is used to simulate the mass transport and conversion processes occurring in the bulk liquid and in the SBR. The biological conversion processes are described using stoichiometric and kinetic parameters from Hao et al. (2001, 2002) and Meijer (2004). The AQUASIM software does not allow the volume of the bulk liquid in the biofilm compartment to vary in time. In order to sort out this problem and simulate the fill and discharge process, two linked compartments have to be defined. A completely mixed liquid compartment with variable volume is connected to the biofilm compartment. A high fluid circulation rate ($Q_{exchange}$) between the two compartments ensures the same bulk liquid concentrations in both compartments (see Fig. 1.6).

The model describes the experimental data well. The effect of process parameters on the nutrient removal rates of aerobic granules could be reliably evaluated using this model. Oxygen penetration depth in combination with the position of the autotrophic biomass plays a crucial role in the conversion rates of the different components and thus on overall nutrient removal efficiency. The ratio between the aerobic and anoxic volumes in the granule strongly determines the N-removal efficiency, as it is shown by model simulations with varying oxygen concentration, temperature, and granule size. The optimum granule diameter for maximum N- and P-removal at DO of 2 mg L^{-1} and 20 °C is found between 1.2 and 1.4 mm, and the optimum COD loading rate is 1.9 kg COD m^{-3} day^{-1}. When all ammonia is oxidized, oxygen diffused to the granule core inhibits the denitrification process. In order to optimize the process, anoxic phases can be implemented in the SBR-cycle configuration, leading to a more efficient overall N-removal. Phosphate removal efficiency mainly depends on the sludge age. At an SRT longer than 30 days, no sufficient sludge is removed from the system to keep effluent phosphate concentrations low.

From the literature, it is concluded that the mathematical modeling of aerobic granular sludge, including the dynamic granulation process and the granule-based SBR performance, is fascinating from a purely scientific perspective as well as engineering application although existing models still have limitations. For further improvements in mathematical modeling of aerobic granular sludge, we need to expand our knowledge of simultaneous increase in density and bioparticle size, detachment due to starvation of inner biomass, which are all important elements in aerobic granulation processes. To acquire this knowledge, a more comprehensive mathematical model to take all these parameters into consideration should be developed in the future to simulate the total number, density, and size in the aerobic granulation process. Another challenge must be to produce more sophisticated model by incorporation of EPS, SMP, or/and X_{STO} production and consumption along the lines of experimental observations and literature information and to address the difficulty of incorporating microbial products formation of both heterotrophs and autotrophs into the modeling of aerobic granular sludge. The long-term challenges are to provide a comprehensive modeling tool for engineers designing and operating all types of wastewater treatment systems whereby, given the feed characteristics, operating conditions, and bacterial populations, one could accurately predict the amount of EPS, SMP and X_{STO} produced even as far as predicting the effluent in terms of individual compounds.

References

Adav, S.S., Lee, D.J., Show, K.Y., Tay, J.H.: Aerobic granular sludge: recent advances. Biotechnol. Adv. **26**, 411–423 (2008)

Allesen-Holm, M., Barken, K.B., Yang, L., Klausen, M., Webb, J.S., Kjelleberg, S., et al.: A characterization of DNA release in *Pseudomonas aeruginosa* cultures and biofilms. Mol. Microbiol. **59**, 1114–1128 (2006)

Aqino, S.F.D.: Formation of soluble microbial products (SMP) in anaerobic digesters during stress conditions. Ph.D. Thesis, Imperial College, London (2004)

Barker, D.J., Stuckey, D.C.: A review of soluble microbial products (SMP) in wastewater treatment systems. Water Res. **33**, 3063–3082 (1999)

Beccari, M., Majone, M., Massanisso, P., Ramadori, R.: A bulking sludge with high storage response selected under intermittent feeding. Water Res. **32**, 3403–3413 (1998)

Beun, J.J., Heijnen, J.J., van Loosdrecht, M.C.M.: N-removal in a granular sludge sequencing batch airlift reactor. Biotechnol. Bioeng. **75**, 82–92 (2001)

Beun, J.J., Hendriks, A., van Loosdrecht, M.C.M., Morgenroth, M., Wilderer, P.A., Heijnen, J.J.:Aerobic granulation in a sequencing batch reactor. Water Res. **33**, 2283–2290 (1999)

Beun, J.J., Paletta, F., van Loosdrecht, M.C.M., Heijnen, J.J.: Stoichiometry and kinetics of poly-b-hydroxybutyrate metabolism in aerobic, slow growing, activated sludge cultures. Biotechnol. Bioeng. **67**, 379–389 (2000)

Boero, V.J., Eckenfelder Jr., W.W., Bowers, A.R.: Soluble microbial product formation in biological systems. Water Sci. Technol. **23**, 1067–1076 (1991)

Boylen, C.W., Ensign, J.C.: Intracellular substrates for endogenous metabolism during long-term starvation of rod and spherical cells of arthrobacter crustallopoietes. J. Bact. **103**, 578–587 (1970)

Burleigh, I.G., Dawes, E.A.: Studies on the endogenous metabolism and senescence of starved sarcin lutea. Biochem. J. **102**, 236–250 (1967)

Chen, M.Y., Lee, D.J., Tay, J.H.: Distribution of extracellular polymeric substances in aerobic granules. Appl. Microbiol. Biotechnol. **73**, 1463–1469 (2007)

Chudoba, J.: Inhibitory effect of refractory organic compounds produced by activated sludge micro-organisms on microbial activity and flocculation. Water Res. **19**, 197–200 (1985)

de Beer, D., O'Flaharty, V., Thaveesri, J., Lens, P., Verstraete, W.: Distribution of extracellular polysaccharides and flotation of anaerobic sludge. Appl. Microbiol. Biotechnol. **46**, 197–201 (1996)

de Kreuk, M.K., Heijnen, J.J., van Loosdrecht, M.C.M.: Simultaneous COD, nitrogen, and phosphate removal by aerobic granular sludge. Biotechnol. Bioeng. **90**, 761–769 (2005)

de Kreuk, M.K., Picioreanu, C., Hosseini, M., Xavier, J.B., van Loosdrecht, M.C.M.: Kinetic model of a granular sludge SBR–Influences on nutrient removal. Biotechnol. Bioeng. **97**, 801–815 (2007)

de Kreuk, M.K., van Loosdrecht, M.C.M. Formation of aerobic granules with domestic sewage. J. Environ. Eng. **132**, 694–697 (2006)

Dignac, M.F., Urbain, V., Rybacki, D., Bruchet, A., Snidaro, D., Scribe, P.: Chemical description of extracellular polymeric substances: implication on activated sludge floc structure. Water Sci. Technol. **38**, 45–53 (1998)

Emery, T.: Iron metabolism in humans and plants. Am. Sci. **70**, 626–632 (1982)

Frolund, B., Griebe, T., Nielsen, P.H.: Enzymatic activity in the activated-sludge floc matrix. Appl. Microbiol. Biotechnol. **43**, 755–761 (1995)

Frolund, B., Palmgren, R., Keiding, K., Nielsen, P.H.: Extraction of extracellular polymers from activated sludge using a cation exchange resin. Water Res. **30**, 1749–1758 (1996)

Gaval, G., Pernelle, J.-J.: Impact of the repetition of oxygen deficiencies on the filamentous bacteria proliferation in activated sludge. Water Res. **37**, 1991–2000 (2003)

Grady Jr, C.P.L., Daigger, G.T., Lim, H.C.: Biological Wastewater Treatment, 2nd edn. pp. 282–284. Marcel Dekker, New York (1999)

Grunheid, S., Amy, G., Jekela, M.: Removal of bulk dissolved organic carbon (DOC) and trace organic compounds by bank filtration and artificial recharge. Water Res. **39**, 3219–3228 (2005)

Harold, F.M.: Conservation and transformation of energy by bacterial membranes. Bact. Rev. **36**, 172–230 (1972)

Holakoo, L., Nakhla, G., Yanful, E.K., Bassi, A.S.: Chelating properties and molecular weight distribution of soluble microbial products from an aerobic membrane bioreactor. Water Res. **40**, 1531–1538 (2006)

Houghton, J., Quarmby, J., Stephenson, T.: Municipal wastewater sludge dewaterability and the presence of microbial extracellular polymer. Water Sci. Technol. **44**, 373–379 (2001)

Hsieh, K.M., Murgel, G.A., Lion, L.W., Shuler, M.L.: Interactions of microbial biofilms with toxic trace metals: 1. Observation and modeling of cell growth, attachment, and production of extracellular polymer. Biotechnol. Bioeng. **44**, 219–231 (1994)

Jarusutthirak, C., Amy, G.: Role of soluble microbial products (SMP) in membrane fouling and flux decline. Environ. Sci. Technol. **40**, 969–974 (2006)

Jendrossek, D., Selchow, O., Hoppert, M.: Poly(3-hydroxybutyrate) granules at the early stages of formation are localized close to the cytoplasmic membrane in *Caryophanon latum*. Appl. Environ. Microbiol. **73**, 586–593 (2007)

Jin, B., Wilen, B.M., Lant, P.A.: Comprehensive insight into floc characteristics and their impact on compressibility and settleability of activated sludge. Chem. Eng. J. **95**, 221–234 (2003)

Kommedal, R., Bakke, R., Dockery, J., Stoodley, P.: Modelling production of extracellular polymeric substances in a *Pseudomonas aeruginosa* chemostat culture. Water Sci. Technol. **43**, 129–134 (2001)

Kuo, W.C.: Production of soluble microbial chelators and their impact on anaerobic treatment. Ph.D. thesis, University of Iowa, Iowa City (1993)

Labbs, C., Amy, G., Jekel, M.: Understanding the size and character of fouling-causing substances from effluent organic matter (EfOM) in low-pressure membrane filtration. Environ. Sci. Technol. **40**, 4495–4499 (2006)

Laspidou, C.S., Rittmann, B.E.: A unified theory for extracellular polymeric substances, soluble microbial products, and active and inert biomass. Water Res. **36**, 2711–2720 (2002a)

Laspidou, C.S., Rittmann, B.E.: Non-steady state modeling of extracellular polymeric substances, soluble microbial products, and active and inert biomass. Water Res. **36**, 1983–1992 (2002b)

Li., J., Zhang, Z.-J., Li, Z.-R., Huang, G.-Y., Abe, N.: Removal of organic matter and nitrogen from distillery wastewater by a combination of methane fermentation and denitrification/ nitrification processes. J. Environ. Sci. **18**, 654–659 (2006)

Li, X.Y., Yang, S.F.: Influence of loosely bound extracellular polymeric substances (EPS) on the flocculation, sedimentation and dewaterability of activated sludge. Water Res. **41**, 1022–1030 (2007)

Liao, B.Q., Allen, D.G., Droppo, I.G., Leppard, G.G., Liss, S.N.: Surface properties of sludge and their role in bioflocculation and settleability. Water Res. **35**, 339–350 (2001)

Liu, Y., Fang, H.H.P.: Influence of extracellular polymeric substances (EPS) on flocculation, settling, and dewatering of activated sludge. Crit. Rev. Environ. Sci. Technol. **33**, 237–273 (2003)

Liu, Y.-Q., Moy, B.Y.-P., Tay, J.-H.: COD removal and nitrification of low-strength domestic wastewater in aerobic granular sludge sequencing batch reactors. Enzym. Microb. Technol. **42**, 23–28 (2007)

Liu, Y., Tay, J.H.: The essential role of hydrodynamic shear force in the formation of biofilm and granular sludge. Water Res. **36**(24), 1653–1665 (2002)

Liu, Y., Tay, J.H.: State of the art of biogranulation technology for wastewater treatment. Biotechnol. Adv. **22**, 533–563 (2004)

Majone, M., Dircks, K., Beun, J.J.: Aerobic storage under dynamic conditions in activated sludge processes—the state of the art. Water Sci. Technol. **39**, 61–73 (1999)

Martins, A.M.P., Heijnen, J.J., Van Loosdrecht, M.C.M.: Effect of dissolved oxygen concentration on sludge settleability. Appl. Microbiol. Biotechnol. **62**, 586–593 (2003)

McSwain, B.S., Irvine, R.L., Wilderer, P.A.: Effect of intermittent feeding on aerobic granule structure. Wat. Sci. Tech. **49**, 19–25 (2004)

McSwain, B.S., Irvine, R.L., Hausner, M., Wilderer, P.A.: Composition and distribution of extracellular polymeric substances in aerobic flocs and granular sludge. Appl. Environ. Microbiol. **71**, 1051–1057 (2005)

Meijer, S.C.F.: Theoretical and practical aspects of modelling activated sludge processes. Ph.D. Thesis, Technical University, Delft, p. 204 (2004)

Meyer, R.L., Saunders, A.M., Zeng, R.J., Keller, J., Blackall, L.L.: Microscale structure and function of anaerobic–aerobic granules containing glycogen accumulating organisms. FEMS Mcrobiol. Ecology. **45**, 253–261 (2003)

Mikkelsen, L.H., Keiding, K.: Physico-chemical characteristics of full scale sewage sludges with implication to dewatering. Water Res. **36**, 2451–2462 (2002)
Morgenroth, E., Sherden, T., van Loosdrecht, M.C.M., Heijnen, J.J., Wilderer, P.A.: Aerobic granular sludge in a sequencing batch reactor. Water Res. **31**, 3191–3194 (1997)
Morel, F.M.M.: Principles of Aquatic Chemistry. John Wiley Interscience, New York (1983)
Morgan, J.W., Forster, C.F., Evison, L.: A comparative study of the nature of biopolymers extracted form anaerobic and activated sludges. Water Res. **24**, 743–753 (1990)
Mosquera-Corral, A., de Kreuk, M.K., Heijnen, J.J., van Loosdrecht, M.C.M.: Effects of oxygen concentration on N-removal in an aerobic granular sludge reactor. Water Res. **39**, 2676–2686 (2005)
Moy, B.Y.P., Tay, J.H., Toh, S.K., Liu, Y., Tay, S.T.L.: High organic loading influences the physical characteristics of aerobic sludge granules. Lett. Appl. Microbio. **34**(6), 407–412 (2002)
Neijssel, O.M., Tempest, D.W.: The role of energy-spilling reactions in the growth of *Klebsiella aerogenes* NCTC 418 in aerobic chemostat culture. Arch. Microbiol. **110**, 305–311 (1976)
Nielsen, P.H., Jahn, A.: Extraction of EPS. In: Wingender, J., Neu, T.R., Flemming, H.C. (eds.). Microbial extracellular polymeric substances: characterization, structure and function. Chapter 3, pp. 49–72, Springer, Heidelberg (1999)
Noguera, D.R., Araki, N., Rittmann, B.E.: Soluble microbial products in anaerobic chemostates. Biotechnol. Bioeng. **44**, 1040–1047 (1994)
Oehmen, A., Yuan, Z., Blackall, L.L., Keller, J.: Comparison of acetate and propionate uptake by polyphosphate accumulating organisms and glycogen accumulating organisms. Biotechnol. Bioeng. **91**, 162–168 (2005)
Owen, W.F., Stuckey, D.C., Healy Jr, J.B., Young, L.Y., McCarty, P.L.: Bioassays for monitoring biochemical methane potential and anaerobic toxicity. Water Res. **13**, 485–492 (1979)
Pan, S., Tay, J.H., He, Y.X., Tay, S.T.L.: The effect of hydraulic retention time on the stability of aerobically grown microbial granules. Lett. Appl. Microbiol. **38**, 158–163 (2004)
Payne, J.W.: Peptides and microorganisms. Adv. Microbiol. Physiol. **13**, 55–113 (1976)
Peng, D., Bernet, N., Delgenes, J.P., Moletta, R.: Aerobic granular sludge-a case study. Water Res. **33**, 890–893 (1999)
Pirt, S.J.: Principles of Microbe and Cell Cultivation. Blackwell Scientific, Oxford (1975)
Pratt, S., Yuan, Z., Keller, J.: Modelling aerobic carbon oxidation and storage by integrating respirometric, titrimetric, and off-gas CO_2 measurements. Biotechnol. Bioeng. **88**, 135–147 (2004)
Quarmby, J., Forster, C.F.: An examination of the structure of UASB granules. Water Res. **11**, 2449–2454 (1995)
Reis, M.A.M., Serafim, L.S., Lemos, P.C., Ramos, A.M., Aguiar, F.R., van Loosdrecht, M.C.M.: Production of polyhydroxyalkanoates by mixed microbial cultures. Bioprocess Biosyst. Eng. **25**, 377–385 (2003)
Rittmann, B.E., McCarty, P.L.: Environmental biotechnology: principles and applications. McGraw Hill, New York (2001)
Robinson, J., Trulear, M.G., Characklis, W.G.: Cellular reproduction and extracellular polymer formation by *Pesudomonas aeruginosa* in continuous culture. Biotechnol. Bioeng. **26**, 1409–1417 (1984)
Rosenberger, S., Laabs, C., Lesjean, B., Gnirss, R., Amy, G., Jekel, M., Schrotter, J.C.: Impact of colloidal and soluble organic material on membrane performance in membrane bioreactors for municipal wastewater treatment. Water Res. **40**, 710–719 (2006)
Saier, M.H., Feucht, B.U., McCaman, M.T.: Regulation of intracellular adenosine cyclic $3'$:$5'$-mono-phosphate levels in *Escherichia coli* and *Salmonella typhimurium*. J. Biol. Chem. **250**, 7593–7601 (1975)
Schwarzenbeck, N., Borges, J.M., Wilderer, P.A.: Treatment of dairy effluents in an aerobic granular sludge sequencing batch reactor. Appl. Microbiol. Biotechnol. **66**, 711–718 (2005)
Sesay, M.L., Ozcengiz, G., Sanin, F.D.: Enzymatic extraction of activated sludge extracellular polymers and implications on bioflocculation. Water Res **40**, 1359–1366 (2006)

Spath, R., Flemming, H.C., Wuertz, S.: Sorption properties of biofilms. Water Sci. Technol. **37**, 207–210 (1998)

Su, K.Z., Yu, H.Q.: Formation and characterization of aerobic granules in a sequencing batch reactor treating soybean-processing wastewater. Environ. Sci. Technol. **39**, 2818–2828 (2005)

Su, K.Z., Yu, H.Q.: A generalized model of aerobic granule-based sequencing batch reactor—I. Model development. Environ. Sci. Technol. **40**, 4703–4708 (2006a)

Su, K.Z., Yu, H.Q.: A generalized model for aerobic granule-based sequencing batch reactor—II. Parametric sensitivity and model verification. Environ. Sci. Technol. **40**, 4709–4713 (2006b)

Tay, J.H., Liu, Q.S., Liu, Y.: The role of cellular polysaccharides in the formation and stability of aerobic granules. Lett. Appl. Microbiol. **33**, 222–226 (2001)

Tsuneda, S., Nagano, T., Hoshino, T., Ejiri, Y., Noda, N., Hirata, A.: Characterization of nitrifying granules produced in an aerobic upflow fluidized bed reactor. Water Res. **37**, 4965–4973 (2003)

Tsuneda, S., Ogiwara, M., Ejiri, Y., Hirata, A.: High-rate nitrification using aerobic granular sludge. Water Sci. Technol. 53, 147–154 (2006)

Turakhia, M.H., Characklis, W.G.: Activity of *Pseudomonas aeruginosa* in biofilms: effect of calcium. Biotechnol. Bioeng. **33**, 406–414 (1989)

van Loosdrecht, M.C.M., Pot, M., Heijnen, J.: Importance of bacterial storage polymers in bioprocesses. Water Res. **35**, 41–47 (1997)

Veiga, M.C., Jain, M.K., Wu, W.M., Hollingsworth, R.I., Zeikus, J.G.: Composition and role of extracellular polymers in methanogenic granules. Appl. Environ. Microbiol. **63**, 403–407 (1997)

Wang, J., Yu, H.Q.: Biosynthesis of polyhydroxybutyrate (PHB) and extracellular polymeric substances (EPS) by Ralstonia eutropha ATCC 17699 in batch cultures. Appl. Microbiol. Biotechnol. **75**, 871–878 (2007)

Washington, D.R., Clesceri, L.S., Young, J.C., Hardt, F.W.: Influence of microbial waste products on the metabolic activity of high solids activated sludge. In: Proceedings of 24th Purdue Industrial Waste Conference, pp. 1103–1117, Purdue University, West Lafayette, IN, (1970)

Wilen, B.M., Keiding, K., Nielsen, P.H.: Anaerobic deflocculation and aerobic reflocculation of activated sludge. Water Res. **34**, 3933–3942 (2000)

Xavier, J.B., de Kreuk, M.K., Picioreanu, C., van Loosdrecht, M.C.M.: Multi-scale individual-based model of microbial and bioconversion dynamics in aerobic granular sludge. Environ. Sci. Technol. **41**, 6410–6417 (2007)

Yang, S.F., Li, X.Y., Yu, H.Q.: Formation and characterisation of fungal and bacterial granules under different feeding alkalinity and pH conditions. Process Biochem. **43**, 8–14 (2008)

Yang, S.F., Liu, Q.S., Tay, J.H., Liu, Y.: Growth kinetics of aerobic granules developed in sequencing batch reactors. Lett. Appl. Microbiol. **38**, 106–112 (2004)

Yang, S.F., Tay, J.H., Liu, Y.: Effect of substrate nitrogen/chemical oxygen demand ratio on the formation of aerobic granules. J. Environ. Eng.-ASCE **131**, 86–92 (2005)

Yang, Z., Peng, X.F., Chen, M.Y., Lee, D.J.: Intra-layer flow in fouling layer on membranes. J. Membrane Sci. **287**, 280–286 (2007)

Zheng, Y.M., Yu, H.Q., Liu, S.J., Liu, X.Z.: Formation and instability of aerobic granules under high organic loading conditions. Chemosphere **63**, 1791–1800 (2006)

Chapter 2
Research Questions and Thesis Overview

2.1 Introduction

This chapter introduces the specific research questions based on the literature review in the previous chapter. These questions encompass the objectives of this PhD study, which focuses on the formation, characterization, and mathematical modeling of the aerobic granular sludge and associated systems. Each of the Chaps. 3–12 will address one of the specific research questions.

2.2 How Can the Dynamic Aerobic Granulation Process be Quantitatively Characterized?

Aerobic granulation is a very complex phenomenon of microbial immobilization. There are numerous internal interactions among process variables, such as growth, storage and endogenous respiration, and sludge characteristics, including biomass detachment, oxygen transfer and diffusion. All of them have significant effects on the overall reactor performance. Therefore, the empirical models are generally descriptive and are not able to simulate the aerobic granulation process in other reactors or cultivated under different conditions. In this work, aerobic granules are cultivated in SBRs fed with both soybean-processing and fatty-acids-rich wastewaters and the aerobic granulation process in terms of mean radius profiles is quantitatively characterized based on experimental observations and formation mechanism analysis in Chap. 3. A new mathematical model incorporating microbial growth, oxygen transfer, substrate diffusion, increased granule size, and biomass detachment is formulated to describe the aerobic granulation process of activated sludge, on the basis of a mixed-culture biofilm model and a simultaneous storage and growth model. The model evaluation results of three different case studies demonstrate that the developed model is applicable to describing the

B.-J. Ni, *Formation, Characterization and Mathematical Modeling of the Aerobic Granular Sludge*, Springer Theses, DOI: 10.1007/978-3-642-31281-6_2,

aerobic granulation process appropriately. With this model, the aerobic granulation process in terms of mean radius profiles could be quantitatively described.

2.3 How Can Autotrophic and Heterotrophic Growth and Competition in Aerobic Granular Sludge be Determined?

The microbial analysis on the aerobic granules reported in Chap. 3 strongly suggests that the architecture of aerobic granules is relevant to their microbial ecology. They consist of two main different microbial groups, i.e., autotrophic and heterotrophic bacteria. Both autotrophs and heterotrophs coexist and interact in aerobic granules. Usually, aerobic granules display considerable heterogeneity, with respect to the microorganisms themselves and their physicochemical microenvironment. The presence of organic matters in wastewater creates competition between them for DO and space in the aerobic granules. Such interspecies competition and mass transfer result in the stratification of microbial species in granules. In Chap. 4, the growth of autotrophs and heterotrophs and their activity in aerobic granules is explored in-depth using experimental and modeling approaches. The fractions of the active biomass (autotrophs and heterotrophs) and inert biomass are determined. It is found that biomass content increases with the increasing SRT, but active biomass ratio decreases. The heterotrophs consume more oxygen than the autotrophs. The autotrophs are mainly located on the outer layers of granules for DO consumption, whereas the heterotrophs occupy the granule center or on the outer layers.

2.4 Can a Thermodynamic Analysis on the Microbial Synthesis of the Aerobic Granular Process be Achieved?

Accurate microbiological modeling requires evaluation of the effects of biological reactions on all important chemical and biological species in the aerobic granular sludge system. The ASM requires a large number of empirical parameters, which are difficult to determine. Moreover, many important chemical and biological species are lumped into one assumed model component or one empirical process. To sort out these problems, a thermodynamic analysis of the biological synthesis in sludge could be performed and the stoichiometrics could be more accurately estimated with the cell yield derived from thermodynamic considerations of the flows of energy and electrons in the catabolic and anabolic pathways. In Chap. 5, the thermodynamic analysis on the microbial synthesis of the aerobic granular process described in Chap. 3 is performed using the bioenergetic methodology established by McCarty after integrating it with a modified ASM1. This approach is able to approximately describe the treatment of soybean wastewater in an SBR in terms of substrate utilization, biomass growth, and the elector acceptor consumption. Such an attempt provides useful information for accurate modeling of the complex aerobic granular process.

2.5 Do Simultaneous Microbial Storage and Growth in Aerobic Granules Occur?

In recent years, many studies refer the production of storage polymers (X_{STO}) by activated sludge when exposed to a transient carbon supply. An aerobic-granule-based SBR is repeatedly subjected to feast and famine conditions. As a result, simultaneous growth and storage processes always occur. The measurements of aerobic granule reactors in Chaps. 3 and 4 indicate that the microorganisms are exposed to significant concentrations of the external substrate only for a relatively short period of time under these dynamic conditions. In excess of external carbon substrate, the uptake is driven to simultaneous growth of biomass and polymer storage, and after substrate exhaustion, stored polymer can be used as energy and carbon sources as well. In Chap. 6, the internal storage mechanisms and electron flow from the external substrate occurring in aerobic granule sludge are explored with storage experiments under different initial conditions. The simultaneous growth and storage processes in aerobic granules are modeled. The results show that simultaneous microbial storage and growth in aerobic granules occur under both aerobic and anoxic conditions.

2.6 How do the Aerobic Granules Produce Extracellular Polymeric Substances?

Aerobic granules are composed of numerous microorganisms, immobilized in EPS and/or matrices constituting polymers of proteins, polysaccharides, humic acids, and lipids. The presence of EPS, a complex high-molecular-weight mixture of polymers in the aerobic granular sludge cultivated in Chaps. 3 and 4, is confirmed and observed using electron microscopy techniques. EPS are sticky solid materials secreted by cells, and they are involved in adhesion phenomena, formation of the matrix structure, controlling the microbial physiology, and the long-term stability of the granules. In Chap. 7, the EPS produced by mixed microbial community are characterized using gel permeating chromatography (GPC) and 3-dimensional excitation emission matrix (EEM) fluorescence spectroscopy measurement, and a novel and convenient approach is also developed to evaluate the EPS production kinetics. Results show that EPS increase rapidly with the substrate consumption, but decrease slightly after the external substrate is completely consumed. Electrons from the external substrate are distributed in the following order: new biomass synthesis of 61 %, oxygen for respiration of 21 %, and EPS of 18 %.

2.7 Can the Formation of Soluble Microbial Products be Fractionized and Determined?

In addition to making different biomass components (active cells, X_{STO}, and EPS), bacteria in aerobic granules also convert a fraction of organic substrate into SMP, which account for the bulk of soluble organic carbon in reactor effluents. Their presence is of particular interest in terms of achieving discharge consent levels for biological wastewater treatment plants. SMP have been found to have a very wide range of MW distribution and different structures/functions. If SMP could be accurately fractionized and quantitatively identified, it is possible to examine how refractory the individual compounds are and to determine which type of SMP is the most difficult to remove from aerobic granular reactor effluents. In Chap. 8, the subfractions of the SMP, i.e., utilization-associated products (UAP) and biomass-associated products (BAP) in terms of formation sequence, MW and chemical nature are characterized, and a new approach for determinating SMP, UAP and BAP, and their production kinetics is developed on the basis of the approach proposed in Chap. 7. The UAP, produced in the substrate utilization process, are found to be carbonaceous compounds with an MW lower than 290 kDa and are quantified separately from BAP. The BAP are mainly cellular macromolecules with an MW in a range of 290–5000 kDa, and could be further classified into the growth-associated BAP (GBAP) with an MW of 1,000 kDa, which are produced in the microbial growth phase, and the endogeny-associated BAP (EBAP) with an MW of 4,500 kDa, which are generated in the endogenous phase.

2.8 How Can a Comprehensive Microbial Products Model (X_{STO}, EPS, and SMP) for Aerobic Granular Sludge be Developed?

The results from Chaps. 6 and 7 indicate that X_{STO} and EPS, as two types of polymers, are produced respectively inside and outside the cell by microorganisms. They could be produced simultaneously. Chapter 8 shows that SMP have been classified into two groups on the basis of the bacterial phase from which they are derived: UAP derived from the original substrate in microbial growth and BAP generated in the endogenous phase. EPS, X_{STO}, and SMP are important sinks for electrons and carbon derived from the original substrate. Phenomena involved with multiple carbon and electron sinks should have significant implications on the performance of aerobic granule systems. Thus, a comprehensive model about these microbial products in aerobic granules is a matter of great interest to improve our understanding about aerobic granular sludge and thereby enhance the efficiency of such system through the optimization of operational parameters. In Chap. 9, the latest representations for the production and consumption of EPS, X_{STO}, and SMP (from Chaps. 6–8) are incorporated into a more comprehensive model to describe the microbial products formation in aerobic granules. The concepts behind this modeling approach quantify the interconnections of EPS, SMP and X_{STO}, and offer a new integrated framework to describe these concepts with mathematical expressions and complete mass balances. The model is able to capture the experimental observations accurately for the soluble and solid components in aerobic granules exposed to dynamic feast and famine conditions. In addition, the model illustrates that X_{STO}, EPS, and the biomass components have different behaviors during feast and famine periods.

2.9 Do the Heterotrophs Utilize the Microbial Products of Autotrophs in Aerobic Granular Sludge for Growth?

Results in Chaps. 4 and 9 indicate that aerobic granular sludge is inherently multi-component as there are electron donors and acceptors, active biomass, EPS, residual inert material, and SMP produced under usual metabolic conditions. In an autotrophic system with the growth of autotrophs, inorganic carbon is converted into organic carbon in cell mass and EPS, and SMP are produced from substrate metabolism. The heterotrophs might utilize the SMP produced by autotrophs for growth, resulting in an existence of heterotrophs and autotrophs. Coexistence of a high level of heterotrophs with autotrophs has been often found in autotrophic nitrifying biofilms cultured without an external organic carbon supply. These microbial products, not contained in the feed solution, are likely to be the energy and carbon sources for the growth of heterotrophs. In Chap. 10, the formation mechanism, component characterization, and mathematical modeling of the microbial products of autotrophs and their exchange with heterotrophs in the nitrifying granular sludge are investigated systemically. Results show that the oxygen diffusion limitation causes anoxic microenvironments in the nitrite oxidizing granules and allows a sequential utilization of nitrate as electron acceptor and microbial products as electron donor for heterotrophic denitrification.

2.10 Is It Feasible to Cultivate Aerobic Granules for the Treatment of Low-Strength Municipal Wastewater in a Pilot-Scale SBR?

The results from Chaps. 3–10 indicate that many factors are involved in the granulation of activated sludge. Contributing factors include carbon source, hydrodynamic shear force, feast–famine regime, feeding strategy, dissolved oxygen, reactor configuration, volume exchange ratio and settling time. It shall be noticed that most of studies concerning aerobic granules have been focused on well-controlled lab-scale reactors with high- or middle-strength synthetic wastewaters.

Operating parameters and experience harvested from Chaps. 3–10 may be applicable to the utilization of aerobic granule process for the treatment of municipal wastewater, which is usually characterized as a low strength wastewater. In China, with the recent rapid economic growth and urbanization, a large number of municipal wastewater treatment plants are being constructed. The increasing population of China results in serious shortage of land area. Thus, aerobic granular process might be the promising technique for municipal wastewater treatment. Chapter 11 reports the cultivation of aerobic granules with an excellent settling ability in a pilot-scale SBR for the treatment of low-strength municipal wastewater for the first time. The volume exchange ratio and settling time are found to be two key factors in the granulation of activated sludge grown on such a low-concentration wastewater in an SBR. The mathematical model developed in Chap. 4 is applied to describe this pilot-scale granular reactor performance successfully.

2.11 Can Aerobic Granular Sludge be Used as Inoculum to Shorten the Startup Period of Anaerobic Ammonium Oxidation Process?

Anaerobic ammonium oxidation (anammox) is a promising new process to treat high-strength nitrogenous wastewater. Due to the slow growth rate of anammox bacteria, efficient biomass retention is essential for reactor operation. The results reported in previous chapters indicate that aerobic granules have a high microbial diversity (Chap. 4) and compact structure (Chap. 3) with very good settling properties (Chap. 11), resulting in an efficient means of biomass retention (Chap. 9). These properties including interspecies competition (Chap. 4) and mass transfer (Chap. 6) result in the stratification of microbial species (Chap. 4) with anoxic zones in the interior of the granules (Chap. 10), which may be suitable to harbor anammox bacteria. In Chap. 12, the feasibility to start up the anammox process by seeding the reactor with aerobic granular sludge is investigated. The anammox granules are successfully cultivated in an upflow anaerobic sludge blanket (UASB) reactor by seeding with aerobic granular sludge. The average total nitrogen removal efficiency exceeds 94 % after the formation of the granules. The heterogeneous composition of the seeding material contributes to a considerable reduce of the startup period to less than 160 days at a loading rate of 0.064 kg N kg VSS^{-1} d^{-1}. The ASM1 extended with a two-step denitrification consideration is incorporated with the anammox process to formulate a new platform to simulate the reactor performance well. Simulation results indicate that the optimum granule diameter for maximum N-removal in the case should be between 1.0 and 1.3 mm and that the optimum N loading rate should be 0.8 kg N m^{-3} d^{-1}.

Chapter 3
Quantitative Description of the Sludge Granulation Process

3.1 Introduction

In recent years, extensive studies have been carried out to cultivate aerobic granular sludge (Beun et al. 1999; Tay et al. 2001; Su and Yu 2005). Aerobic granules, compared with conventional activated sludge flocs, are well known for their regular, dense, and strong microbial structure, good settling ability, high biomass retention, and great ability to withstand shock loading rates (Tay et al. 2001). Studies have demonstrated that the aerobic granules could be applied for the treatment of high-strength wastewaters (Su and Yu 2005), simultaneous removal of organics, nitrogen and phosphorus (de Kreuk et al. 2005), and decomposition of toxic wastewaters (Jiang et al. 2002). Thus, this new kind of activate sludge, like the anaerobic granular sludge, could be employed for the treatment of municipal and industrial wastewaters in the near future. Aerobic granules have been cultivated in SBR fed with synthetic wastewaters composed of various substances, such as acetate (Tay et al. 2001), molasses (Morgenroth et al. 1997), and ethanol (Beun et al. 1999). SBR with aerobic granular sludge is recognized as a very promising process for wastewater treatment.

Microscopic observation shows that the formation of aerobic granules is a gradual process from seed sludge to compact aggregates, further to granular sludge, and finally to mature granules (Su and Yu 2005). Previous work on aerobic granulation has focused mainly on the factors involved in the formation of aerobic granules, but little information is available for the quantitative description of such a granule growth course yet. A modified Logistic model was adapted by Su and Yu to describe the granulation process in terms of diameter (Su and Yu 2005). A linear phenomenological equation was applied to model the aerobic granulation process (Yang et al. 2004). These models are generally empirical and descriptive, and are not able to simulate the aerobic granulation process in other reactors or cultivated under different conditions. Recently, a mathematical model developed by de Kreuk et al. (2007) is able to describe an aerobic-granule-based reactor effectively.

B.-J. Ni, *Formation, Characterization and Mathematical Modeling of the Aerobic Granular Sludge*, Springer Theses, DOI: 10.1007/978-3-642-31281-6_3,

However, it is a steady-state model and only granules with constant size are modeled. Thus, aerobic granulation process with a varied sludge size cannot be simulated by this model.

Aerobic granulation is a very complex phenomenon of microbial immobilization. There are numerous internal interactions among process variables, such as growth, storage and endogenous respiration, and sludge characteristics, including biomass detachment, oxygen transfer, and diffusion. All of them have significant effects on the overall reactor performance. However, a mathematic structure model to describe aerobic granulation process with a full consideration of all the factors above is not available yet. Thus, a unified mathematic model is highly desirable to describe the aerobic granulation process.

Biofilm thickness has been widely used as an index to describe the microbial growth on carrier surface, and a number of comprehensive growth models for biofilm formation can be found in the literature (Wanner and Reichert 1996; Rauch et al. 1999). Therefore, this study aims at formulating a mathematical model to describe the growth process of aerobic granules in SBRs. Using the model established, the aerobic granulation process in terms of mean radius could be characterized and the granulation mechanisms could be understood better.

3.2 Model Development

3.2.1 General Description

The growth of aerobic granules after the initial cell-to-cell self-attachment is similar to the growth of biofilm, and is the net result of interactions between bacterial growth and detachment. The balance between the growth and detachment processes in turn might lead to an equilibrium granule size (Yang et al. 2004). Compared with biofilm process, aerobic granulation is a self-immobilization process, rather than an initial cell attachment to a solid surface. Microbial growth results in an increase in granule size or volume. Thus, the sludge size evolution can be used to describe the growth process of aerobic granules. A mixed-culture biofilm model (Wanner and Reichert 1996), which has been implemented in AQUASIM (Reichert 1998), is combined with a simultaneous storage and growth model to formulate a new model to describe the growth process of aerobic granules in SBRs.

3.2.2 Biomass Growth

The simultaneous storage and growth model developed by Karahan et al. (2006) is adopted to describe the microbial growth kinetics. In this model two distinct phases, feast (external carbon is present) and famine (external carbon is depleted) phases, are taken into account. Thus, microbial growth takes place under non-steady-state conditions: presence (feast) and absence (famine) of external carbon. Based on a conventional structure of activated sludge model (ASM), three distinctive yield coefficients independent from each other are used for storage (Y_{STO}), direct growth on external substrate ($Y_{H,S}$), and growth on internal storage products ($Y_{H,STO}$), respectively (Karahan et al. 2006).

At feast phase, because of a pulse dosage of substrate (S_S), the biomass primarily utilizes the substrate available through storage and growth. With the consumption of oxygen (S_O), microbial growth takes place through the degradation of S_S when the external carbon is present, or through the degradation of storage products (X_{STO}) as the external carbon becomes depleted. Furthermore, energy is required for growth, and also for maintenance and lysis, which are described as endogenous respiration following first-order reaction kinetics.

The change in active biomass attributed to growth and endogenous respiration is described in Eq. 3.1. Heterotrophic growth occurs through the degradation of S_S (first term), and the degradation of X_{STO} (second term), which represents the loss of active biomass associated with the endogenous respiration.

$$\frac{dX_H(t)}{dt} = \mu_{H,S} M_S(t) M_O(t) X_H(t) + \mu_{H,STO} M_{STO}(t) I_S(t) M_O(t) X_H(t) - b_H M_O(t) X_H(t) \tag{3.1}$$

$$M_{STO}(t) = \frac{X_{STO}(t)/X_H(t)}{K_{STO} + X_{STO}(t)/X_H(t)} \tag{3.2}$$

$$M_S(t) = \frac{S_S(t)}{K_S + S_S(t)} \tag{3.3}$$

$$M_O(t) = \frac{S_O(t)}{K_{O_2} + S_O(t)} \tag{3.4}$$

$$I_S(t) = \frac{K_S}{K_S + S_S(t)} \tag{3.5}$$

where $\mu_{H,S}$ is the maximum microbial growth rate on external substrate, $\mu_{H,STO}$ is the maximum microbial growth rate on storage products, b_H is the microbial endogenous decay coefficient. M_S stands for a Monod kinetic function of S_S, and M_O is a Monod kinetic function of DO. M_{STO} is a surface saturation-type kinetic function of X_{STO}, while I_S is a Monod inhibition function for S_S. The first term on the right side of Eq. 3.1 is for describing the biomass growth through the degradation of S_S and S_O, while the second term is for the biomass growth through the degradation of X_{STO} and S_O when S_S is depleted. The last term represents the biomass decrease due to the endogenous respiration with consumption of S_O.

This simultaneous storage and growth model involves five microbial processes: heterotrophic growth on S_S, heterotrophic storage, heterotrophic growth on X_{STO}, heterotrophic endogenous respiration, and respiration of X_{STO}.

3.2.3 Oxygen Transfer

Before reaching the granule surface for diffusion and reaction, oxygen must transfer from gas phase to solid phase. The gas–liquid oxygen transfer rate is assumed to be proportional to the difference in oxygen concentrations between gas and liquid interface, and the proportionality factor is the volumetric oxygen transfer coefficient k_La (Nicolella et al. 1998). On the granule surface, oxygen transferred from the gas phase is equal to that diffused into granules. It follows a mass balance equation below:

$$\frac{dS_O(t)}{dt} = k_La\left(S_O^* - S_O(t)\right) - r_O(t) \tag{3.6}$$

$$\begin{aligned} r_O(t) = {} & \frac{(1 - Y_{\mathrm{H,S}})}{Y_{\mathrm{H,S}}}\mu_{\mathrm{H,S}}M_{\mathrm{S}}(t)M_{\mathrm{O}}(t)X_{\mathrm{H}}(t) + \frac{1 - Y_{\mathrm{H,STO}}}{Y_{\mathrm{H,STO}}}\mu_{\mathrm{H,STO}} \\ & *M_{\mathrm{STO}}(t)I_{\mathrm{S}}(t)M_{\mathrm{O}}(t)X_{\mathrm{H}}(t) + (1 - f_{\mathrm{I}})b_{\mathrm{H}}M_{\mathrm{O}}(t)X_{\mathrm{H}}(t) \\ & + \frac{1 - Y_{\mathrm{STO}}}{Y_{\mathrm{STO}}}k_{\mathrm{STO}}M_{\mathrm{S}}(t)M_{\mathrm{O}}(t)X_{\mathrm{H}}(t) + b_{\mathrm{STO}}M_{\mathrm{O}}(t)X_{\mathrm{STO}}(t) \end{aligned} \tag{3.7}$$

where b_{STO} is the endogenous decay coefficient of storage products; S_{O}^* is the maximum oxygen solubility in liquid phase; S_{O} is the oxygen concentration on the granule surface, equal to that in the bulk liquid when the liquid–solid oxygen transfer resistance is ignored; $r_{\mathrm{O}}(t)$ is oxygen uptake dynamics including growth on S_{S}, aerobic storage, growth on X_{STO}, endogenous respiration of X_{H}, and endogenous respiration of X_{STO}. a is gas–liquid interfacial area per unit liquid volume. The five terms in the right side of Eq. 3.7 represent oxygen uptake by all respiration processes: utilization of S_{S} for biomass growth (term 1), utilization of X_{STO} for biomass growth (term 2), endogenous respiration of X_{H} (term 3), storage of X_{STO} (term 4), and endogenous respiration of X_{STO} (term 5).

3.2.4 Mixed-Culture Biofilm Model

The mixed-culture biofilm model established by Wanner and Reichert is adopted to describe the aerobic granules in this work (Wanner and Reichert 1996). Similar to biofilm, aerobic granules consist of a solid matrix with pore water that contains dissolved substances and suspended solids (Reichert 1998). The microbial growth forms the solid matrix and leads to the expansion of granules. The microbial consumption of substrate at a high concentration in the granule solid matrix results in a growth limitation by the diffusive mass transfer into the granule depth.

Granules with a wide size range should be simulated with a consideration of component diffusion limitation within granules. Transport of dissolved components in the liquid phase of granules is described by Fick's first law as a diffusive flux:

$$J_C = -AD_C \frac{\partial C}{\partial z} \tag{3.8}$$

where D_C is the diffusion coefficient of the dissolved component in the liquid phase of the granule, J_C is amount of the conserved quantity transported per unit time, and A is the granule surface area. Equation 3.8 is used for all the dissolved components, including the dissolved oxygen in the granular system.

In the present study, detachment is also taken into account, as it plays a crucial rule in the granulation process. In SBRs, the shear force applied to granules is attributed to the relative velocities of gas and liquid as well as collisions among granules. The progression of the granule radius (*GR*) is given by the following equation (Wanner and Reichert 1996):

$$\frac{\mathrm{d}GR}{\mathrm{d}t} = u_L \tag{3.9}$$

where u_L is the velocity of the granule surface. This velocity can be calculated from Eq. 3.10 (Rauch et al. 1999):

$$u_L = u_F(GR) - u_{\mathrm{de}} + u_{\mathrm{at}} \tag{3.10}$$

where u_{de} is the velocity by which particulate components are detached from the granule surface, u_{at} is the velocity by which cells and particles suspended in the bulk fluid are attached to the granule surface, u_F is the advective velocity. The detachment velocity is the global function of the granule size. The solids are detached from the granules according to their relative occurrence at the granule surface.

3.2.5 Simulation Methods

The model established is calculated with AQUASIM 2.0 (Reichert 1998), a program designed for parameter estimation and sensitivity analysis. This software has been widely used in model calculation and simulation for biological wastewater treatment processes. In AQUASIM, as a first step, the partial differential equations are discretized in space. Then, the spatially discretized partial differential equations together with the ordinary differential equations and the algebraic equations are integrated numerically in time with the algorithm DASSL (Reichert 1998), which is based on the implicit (backward differencing) variable step, variable-order Gear integration technique. Model parameters represented by constant variables can be estimated with AQUASIM through minimizing the sum of the squares of the weighted deviations between the measurements and calculated model results. The reactor in modeling is described with a biofilm compartment connected to a mixed compartment with an advective link, and a high recirculation flow rate is incorporated from the biofilm compartment to the completely mixed one. The high recirculation rate is chosen to ensure the same concentrations in the liquor of biofilm compartment and the mixed compartment. The biofilm compartment with a volume of 1 L contains the biomass granules and bulk liquid volume. The mixed compartment of 1 L includes the remaining liquor volume. The number of aerobic granules and their diameter has an effect on the simulation results, because they influence the overall liquid/granule interfacial area. The granules grown in the SBR reactor had a size distribution with a mean granule size. However, the use of a granule size distribution was avoided in this model, since it will significantly increase the complexity of the numerical computations and have no substantial contribution to a better understanding of the system. Therefore, in the simulations the diameter of the aerobic granules was chosen to be the mean granule size, which was the most representative for the granules in the SBR reactor. The mean granule size was obtained through calculating the mean value of all the radius of the existing granules in the reactor. The mean radius and number of granules could be used to yield a given total biomass volume, which was the comparable amount of biomass present in the reactor.

3.3 Materials and Methods

3.3.1 Experimental Set-Up and Operation

Aerobic granules were cultivated in two identical lab-scale SBRs, which had a working volume of 2 L, an internal diameter of 7.0 cm, and a height of 100 cm. In this work, one reactor (designated as R1) was fed with a soybean-processing wastewater as the influent, while the other reactor (designated as R2) was fed with a fatty-acids-rich wastewater. Other operating conditions were kept identical for both reactors. Effluent was drawn from the middle of the reactors, resulting in 1.0 L of mixed liquor left in each reactor after effluent withdrawal.

The two reactors were operated for 4 h per cycle with a hydraulic retention time (HRT) of 8 h and an influent chemical oxygen demand (COD) of 800 mg L^{-1}. Both reactors were operated at 20 °C in a sequential mode: 3 min of influent filling, 227 min of aeration, 5 min of settling, and 5 min of effluent withdrawal. Air was introduced through an air diffuser at the reactor bottom by an air pump. The airflow rate was controlled via a gas-flow controller to keep the dissolved oxygen (DO) level over 2.0 mg L^{-1} in each aeration cycle. An air velocity of 0.4 $m^3 h^{-1}$ was applied to each reactor, equivalent to a superficial upflow velocity of 2.8 cm s^{-1}. Aeration was started along with influent dose and was stopped before settling and effluent withdrawal. Meanwhile, in order to prevent the accumulation of suspended sludge and to accelerate the granule formation, the settling time was chosen to create a selective pressure such that only sludge with a higher settling velocity than the corresponding minimal settling velocity was efficiently retained in the reactor.

Table 3.1 Characteristics of the two wastewaters used in this chapter

Characteristics	Soybean-processing wastewater	Fatty-acids-rich wastewater
pH	4.2	4.4
COD (mg L^{-1})	21,100	–
Total N (mg L^{-1})	974	–
Total P (mg L^{-1})	15	–
VSS (mg L^{-1})	305	–
Butyrate (mM)	–	11.2
Acetate (mM)	–	3.97
Ethanol (mM)	–	5.18

3.3.2 Wastewater and Seed Sludge

The soybean-processing wastewater obtained from a local plant contained soluble proteins of 5.5 g L^{-1} and carbohydrates of 7.4 g L^{-1}. Because it had a sufficient amount of nitrogen, only phosphorus as Na_2HPO_4 was added to ensure the ratio of COD to P to be 100:1. In addition, a microelement solution of 1.0 mL L^{-1} was added, which contained (in mg L^{-1}): H_3BO_3, 50; $ZnCl_2$, 50; $CuCl_2$, 30; $MnSO_4{\cdot}H_2O$, 50; $(NH_4)_6Mo_7O_{24}{\cdot}4H_2O$, 50; $AlCl_3$, 50; $CoCl_2{\cdot}6H_2O$, 50; and $NiCl_2$, 50. The influent pH value was adjusted to 7.0 through the dose of $NaHCO_3$ or HCl. The raw wastewater was diluted by 10 times using tap water to get the influent to R1.

The fatty-acids-rich wastewater was the effluent of a lab-scale acidogenic reactor fed with a sucrose-rich wastewater (Mu and Yu 2006). Butyrate, acetate, and ethanol were its main constituents. The raw wastewater was diluted by seven times using tap water to get the influent to R2. The characteristics of both wastewaters used in this work and key reactor constituents are listed in Table 3.1.

Activated sludge was collected from an aeration tank in the Wangxiaoying Municipal Wastewater Treatment Plant, Hefei, China, as inoculum for the two SBRs. The seed sludge had a mixed liquor suspended solids (MLSS) concentration of 8 g L^{-1} and a sludge volume index (SVI) of 51.2 mL g^{-1}. One thousand milliliter of inoculum was seeded to each reactor, resulting in an initial mixed liquor volatile suspended solid (MLVSS) of 3,000 mg L^{-1} in each reactor.

3.3.3 Parameter Determination

For the aerobic granules, diffusion resistance could mask the intrinsic properties of substrate utilization. Therefore, determination of kinetic and stoichiometric parameters was performed using fine floc particles (<100 μm) obtained through disrupting the granules into fragments at 4 °C using a mortar and pestle as described by Wu et al. (1995). The fragments were further fractionated to flocs through magnetic stirring at 300 rpm for 6 h. The resultant flocs were then used for the determination of kinetic and stoichiometric parameters. Parameters were determined at 20 °C and pH 7.0. Heterotrophic storage yield coefficient (Y_{STO}) was measured using the method proposed by Karahan et al. Biomass yield on substrate ($Y_{H,S}$), maximum growth rate ($u_{H,S}$), substrate half-saturation coefficient (K_S), and biomass decay coefficient (b_H) were determined as described by Henze et al. (1987). The increase in substrate utilization rate with the increasing S_S concentration in independent batch tests was fitted with Monod kinetics to find the best-fit $\mu_{H,S}$-K_S values. For a batch test, granular sludge samples were collected in an amber glass bottle at the end of one operating cycle of the SBR when substrate became depleted in the medium, and a 1 L batch reactor was inoculated with the sludge by disrupting these granules to conduct the batch tests. Five batch tests with different initial S_S concentrations were conducted for parameters determination. To estimate b_H, the sludge was drawn from a completely mixed reactor and put into an aerated and non-fed 1 L batch reactor. The oxygen respiration rate profiles were measured over a period of several days. A plot of the logarithm of the respiration rate versus time gave a straight line with a slope of b_H. The storage process parameters were assessed in batch respirometric experiments. Prior to a respirometric test, the granules were washed twice with distilled water to remove carry-over matters, and the wastewater was then added to reactors with the granules. The DO concentration was monitored, and the calculated oxygen uptake rate (OUR) was used for parameter estimation with the methods proposed by Avcioglu et al. (2003). The global detachment rate coefficient was determined through fitting the experimental data.

3.3.4 Analytical Methods

The granule size was measured using an image analysis system (Image-pro Express 4.0, Media Cybernetics) with an Olympus CX41 microscope and a digital camera (Olympus C5050). Approximately 10 mL of biomass granules were randomly removed from the lower-, middle-, and upper-part of the sludge-bed zone of each SBR. The granules were then washed with distilled water for 3 times. Their average diameter was measured with a representative sample, in which at least 100 particles were analyzed for granule radius with the image analyzer. The DO concentration in the reactors was measured with a DO electrode (MO128, Mettler-Toledo Gmbh, Switzerland). Measurement of COD, MLSS, and MLVSS was performed according to the Standard Methods (APHA 1995).

OUR was measured using a respirometer that consisted of a 300 mL glass vessel with a pot at the top for insertion of a DO probe (MO128, Mettler-Toledo Gmbh, Switzerland). A magnetic stirring bar and a stirring plate provided internal mixing of the liquor and sludge. The respirometer was kept at pH of 7.0 ± 0.1 using NaOH or HCl solutions and an auto-titrator, which enabled the pH to be maintained at 7.0. For each respirometric test, 250 ml of sludge was collected from the batch reactors and added to the respirometer. The slope of the DO concentration decline was measured over 15 min. The OUR, calculated from the slope though linear regression, was determined twice for each test, and the averages are reported here.

3.4 Results and Discussion

3.4.1 Experimental Observations

Granules with a clear and spherical outer shape and stable size were successfully cultured after 5 weeks of operation in the two SBRs. Figure 3.1 shows the evolution of bioparticles from dispersed sludge to mature granules in terms of mean size in the two reactors. The particle size gradually increased over operating time, and finally reached a stable value. There was no significant change for the sludge in the first week (day 1–7). After 7-day operation, tiny granules were first observed in R2. On day 10, granules with small size also appeared in R1 fed with the soybean-processing wastewater. Fifteen days later, aerobic granules became obviously visible in both reactors. After that, the granule sizes in R1 and R2 increased gradually. After a 5-week operation, granular sludge became matured. The granule mean radius was 1.25 mm for R1 and 1.5 mm for R2, respectively, on day 35.

Figure 3.2 shows the morphology of the seed sludge and the aerobic granules cultured in R1 and R2. The sludge color changed from brown to yellow gradually. The seed sludge consisted of flocs only, and had a fluffy, irregular, and loose-structure morphology. On the other hand, the matured granules had a compact structure with a regular and spherical outer shape.

In the granulation process, the reactor performance was improved continuously in terms of COD removal efficiency. For R2 the COD removal efficiency was kept at 95–98 % after 10 day of operation (Fig. 3.3a). Similarly, the MLVSS concentration kept increasing. After day 30, the MLVSS was stabilized.

After 60-day operation, a constant DO and OUR variation, as well as COD removal in the two reactors were achieved, indicating a steady-state operation of the reactors. As shown in Fig. 3.5b, the OUR increased, attributed to the rapid oxidation of the external substrate. The sharp bending point in the OUR curve corresponded with the complete removal of the external substrate added (Fig. 3.5c). Complete oxidation of COD occurred within 1.5 h.

3.4.2 Stoichiometric and Kinetic Parameters

The stoichiometric and kinetic coefficients $Y_{H,\ S}$, $\mu_{H,S}$, K_S and b_H were independently measured from the experimental results, and were used in the model simulation later. The kinetic coefficients for the storage of substrate and the utilization

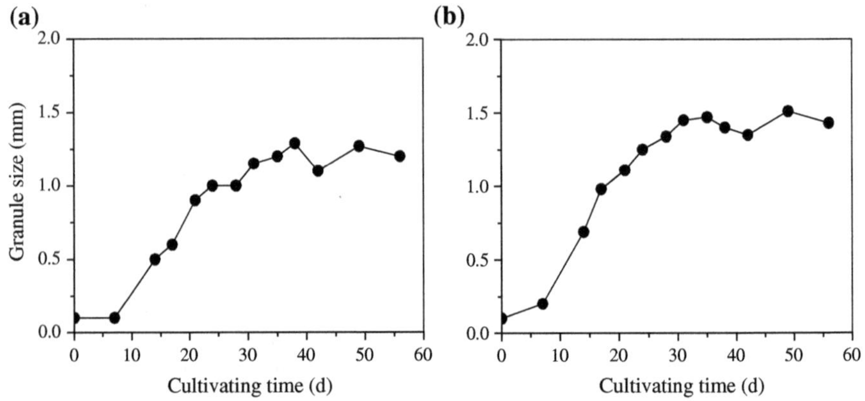

Fig. 3.1 Granule mean diameter profiles in the granulation process of: **a** R1; and **b** R2 (Reprinted from Ni et al. (2010) with permission of American Chemical Society)

of storage polymer were determined by fitting the model to the batch experimental data using the estimation methods provided in AQUASIM. These values were estimated by minimizing the sum of squares of the deviations between the measured OUR data and the model predictions. In this work, *fI* is fixed to the value mentioned in ASM3, 0.2 g COD g^{-1} COD. The endogenous rate of X_{STO}, b_{STO}, is taken as same as the microbial endogenous rate, b_H (Gujer et al. 1999).

To determine the Monod parameter ($Y_{H,S}$, $\mu_{H,S}$, and K_S) using batch tests, parameter identification is important as it is possible that different parameter combinations can give similar simulation accuracy in an aerobic system, in which biomass grows significantly. Oxygen was continuously provided to ensure a high DO level in these batch reactors. Thus, the DO was assumed not to be a limiting factor for the Monod parameter determination. The observed correlation between parameter sets follows the Monod-based equation.

In our work, a least-squared analysis and evaluated parameter confidence intervals were applied to analyze the parameter identification. Overall, the 95 % confidence regions for the parameters combinations of $Y_{H,S}$, $\mu_{H,S}$, and K_S in the two cases were bounded by ellipsoids with mean value for the parameter estimates approximately at the center (Fig. 3.4a, b). The shape of parameter response surface for $\mu_{H,S}$, and K_S (Fig. 3.4a, b) plots shows no evidence of cross-correlation. The fact that $\mu_{H,S}$, and K_S shows no cross-correlation is an especially significant sign of uniqueness for the parameter values. In addition, the calculated 95 % confidence intervals of $Y_{H,S}$, $\mu_{H,S}$, and K_S presented as an absolute percentage of the determined parameters are 8.6, 9.5 and 4.5 % for R1 and 9.2, 10.3, and 4.9 % for R2, respectively, which are not very higher compared to the best-determine values, suggesting a good identification of these parameters.

The kinetic coefficients for the storage of substrate and the utilization of storage polymer were determined through minimizing the sum of squares of the deviations between the measured data and the model predictions using AQUASIM. As shown in Fig. 3.4c, d the 95 % confidence regions for the parameter combinations of

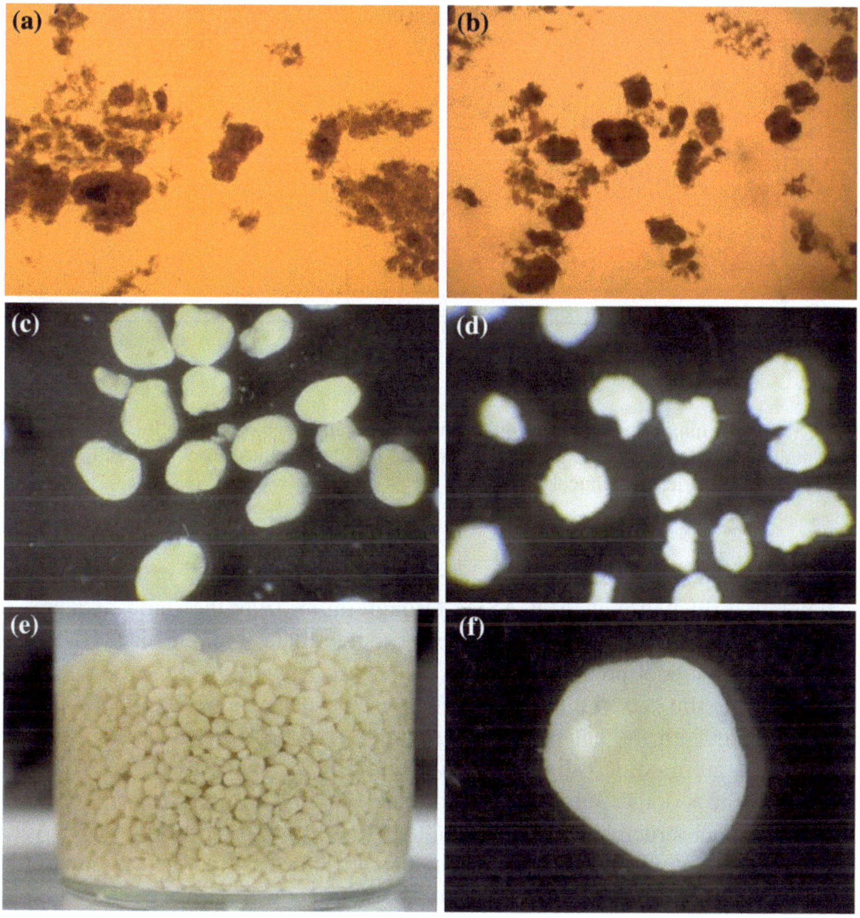

Fig. 3.2 Images of sludge in the granulation process: **a** seed sludge; **b** seed sludge; **c** matured granules in R1; **d** matured granules in R2; **e** granule compact structure; and **f** regular and spherical outer shape (Reprinted from Ni et al. (2010) with permission of American Chemical Society)

storage kinetics for both cases were also bounded by ellipsoids with mean value for the parameter estimates approximately at the center, which showed no cross-correlation. The calculated 95 % confidence intervals were all lower than 15 % (most are $\leq$10 %) for the two cases compared to the best determine values, suggesting a good identification of these parameters. In addition, to the analysis of the response surface, three different experimental parameters (OUR, COD, and granule size), which reflect different aspects of the kinetics of the microbial process, were used to further support the uniqueness of the parameter values.

Table 3.2 summarizes the parameter values in the model, which are used to generate the model outputs. Parameters determined with the experimental data of the substrate conversion batch tests ($Y_{H,S}$, $\mu_{H,S}$, and K_S) with good parameter identification and b_H which was independently determined with non-fed batch

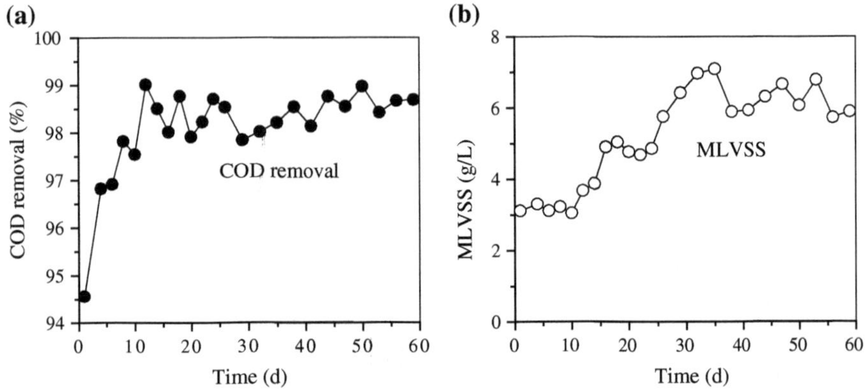

Fig. 3.3 **a** COD removal efficiency profiles during the granulation process in R2; and **b** MLVSS concentration profiles during the granulation process in R2 (Reprinted from Ni et al. (2010) with permission of American Chemical Society)

reactor over a period of several days are marked as "Measured" in Table 3.2. The parameters estimated with AQUASIM through minimizing the sum of squares of the deviations between the measured experimental data (granule size, OUR, and COD) and the model predictions (Y_{STO}, $Y_{H,STO}$, k_{STO}, $\mu_{H,STO}$, K_{STO}, and k_{d1}) are marked as "Estimated" in Table 3.2. With a consideration of the kinetic parameters which govern the heterotrophic growth, the values recommended by Gujer et al. (1999) are modified slightly. This is partially attributed to the heterotrophic growth on the storage polymers, rather than directly on the soluble substrates. Different model structures might be another reason for such a difference. Additionally, different wastewater characteristics, reactor conditions, and microbial communities may also partially explain the difference.

A high level of readily biodegradable substrate results in a relatively rapid storage process with a maximum rate, k_{STO} (Table 3.2). This favors a rapid substrate storage accompanied by an optimized primary growth on substrate available, when the excess substrate is present in reactor (Avcioglu et al. 2003). The heterotrophs utilize the substrate storage as a tool for maximizing substrate uptake when retaining an optimum growth rate (Avcioglu et al. 2003). The relatively high yield coefficients for storage polymer and biomass are associated with the high biomass growth rate at the early stage of aerobic granulation process. The kinetic and stoichiometric coefficients used in this work, are comparable to those reported in the literature (Avcioglu et al. 2003).

Comparison of model parameter values provides insight into the aerobic granulation processes in different cases. The obtained kinetic and stoichiometric coefficients for R1, have different parameter values from those of R2 (Table 3.2). A slightly lower true microbial yield in R1 is obtained (0.63 vs. 0.65 g COD_X g^{-1} COD_S), which is partly attributable to the fact that the soybean-processing wastewater is a complex mixture of carbohydrates and proteins, while the fatty-

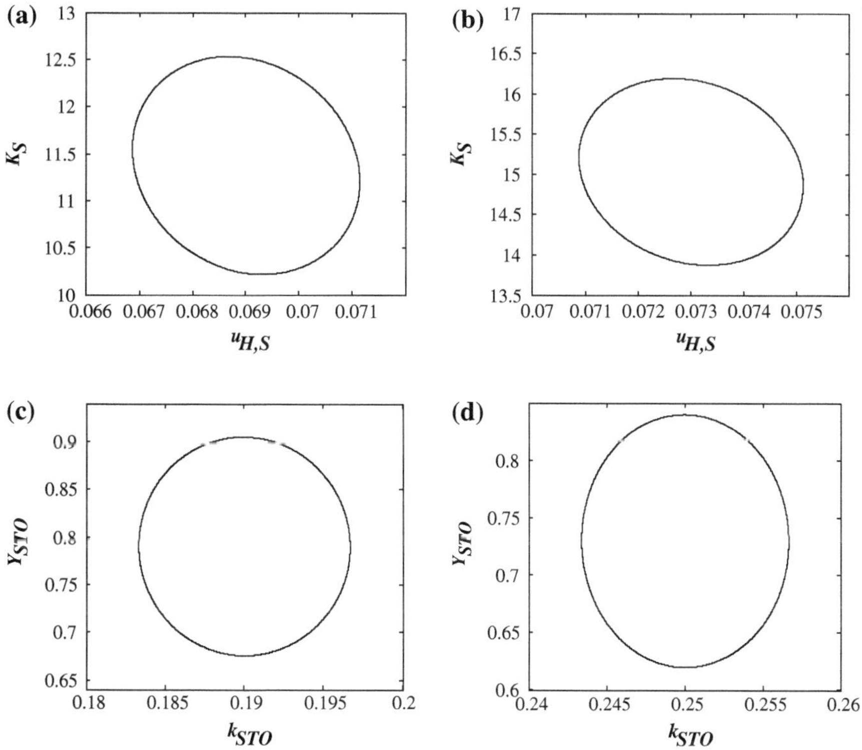

Fig. 3.4 The 95 % confidence ellipsoids for the kinetic parameters: **a** $\mu_{H,S}$ versus K_S for R1; **b** $\mu_{H,S}$ versus K_S for R2; **c** k_{STO} versus Y_{STO} for R1; and **d** k_{STO} versus Y_{STO} for R2 (Reprinted from Ni et al. (2010) with permission of American Chemical Society)

acids-rich wastewater is composed of acetate, butryrate, and ethanol. The different kinetic parameters of X_{STO} formation and consumption (Y_{STO}, k_{STO}, $\mu_{H,STO}$, $Y_{H,STO}$, K_{STO}) reflect the different microbial communities in the two reactors because of different feeding wastewaters. One parameter has great difference for the two reactors: the half-saturation constant, K_S. The half-saturation coefficients for R2 are much larger than that of R1, ensuring slow biodegradation kinetics under most conditions. The decay rate for R1 is 0.0093 h^{-1}, which is larger than the decay rate for R2 of 0.0081 h^{-1}. This difference means that biomass in R1 are more easily detached from the granules which might result in the smaller granule size in R1. The maximum specific growth rate for the external substrate ($\mu_{H,S}$) of R1 is 0.069 h^{-1}, which is only slightly slower than that of R2 of 0.073 h^{-1}. Thus, growth through utilization of soybean-processing wastewater is commensurate with growth on fatty-acids-rich wastewater.

Generally, the obtained parameters values for R1 can be fitted to use the model in the systems feeding with comparatively complex wastewater or carbohydrates such as starch, glucose, molasses, sucrose, malting wastewater, dairy effluents, and

Table 3.2 Parameters for the model simulation (20 °C)

Parameter	Unit	Values for R1	Values for R2	Reference
Stoichiometry				
Y_{STO}	g STO g^{-1} S	0.79	0.73	Estimated
$Y_{H,S}$	g X g^{-1} S	0.63	0.65	Measured
$Y_{H,STO}$	g X g^{-1} STO	0.53	0.63	Estimated
f_I	g X g^{-1} X	0.2	0.2	Gujer et al.
Kinetics				
k_{STO}	h^{-1}	0.19	0.25	Estimated
$\mu_{H,S}$	h^{-1}	0.069	0.073	Measured
$\mu_{H,STO}$	h^{-1}	0.061	0.064	Estimated
b_H	h^{-1}	0.0093	0.0081	Measured
b_{STO}	h^{-1}	0.0093	0.0081	Measured
K_S	g S m^{-3}	11.38	15.03	Measured
K_{STO}	g STO g^{-1} X	3.5	2.8	Estimated
K_{O_2}	g O_2 m^{-3}	0.2	0.2	Gujer et al.
Diffusion coefficients				
D_S	cm h^{-1}	0.021	0.021	Horn et al.
D_{O_2}	cm h^{-1}	0.087	0.087	Horn et al.
Density of solid phase				
ρ_{XH}	g m^{-3}	24,000	24,000	Horn et al.
ρ_{XI}	g m^{-3}	24,000	24,000	Horn et al.
ρ_{STO}	g m^{-3}	24,00,000	24,00,000	Beun et al.
Detachment				
k_{d1}	–	0.25	0.3	Estimated
GR_{eq}	mm	1.25	1.4	Measured

other similar industrial wastewater. On the other hand, the obtained parameters values for R2 can be fitted to use the model in the systems feeding with acetate, ethanol, and other volatile fatty acids-containing wastewater. However, it should be noted that the substrate type is an important factor affecting the aerobic granulation processes. The parameters reported here are partially dependent on wastewater characteristics. To more accurately describe the activated sludge granulation process in other cases, more experimental work is needed to calibrate model parameters with a consideration of wastewater characteristics and reactor operating conditions.

The parameters of biomass yield on substrate ($Y_{H,S}$), maximum growth rate ($\mu_{H,S}$), substrate half-saturation coefficient (K_S) should be specifically estimated in each aerobic granule system with different wastewater characteristics and reactor operating conditions based on the substrate conversion batch experiments. Most importantly, parameter identification is important for this determination procedure as it is possible that different parameter combinations can give similar simulation accuracy in these systems in which biomass grows very significantly. For the different aerobic granule systems with different microbial communities, the parameters for storage process and the utilization of storage polymer (Y_{STO}, $Y_{H,STO}$, k_{STO}, and $\mu_{H,STO}$) should be assessed with batch respirometric experiments under feast and famine conditions, in which the external carbon could be converted to storage products (X_{STO}) and the degradation of storage products (X_{STO}) can take place as the external carbon becomes depleted. Other parameters, such as f_I, the half saturation constant for dissolved oxygen K_{O_2}, diffusion coefficients for dissolved components, and biomass density, are probably more generally applicable in different activated sludge systems.

3.4.3 Detachment

In the biofilm models established by Wanner and Reichert (1996) a detachment equation is included. However, these equations are not related to any fundamental process. Such an inclusion allows the occurrence of a steady-state biofilm thickness. Since information about the detachment of aerobic granules is very limited, the method established by Wanner and Reichert (1996) is applied to model the granule detachment in our work. Solids are detached from the granules, depending on their relative occurrence at the granule surface. In this model, detachment is described using a global detachment velocity, u_{de}, i.e., the rate at which the granule retracts as a consequence of biomass detachment, allowing detachment kinetics to be a function of time, or other relevant physical quantities. It should be noticed that the specific detachment rate is not constant in the aerobic granulation process.

Two approaches are required to model the detachment and to describe the two distinct phases of the aerobic granulation. The first phase is the biomass growth accompanied with a stable increase in granule size. The second phase is the quasi-steady-state with spontaneous detachment and a relatively constant mean granule size. At the first phase, the biomass detachment rate coefficient k_{d1} is considered to be dependent on the advective velocity u_F, and is described as follows:

$$\frac{dGR}{dt} = u_F * k_{d1} \tag{3.11}$$

where k_{d1} ranges between 0 and 1, and couples the detachment velocity to the current growth rate. This allows the description of a continuous detachment rate at the granule growth phase. k_{d1} is determined as 0.25 for R1 and 0.3 for R2, respectively, through data fitting in the model simulation.

After the biomass growth phase, the granule size GR in both R1 and R2 reach quasi-steady-state with a slight fluctuation (Fig. 3.1). In this case, the fluctuation is modeled using a fixed equilibrium granule size to formulate a random detachment process:

$$\frac{dGR}{dt} = \mu_F * k_{d2} = \mu_F * \frac{GR}{GR_{eq}} \tag{3.12}$$

where k_{d2} is random detachment coefficient, and GR_{eq} is the equilibrium granule size.

3.4.4 Sensitivity Analysis

A sensitivity analysis is performed for the model established in this work through independently altering the parameters used in the model and assessing how much the variation affects the modeled components. This is completed with AQUASIM, and the "absolute–relative" sensitivity function is used as follows:

$$\delta_{yp}^{ar} = p\frac{\partial y}{\partial p} \tag{3.13}$$

where $\delta_{yp}^{\alpha r}$ is the sensitivity analysis output to a model parameter, y is an arbitrary value (set to the same to the actual operation conditions) related to a model variable (e.g., S_S or X_H) calculated with AQUASIM, and p is a model parameter (e.g., $Y_{H,S}$, or K_{STO}). The function measures the absolute change of y for a 100 % change in p.

The sensitivity analysis gives a ranking of the averages of the absolute values of the absolute-relative sensitivity functions $\delta_{yp}^{\alpha r}$ for all model components related to the parameters. The units of $\delta_{yp}^{\alpha r}$ do not depend on the unit of the parameter. This makes quantitative comparisons of the effect of different parameters p on a common model component y possible. For models with many parameters, this ranking is very useful to quickly find out which are the most important parameters influencing the model components in the system.

The analytical results summarized in Table 2.3 indicate that the substrate affinity constant, i.e., biomass affinity constant for S_S and biomass affinity constant for X_{STO}, is the most influential factor. Both kinetic and stoichiometric parameters related to the biomass growth have a considerable effect on the variability of the model, especially on the simulation of biomass content and substrate consumption. The parameters governing the growth of heterotrophs and the production of polymers affect the modeling of total VSS content in the granules. The detachment also has a significant effect on the variability of the model components, but the diffusivity of soluble components has no marked impact.

3.4.5 Appropriateness of the Model with Experimental Results

The model evaluation is performed through comparing the measured and simulated data of the aerobic granulation processes. Various matured granule sizes (0.85–1.5 mm in radius) are used for the model evaluation. In addition, the applied ASM-based model to describe biological reactions occurring within granules in our established model could deal with various substrates with complex compositions. Such an extension enables this developed model to evaluate the granulation processes in different cases. The characterizing parameter of aerobic granules, i.e., mean radius, is used for model evaluation. Stabilization of this parameter value is regarded as the indication of the maturation of aerobic granules.

The model parameters are calibrated according to a logical step-wise procedure at two levels. The calibration procedure includes the long-term dynamic simulations of the evolvement of granule size in the SBR and the reactor performance at steady-state after the mature of the granules. The calibration at each level is not independent, because the processes incorporated in the model are coupled together. Consequently, several iterations with loops to the earlier stages are needed to complete the entire calibration procedure. The standard deviation in parameter calibration is required to be lower than 50 % to ensure the validity and identifiability of the values of the obtained parameters.

Case 1: Aerobic granulation in R1. Experimental results of R1 are used for model evaluation. Figure 3.5a shows both measured and simulated granule mean radius profiles. The measured radius gradually increased from 0.1 to 1.25 mm in the granulation process. The formation of granules was found to experience shaping, developing, and maturating phases. The granule size slightly changed in the initial 7 day (shaping period) and then significantly increased from day 8 to day 25. Attributed to the effects of microbial growth and detachment, the granule size became stabilized with their maturation on day 30. Figure 3.5a shows that the model provides a good description of the granulation process in terms of variation of the mean radius in R1.

The measured OUR and COD concentrations of R1 at steady-state are used for model evaluation. The simulating results are shown in Fig. 3.5a. The OUR and COD profiles resulting from the model simulation under steady-state have less than 15 % difference from the values measured in the experiments. The model is able to appropriately predict the OUR and COD profiles.

Case 2: Aerobic granulation in R2. Experimental results of granule size profiles in R2 are also used for model evaluation (Fig. 3.5b). The stoichiometric and kinetic parameters for the simulation of R2 are given in Table 2.2. In R2 the measured granule mean radius gradually increased from 0.1 to 1.5 mm after 35-day operation. There is a good agreement between the observed and simulated results of the aerobic granulation process in R2. This further confirms that the model developed is able to quantitatively characterize the aerobic granulation process in SBRs.

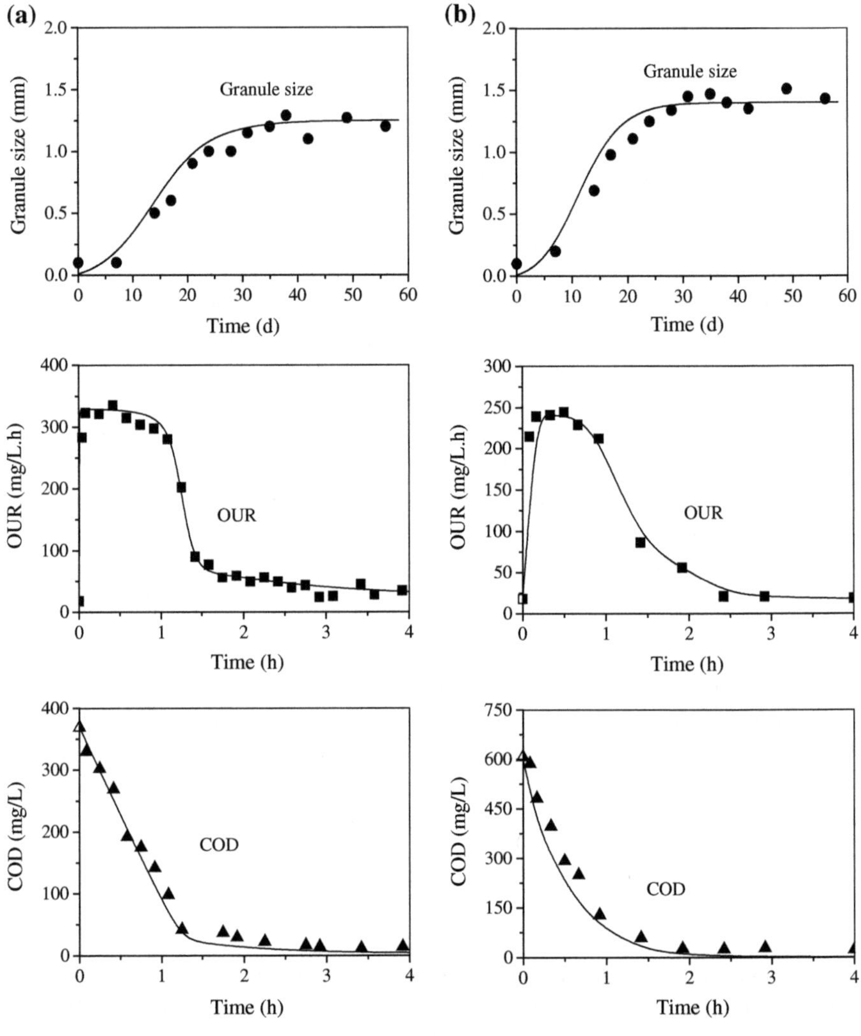

Fig. 3.5 Size evolution of the granules formed, measured concentration profiles of OUR in a steady-state cycle and COD variation in a steady-state of **a** R1 and **b** R2 (Reprinted from Ni et al. (2010) with permission of American Chemical Society)

Similarly, the measured OUR and COD concentrations of R2 at steady-state are also used for model evaluation. Figure 3.5b compares the simulated and experimental results for both OUR and COD profiles. The model simulations match all the experimental measurements well, suggesting the validity of the model developed in this work.

Case 3: Aerobic granulation reported by Yang et al. *(2004).* A series of experiments were conducted by Yang et al. (2004) to evaluate the factors governing the granulation process in an SBR, which was supplied with a synthetic

Table 3.3 Sensitivity analysis: Sens AR represents the sensitivity analysis output to a model parameter to the model variable)

p	Sens AR S_O (mg L^{-1})	Sens AR S_S (mg L^{-1})	Sens AR X_{STO} (mg L^{-1})	Sens AR X_H (mg L^{-1})	Sens AR X_I (mg L^{-1})
Y_{STO}	0.5513	203	387.4	314.1	92.18
$Y_{H,S}$	0.2319	80.14	225.4	131.8	67.4
$Y_{H,STO}$	0.3756	86.17	208	173.2	57.57
k_{STO}	0.2073	10.85	479.5	168.8	117.6
$\mu_{H,S}$	0.5358	32.62	530.9	388.9	288.6
$\mu_{H,STO}$	0.04471	6.286	79.55	27.44	26.91
b_H	0.1441	6.206	109.6	172.2	125.9
b_{STO}	0.03084	0.4107	44.52	7.691	13.54
K_S	25.29	19,850	18,440	15,030	13,760
K_{STO}	0.7943	406.1	529.6	571.2	636
K_{O2}	0.1545	18.18	85.77	87.77	146.7
D_S	0.0004167	0.009949	0.2586	0.2614	0.2949
D_{O2}	0.003769	0.1423	0.782	0.7237	1.086
k_{d1}	0.08509	5.961	51.92	52.49	38.09

sodium acetate-rich wastewater of 1,000 mg COD L^{-1} and had a superficial gas velocity of 2.4 cm s^{-1}. The settling time was gradually decreased from 10 to 2 min. One set of experimental results is used to evaluate the model developed in our work. Since the characteristics of the synthetic wastewater were similar to those of the fatty-acids-rich wastewater used in our work. Thus, the stoichiometric and kinetic parameters for the simulation of R2 shown in Table 3.3 are also used for the simulation of this case study. Figure 3.6a illustrates the evolution of bioparticles from dispersed sludge to mature granules in terms of mean size. The bioparticle size gradually increased over operating time, and finally stabilized. There is a satisfactory agreement between the measured and simulated results (Fig. 3.6a). Again, the model provides a good simulation to the experimental data.

3.4.6 *Aerobic Granules Evolution in Time*

Microbial granulation can be regarded as a self-immobilization community of bacteria. The formation of granules is a multiple-step process, to which physico-chemical and biological forces make a significant contribution. It is proposed that cell immobilization can be roughly described as follows: physical movement initiates bacterium-to-bacterium contact; initial attractive forces keep stable bacteria solid surface and multicellular contact; microbial forces make attached bacteria or aggregated bacteria mature: production of EPS; steady state 3D structure of granule shaped.

The evolution of the three simulated granules from R1, R2, and Yang et al. is presented in Fig. 3.6b. Three phases can be clearly distinguished in the granulation

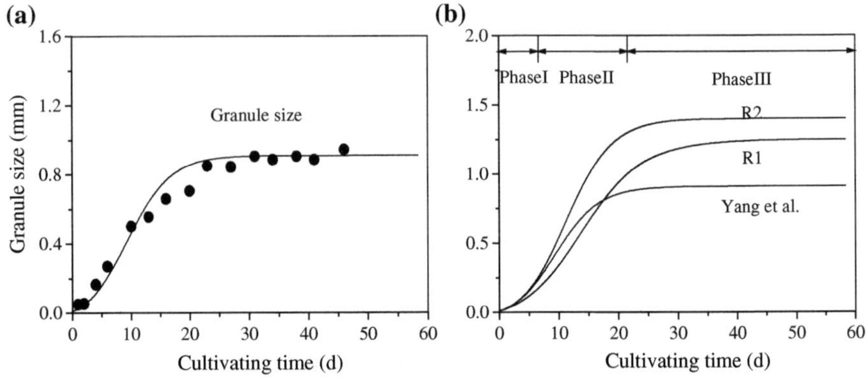

Fig. 3.6 **a** Size evolution of the granules formed in Yang et al. (2004); and **b** three phases in the evolution process of granule size. Model simulation: *solid line* (Reprinted from Ni et al. (2010) with permission of American Chemical Society)

Fig. 3.7 Model simulation of the oxygen penetration profiles as a function of granule radius (r), numbers in the figure indicate cycle time in hours (Reprinted from Ni et al. (2010) with permission of American Chemical Society)

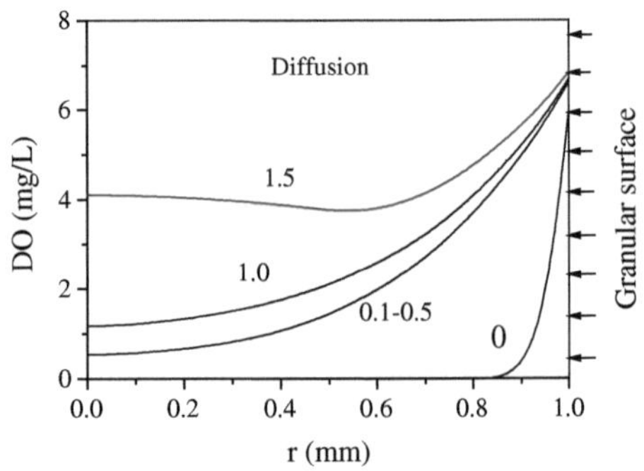

process. At the initial phase, there is no substrate limitation because the granules are small, and microbial growth occurs in the entire granule at a rate close to its maximum. The biomass growth follows an exponential curve (Phase I). There is a lag phase, as seen in Fig. 3.6b. When the granules become larger, there is substrate limitation in the interior granule layers (Ni et al. 2008). The oxygen penetration profiles in the granules are shown in Fig 3.7. In the feast period, the oxygen penetration concentration is very low. As substrate is depleted, the oxygen penetration concentrationre incases. In this case, microbial growth occurs only at an active layer close to the granule–liquid interface. In this mass-transfer-limited regime, the biomass grows linearly (Phase II), and is produced only at a layer with constant thickness. At Phase III, the microbial growth slows down, because biomass decays at the layers below the substrate penetration depth. At all three phases, granule growth is balanced only by superficial erosion. When the balance between growth and detachment processes is achieved, the resultant radius is the equilibrium granule size.

3.5 Conclusions

The sludge particle size gradually increased over the operating time until reaching a stable value during the aerobic granulation process in two SBRs fed with two different wastewaters. The good agreement between the measured and simulated results is obtained for three case studies concerning aerobic granulation in SBRs, confirming that the developed model is able to quantitatively characterize the aerobic granulation process in SBRs. Parameter estimation results of no evident cross-correlation and low 95 % confidence intervals indicate a good identification of the obtained parameter values. Three phases can be clearly distinguished in the aerobic granule formation: initial exponential growth, linear growth afterwards, and final stable phase (equilibrium granule size). The parameters of biomass yield on substrate, maximum growth rate, and substrate half-saturation coefficient should be specifically estimated for different aerobic granular systems when the model is used.

References

APHA: Standard Methods for the Examination of Water and Wastewater, 19th edn. American Public Health Association, Washington (1995)

Avcioglu, E., Karahan, G., Orhon, D.: Estimation of stoichiometric and kinetic coefficients of ASM3 under aerobic and anoxic conditions via respirometry. Water Sci. Technol. **48**, 185–194 (2003)

Beun, J.J., Hendriks, A., van Loosdrecht, M.C.M., Morgenroth, E., Wilderer, P.A., Heijnen, J.J.: Aerobic granulation in a sequencing batch reactor. Water Res. **33**, 2283–2290 (1999)

de Kreuk, M.K., Picioreanu, C., Hosseini, M., Xavier, J.B., van Loosdrecht, M.C.M.: Kinetic model of a granular sludge SBR—Influences on nutrient removal. Biotechnol. Bioeng. **97**, 801–815 (2007)

de Kreuk, M.K., Heijnen, J.J., van Loosdrecht, M.C.M.: Simultaneous COD, nitrogen, and phosphate removal by aerobic granular sludge. Biotechnol. Bioeng. **90**, 761–769 (2005)

Gujer, W., Henze, M., Mino, T., van Loosdrecht, M.C.M.: Activated sludge model NO. 3. Water Sci. Technol. **39**, 183–193 (1999)

Henze, M., Grady, C.P.L. Jr, Gujer, W., Marais, G.V.R., Matsuo, T.: Activated sludge model No. 1. Scientific and Technical Report No. 1. IAWPRC, London (1987)

Jiang, H.L., Tay, J.H., Tay, S.T.L.: Aggregation of immobilized activated sludge cells into aerobically grown microbial granules for the aerobic biodegradation of phenol. Lett. Appl. Microbiol. **35**, 439–445 (2002)

Karahan, O., van Loosdrecht, M.C.M., Orhon, D.: Modeling the utilization of starch by activated sludge for simultaneous substrate storage and microbial growth. Biotechnol. Bioeng. **94**, 43–53 (2006)

Morgenroth, E., Sherden, T., van Loosdrecht, M.C.M., Heijnen, J.J., Wilderer, P.A.: Aerobic granular sludge in a sequencing batch reactor. Water Res. **31**, 3191–3194 (1997)

Mu, Y., Yu, H.Q.: Biological hydrogen production in a UASB reactor with granules I: physicochemical characteristics of hydrogen-producing granules. Biotechnol. Bioeng. **94**, 980–987 (2006)

Ni, B.J., Yu, H.Q., Sun, Y.J.: Modeling simultaneous autotrophic and heterotrophic growth in aerobic granules. Water Res. **42**, 1583–1594 (2008)

Ni, B.J., Sheng, G.P., Li, X.Y., Yu, H.Q.: Quantitative simulation of the granulation process of activated sludge for wastewater treatment. Ind. Eng. Chem. Res. **49**, 2864–2873 (2010)

Nicolella, C., van Loosdrecht, M.C.M., Heijnen, J.J.: Mass transfer and reaction in a biofilm airlift suspension reactor. Chem. Eng. Sci. **53**, 2743–2753 (1998)

Rauch, W., Vanhooren, H., Vanrolleghem, P.A.: A simplified mixed-culture biofilm model. Water Res. **33**, 2148–2162 (1999)

Reichert, P.: Aquasim 2.0-User Manual, Computer Program for the Identification and Simulation of Aquatic Systems. EAWAG, Dübendorf (ISBN 3 906484 16 5) (1998)

Su, K.Z., Yu, H.Q.: Formation and characterization of aerobic granules in a sequencing batch reactor treating soybean-processing wastewater. Environ. Sci. Technol. **39**, 2818–2827 (2005)

Tay, J.H., Liu, Q.S., Liu, Y.: The effects of shear force on the formation, structure and metabolism of aerobic granules. Appl. Microbiol. Biotechnol. **57**, 227–233 (2001)

Wanner, O., Reichert, P.: Mathematical modeling of mixed-culture biofilms. Biotechnol. Bioeng. **49**, 172–184 (1996)

Wu, M.M., Criddle, C.S., Hickey, R.F.: Mass transfer and temperature effects on substrate utilization in brewery granules. Biotechnol. Bioeng. **46**, 465–475 (1995)

Yang, S.F., Liu, Q.S., Tay, J.H., Liu, Y.: Growth kinetics of aerobic granules developed in sequencing batch reactors. Lett. Appl. Microbiol. **38**, 106–112 (2004)

Chapter 4
Autotrophic and Heterotrophic Growth in Aerobic Granular Sludge

A mathematical model has been developed to describe the simultaneous autotrophic and heterotrophic growth in granule-based sequencing batch reactor (SBR). Experimental results of a laboratory-scale granule-based SBR are used to calibrate and validate the model. The model is able to simulate the reactor performance and get insight in autotrophic and heterotrophic growth in the granules. With the established model, the fractions of the active biomass (autotrophs and heterotrophs) and inert biomass are predicted to be 55.6 % and 44.4 % of the total mixed liquid volatile suspended solid, respectively, at a solids retention time (SRT) of 20 days. Biomass content increases with the increasing SRT, but active biomass ratio decreases. Autotrophs have no significant effect on the total biomass content, although they play an important role in nitrogen removal. Simulation results also demonstrate the key role of the influent substrate and $NH4^+$ –N in governing the composition of the heterotrophic and autotrophic biomass in the granule-based SBR. The autotrophs are mainly located on the outer layer of granules, whereas the heterotrophs are present in the center of granules, or on the outer layer of granules.

4.1 Introduction

Aerobic granule has been extensively investigated (Peng et al. 1999; de Kreuk et al. 2005; Su and Yu 2005a). Compared with conventional sludge flocs, aerobic granules have more compact structure, better settleability, and greater biomass retention. Studies show that aerobic granules could be applied for the treatment of high-strength wastewaters, simultaneous removal of organics matters, nitrogen and phosphorus (de Kreuk et al. 2005), and decomposition of toxic wastewaters (Jiang et al. 2002). SBR has been used for granulation of activated sludge (Su and Yu 2005a; de Kreuk and van Loosdrecht 2006). Aerobic-granule-based SBR has been

B.-J. Ni, *Formation, Characterization and Mathematical Modeling of the Aerobic Granular Sludge*, Springer Theses, DOI: 10.1007/978-3-642-31281-6_4,

proven to be applicable for treating wastewaters from various industries, such as malting (Schwarzenbeck et al. 2004), dairy (Arrojo et al. 2004), and soybean processing (Su and Yu 2005a), as well as municipal wastewater (de Kreuk and van Loosdrecht 2006). These results demonstrate that aerobic-granule-based SBR has a great potential in municipal and industrial wastewater treatment.

The architecture of aerobic granules is relevant to their microbial ecology, as they are macroscopic microbial consortia (Arrojo et al. 2004; Su and Yu 2005a). They consist of two main different microbial groups, i.e., autotrophic and heterotrophic bacteria (Su and Yu 2006). Both autotrophs and heterotrophs coexist and interact in aerobic granules. The autotrophs are responsible for nitrification, whereas the heterotrophs are responsible for organic carbon oxidation. Both of them play a vital role in the removal of nitrogen and conversion of a wide diversity of organic matters present in wastewater. The dynamics, composition, and fraction of autotrophs and heterotrophs growing in a bioreactor are governed by its SRT and other factors (Okabe et al. 1996; Grady et al. 1999; Moussa et al. 2005). SRT is an appropriate and powerful design and operating parameter, and regulates the dynamics and activities of autotrophic and heterotrophic biomass population (Okabe et al. 1996; Liao et al. 2006).

Usually, aerobic granules display considerable heterogeneity, with respect to the microorganisms themselves and their physicochemical microenvironment (De Beer et al. 1994; Okabe et al. 1996). The presence of organic matters in wastewater creates competition between the autotrophs and heterotrophs for dissolved oxygen (DO) and space in the aerobic granules (Okabe et al. 1996). In large-sized granules with mixed population, oxygen diffusion limitation creates both aerobic and anoxic microenvironments, allowing sequential utilization of electron acceptors such as oxygen and nitrate. Such interspecies competition and mass transfer result in the stratification of microbial species in granules. The growth rate of autotrophs is considerably lower than that of heterotrophs. Inhibition or elimination of the autotrophs by these interspecies competitions usually leads to a decrease in nitrification efficiency or even to a process failure (Okabe et al. 1996). Thus, an in-depth understanding of the growth of autotrophs and heterotrophs and their activity in aerobic granules is essential for optimizing nutrient removal from wastewater in granule-based reactors.

For biofilms, the competition between autotrophs and heterotrophs for substrates (oxygen and ammonia) and space in biofilm is of major practical importance and thus has been previously studied (Tijhuis et al. 1994; van Benthum et al. 1997; Nogueira et al. 2002a). Tijhuis et al. (1994) and van Benthum et al. (1997) observed that competition in biofilm results in a stratified biofilm structure. The fast growing heterotrophs are located on the outer layers, where both substrate concentration and detachment rate are high, while the slow-growing autotrophs (nitrifying bacteria) stay deeper inside the biofilm. Nogueira et al. (2002b) operated two biofilm reactors at HRTs of 0.8 and 5.0 h, to study the links between population dynamics and reactor operation performance during a shift in process operating from pure nitrification to combined nitrification and organic carbon removal. They found that an increase in HRT resulted in a decrease in the nitrification activity, and these results are not in agreement with those of van Benthum et al. (1997). However, so far little information about the autotrophic and heterotrophic growth in aerobic-granule-based SBRs is available. In addition, the effect of SRT on the active biomass fraction and roles of autotrophs in biomass remain unknown.

Mathematical simulation is a useful tool to evaluate the growth of autotrophs and heterotrophs in activated sludge. Therefore, the main objective of this study is to develop a mathematical model to describe the autotrophic and heterotrophic growth in an aerobic-granule-based SBR. This model is used to simulate the conversion processes occurring in granules, and to elucidate the fraction of autotrophic and heterotrophic biomass in granules. The effect of SRT and influent substrate concentration on the biomass fraction in the systems is also explored. In addition, autotrophic and heterotrophic population distributions in the aerobic granules are also elucidated.

4.2 Materials and Methods

4.2.1 Reactor Set-Up and Operation

Aerobic granules were cultivated in a laboratory-scale SBR, as described in Chap. 3. The desired SRT was set by controlling the amount of sludge wasted from the reactor in each cycle. The reactor was extensively monitored for pH and DO, and sampled for the measurement of COD, storage polymers (X_{STO}), MLSS, MLVSS, NH_4^+–N, and OUR. The criteria for reaching steady state were constancy of the MLSS concentration, and of the dynamic patterns of OUR and pH in repetitive cycles.

4.2.2 Seed Sludge and Wastewater

Sludge from a conventional activated sludge plant in Hefei, China, was used as inoculum. The SBR was fed with a fatty-acids-rich wastewater at COD of approximately 800 mg L^{-1}. The fatty-acids-rich wastewater was the effluent of a lab-scale acidogenic reactor fed with sucrose-rich wastewater (Mu and Yu 2006), as described in Chap. 3.

4.2.3 Parameter Determination

The substrate half-saturation coefficient (K_S) was estimated as described by Henze et al. (1987). An increase in the growth rate with increasing Ss concentration could be achieved in batch experiments. Since that the concentration of substrate in the wastewater was known, the parameter K_S could then be estimated by fitting data with the Monod model. The heterotrophic endogenous respiration coefficient (b_H) and anoxic reduction factor (η_{NOx}) were also estimated following the approach of Henze et al. (1987). Sludge was removed from a completely mixed reactor and put into an aerated and non-fed batch reactor. The endogenous respiration rate was measured at a given time intervals over a period of several days. The plot of the logarithm of the respiration rate versus time visualized the exponential decrease in the biomass as a straight line with a slope of b_H. Moreover, the endogenous utilization rate of X_{STO}, b_{STO}, was taken as the same as the endogenous decay rate of biomass, b_H, using a similar approach in ASM3 (Gujer et al. 1999). The respirometric test, in which two batch reactors (anoxic/aerobic) with the same wastewater and aerobic granule samples were operated in parallel, was used to determine the anoxic reduction factor.

4.2.4 Analytical Procedures

The analysis of DO, OUR, and sludge image was described in Chap. 3. Measurement of COD, NH_4^+–N, MLSS, and MLVSS was conducted according to the Standard Methods (APHA 1995). The internal storage polymers, i.e., polyhydroxyalkanoate (PHA), were measured according to Pratt et al. (2004).

4.3 Results

4.3.1 Reactor Operation

In the initial operating days, the influent COD was stepwise increased from initial 200 to 800 mg L^{-1} on day 15, and correspondingly the loading rate was increased from 0.6 to 2.4 kg COD $m^{-3} \cdot d^{-1}$. The reactor performance was improved continuously in terms of COD removal efficiency (data not shown). Such an improvement became significant after 10 days of operation, and the COD removal efficiency was kept at 95–98 % afterwards. A settling period of 10 min was applied in the first week to prevent severe sludge wash out and was then reduced to 5 min in the subsequent operating period. As a result of the improvement in settling ability, the MLVSS concentration kept increasing despite of the sludge wash out, even after the settling period was reduced to 1 min on day 20. After day 30, the MLVSS became stabilized. More ten days later, sphere- or ellipse-shaped granules became physiologically stabilized. The particle size gradually increased over operating time, and it finally reached its plateau after 40 days of operation.

After 60-day operation, the SRT of the reactor was switched to 20 d. A constant DO and MLVSS concentration, and COD removal in the reactor were the indication of a steady-state operation. Experimental assessment of the autotrophic and heterotrophic growth in the granules was performed with the measurement of concentration profiles of several components, i.e., NH_4^+–N, COD, and other parameters, such as MLVSS and OUR etc., throughout a representative complete cycle in the steady-state operation of each run. Full oxidation of ammonia and COD occurred within 1.5 h, and in the rest of one cycle the storage polymer (X_{STO}) was consumed by the heterotrophs. The total OUR measured was related with the heterotrophic and autotrophic growth activities.

4.3.2 Model Development

The mathematical model for describing the reactor performance is implemented in the well-established AQUASIM simulation software (Reichert 1998). A combination of completely mixed reactor and biofilm reactor compartments (Wanner and Reichert 1996), provided by AQUASIM, is used to simulate the mass transfer and conversion processes occurring in the bulk liquid and granules (Beun et al. 2001; Moussa et al. 2005; de Kreuk et al. 2007). The biological conversion processes are modeled using a modified ASM3 with a consideration of carbon removal, nitrification, and denitrification (Gujer et al. 1999). The established model is calibrated and used to simulate the biological reactions that occur in the SBR, and the simulation results are compared with the experimental data obtained.

4.3.2.1 Biological Reactions

Because of the characteristics of an SBR and the difference between granules and flocs, several modifications are made to the ASM3 to describe the biological reactions occurring in the granules. In ASM3 all readily biodegradable substrate is assumed to be initially stored as internal storage products before it is used for growth at famine phase. However, this is not true in real case, because simultaneous aerobic/anoxic storage and growth on organic substrates by the heterotrophs occur under feast conditions (Krishna and van Loosdrecht 1999). It is especially true for the case of an aerobic-granule-based SBR. Such a system is repeatedly subject to feast and famine conditions (Arrojo et al. 2004; Su and Yu, 2005a). As a result, the simultaneous aerobic/anoxic storage and growth processes always occur in such an aerobic-granule-based SBR (Beun et al. 2001; Su and Yu 2006).

The model proposed to describe the biological reactions in aerobic granules has nine model components or state variables, i.e., heterotrophic microorganisms (X_H), autotrophic microorganisms (X_A), storage products of heterotrophic microorganisms (X_{STO}), readily biodegradable substrate (S_S), residual inert biomass (X_I), ammonia–N (S_{NH4}), nitrate–N (S_{NO3}), nitrogen (S_{N2}), and dissolved oxygen (S_O).

This model basically has 13 microbial processes: heterotrophic processes including aerobic storage, anoxic storage, aerobic growth on S_S, anoxic growth on S_S, aerobic growth on X_{STO}, anoxic growth on X_{STO}, aerobic endogenous respiration, anoxic endogenous respiration, aerobic respiration of X_{STO}, and anoxic respiration of X_{STO}; autotrophic processes including aerobic growth, aerobic endogenous respiration, and anoxic endogenous respiration. Related process kinetics and stoichiometrics to describe the interactions and transformations among model components are expressed compatibly with other activated sludge models proposed previously (Gujer et al. 1999). The structure of the proposed model is presented in a matrix format, reflecting the basic stoichiometric relationships constituting the model backbone. This matrix format is outlined in Tables 4.1 and 4.2, where model components are listed in the upper row of Table 4.1, and the relevant stoichiometric coefficients are also incorporated in appropriate matrix cells of Table 4.1; the process rate expressions are given in Table 4.2. In this way, the changing rate (generation or utilization) of a model component for a given biochemical process can be obtained through multiplication of related process stoichiometrics and kinetics (Gujer and Larsen 1995; Insel et al. 2007).

At the feast phase, because of a pulse substrate dose, the heterotrophs are likely to primarily utilize the readily biodegradable substrate (S_S) available via storage and growth under both aerobic and anoxic conditions. The autotrophs utilize the influent S_{NH_4} for their aerobic growth. Oxygen utilizations associated with the primary aerobic heterotrophic consumption of substrate and aerobic storage of substrate (X_{STO}) are defined by the following process kinetics:

$$\mathrm{OUR}_{\mathrm{grow}S_S} = \mu_{H,S_S} X_H \frac{1 - Y_{H}, S_S}{Y_{H}, S_S} \frac{S_S}{K_S + S_S} \frac{S_O}{K_{O_2} + S_O} \frac{S_{NH_4}}{K_{NH_4} + S_{NH_4}} \tag{4.1}$$

$$\mathrm{OUR}_{STO} = k_{STO} X_H \frac{1 - Y_{STO}}{Y_{STO}} \frac{S_S}{K_S + S_S} \frac{S_O}{K_{O_2} + S_O} \tag{4.2}$$

Utilization of nitrate associated with the primary anoxic heterotrophic consumption of substrate (denitrification) and anoxic storage of substrate (X_{STO}) are described by the following equations:

$$\mathrm{NUR}_{\mathrm{grow}S_S} = \mu_{H,S} \eta_{NOx} X_H \frac{1 - Y_{H,S}}{2.86\, Y_{H,S}} \frac{S_S}{K_S + S_S} \frac{K_{O_2}}{K_{O_2} + S_O} \frac{S_{NO_3}}{K_{NO_3} + S_{NO_3}} \frac{S_{NH_4}}{K_{NH_4} + S_{NH_4}} \tag{4.3}$$

$$\mathrm{NUR}_{STO} = k_{STO} \eta_{NOx} X_H \frac{1 - Y_{STO}}{2.86\, Y_{STO}} \frac{S_S}{K_S + S_S} \frac{K_{O_2}}{K_{O_2} + S_O} \frac{S_{NO_3}}{K_{NO_3} + S_{NO_3}} \tag{4.4}$$

Oxygen utilization attributed to the autotrophic NH_3–N consumption is defined by the equation below:

$$\mathrm{OUR}_{\mathrm{grow}X_A} = \mu_A X_A \frac{1 - Y_A}{Y_A} \frac{S_{NH_4}}{K_{A,NH_4} + S_{NH_4}} \frac{S_O}{K_{A,O_2} + S_O} \tag{4.5}$$

Table 4.1 Stoichiometric matrix for the heterotrophs and autotrophs of the biological models

Component	S_O	S_S	S_{NH_4}	S_{NO_3}	S_{N_2}	X_A	X_H	X_{STO}	X_I
Process	O_2	COD	N	COD	N	COD	COD	COD	COD
Heterotrophic organisms									
Aerobic storage	$-\frac{1-Y_{STO}}{Y_{STO}}$	$-\frac{1}{Y_{STO}}$						1	
Anoxic storage		$-\frac{1}{Y_{STO}}$		$-\frac{1-Y_{STO}}{2.86\,Y_{STO}}$	$\frac{1-Y_{STO}}{2.86\,Y_{STO}}$			1	
Aerobic growth on S_S	$-\frac{1-Y_{H,S}}{Y_{H,S}}$	$-\frac{1}{Y_{H,S}}$	i_{NBM}				1		
Anoxic growth on S_S		$-\frac{1}{Y_{H,S}}$	$-i_{NBM}$	$-\frac{1-Y_{H,S}}{2.86\,Y_{H,S}}$	$\frac{1-Y_{H,S}}{2.86\,Y_{H,S}}$		1		
Aerobic growth on X_{STO}	$-\frac{1-Y_{H,STO}}{Y_{H,STO}}$		$-i_{NBM}$				1	$-\frac{1}{Y_{H,STO}}$	
Anoxic growth on X_{STO}			$-i_{NBM}$	$-\frac{1-Y_{H,STO}}{2.86\,Y_{H,STO}}$	$\frac{1-Y_{H,STO}}{2.86\,Y_{H,STO}}$		1	$-\frac{1}{Y_{H,STO}}$	
Aerobic endogenous respiration	$-(1-f_I)$		$i_{NBM}-f_I i_{NXI}$				-1		f_I
Anoxic endogenous respiration			$i_{NBM}-f_I i_{NXI}$	$-\frac{(1-f_I)}{2.86}$	$\frac{(1-f_I)}{2.86}$		-1		f_I
Aerobic respiration of X_{STO}	-1							-1	
Anoxic respiration of X_{STO}				$-\frac{1}{2.86}$	$\frac{1}{2.86}$			-1	
Autotrophic organisms									
Aerobic growth	$-\frac{4.57-Y_A}{Y_A}$		$-i_{NBM}-\frac{1}{Y_A}$	$\frac{1}{Y_A}$		1			
Aerobic endogenous respiration	$-(1-f_I)$		$i_{NBM}-f_I i_{NXI}$			-1			f_I
Anoxic endogenous respiration			$i_{NBM}-f_I i_{NXI}$	$-\frac{(1-f_I)}{2.86}$	$\frac{(1-f_I)}{2.86}$	-1			f_I

Table 4.2 Kinetics rate expressions for the biological reaction processes

Heterotrophic organisms	
1	$k_{STO}\frac{S_S}{k_S+S_S}\frac{S_O}{k_{O_2}+S_O}X_H$
2	$k_{STO}\eta_{NOx}\frac{S_S}{K_S+S_S}\frac{K_{O_2}}{K_{O_2}+S_O}\frac{S_{NO3}}{K_{NO3}+S_{NO3}}X_H$
3	$\mu_{H,S}\frac{S_S}{K_S+S_S}\frac{S_O}{K_{O_2}+S_O}\frac{S_{NH_4}}{K_{NH_4}+S_{NH_4}}X_H$
4	$\mu_{H,S}\eta_{NOx}\frac{S_S}{K_S+S_S}\frac{K_{O_2}}{K_{O_2}+S_O}\frac{S_{NO3}}{K_{NO3}+S_{NO3}}\frac{S_{NH_4}}{K_{NH_4}+S_{NH_4}}X_H$
5	$\mu_{H,STO}\frac{X_{STO}/X_H}{K_{STO}+X_{STO}/X_H}\frac{K_S}{K_S+S_S}\frac{S_O}{K_{O_2}+S_O}\frac{S_{NH_4}}{K_{NH_4}+S_{NH_4}}X_H$
6	$\mu_{H,STO}\eta_{NOx}\frac{X_{STO}/X_H}{K_{STO}+X_{STO}/X_H}\frac{K_S}{K_S+S_S}\frac{K_{O_2}}{K_{O_2}+S_O}\frac{S_{NO3}}{K_{NO3}+S_{NO3}}\frac{S_{NH_4}}{K_{NH_4}+S_{NH_4}}X_H$
7	$b_H\frac{S_O}{K_{O_2}+S_O}X_H$
8	$b_H\eta_{NOx}\frac{K_{O_2}}{K_{O_2}+S_O}\frac{S_{NO3}}{K_{NO3}+S_{NO3}}X_H$
9	$b_{STO}\frac{S_O}{K_{O_2}+S_O}X_{STO}$
10	$b_{STO}\eta_{NOx}\frac{K_{O_2}}{K_{O_2}+S_O}\frac{S_{NO3}}{K_{NO3}+S_{NO3}}X_{STO}$
Autotrophic organisms	
11	$\mu_A\frac{S_O}{K_{A,O_2}+S_O}\frac{S_{NH_4}}{K_{A,NH_4}+S_{NH_4}}X_A$
12	$b_A\frac{S_O}{K_{A,O_2}+S_O}X_A$
13	$b_A\eta_{NOx}\frac{K_{A,O_2}}{K_{A,O_2}+S_O}\frac{S_{NO3}}{K_{NO3}+S_{NO3}}X_A$

After the consumption of the primary S_S, the secondary heterotrophic growth process on the stored X_{STO} occurs at the famine phase. When one unit of X_{STO} is consumed for growth, its $Y_{H,STO}$ fraction is converted into X_H and the remaining stoichiometric fraction ($1-Y_{H,STO}$) is used for consuming an equivalent amount of oxygen under aerobic conditions or nitrate under anoxic conditions. Oxygen utilization attributed to the secondary aerobic heterotrophic consumption of substrate (X_{STO}) is defined by Eq. 4.6, while nitrate utilization attributed to the secondary anoxic heterotrophic consumption of substrate (X_{STO}) is described by Eq. 4.7.

$$\mathrm{OUR}_{\mathrm{growSTO}} = \mu_{H,STO}X_H\frac{1-Y_{H,STO}}{Y_{H,STO}}\frac{K_S}{K_S+S_S}\frac{X_{STO}/X_D}{K_{STO}+X_{STO}/X_D}\frac{S_{NH_4}}{K_{NH_4}+S_{NH_4}}\frac{S_O}{K_{O_2}+S_O} \tag{4.6}$$

$$\mathrm{NUR}_{\mathrm{growSTO}} = \mu_{H,STO}\eta_{NOx}X_H\frac{1-Y_{H,STO}}{2.86\,Y_{H,STO}}\frac{K_S}{K_S+S_S}\frac{X_{STO}/X_H}{K_{STO}+X_{STO}/X_H}\frac{S_{NH_4}}{K_{NH_4}+S_{NH_4}}\frac{K_{O_2}}{K_{O_2}+S_O}\frac{S_{NO_3}}{K_{NO_3}+S_{NO_3}} \tag{4.7}$$

4.3.2.2 Oxygen transfer

Before reaching the granule surface for diffusion and reaction, oxygen must transfer from gas phase to solid phase. The gas–liquid oxygen transfer rate is assumed to be proportional to the difference in oxygen concentrations between the gas–liquid interface, and the proportionality factor is volumetric oxygen transfer coefficient k_La (Nicolella et al. 1998). On granule surface, oxygen transferred from the gas phase is equal to that diffused into granules, following a mass balance equation below:

$$\frac{\mathrm{d}S_O(t)}{\mathrm{d}t} = k_La\left(S_O^* - S_O(t)\right) \tag{4.8}$$

where S_O^* is the maximum oxygen solubility in liquid phase; S_O is the oxygen concentration on granule surface, equal to that in the bulk liquid, because the liquid–solid oxygen transfer resistance is ignored (Pras 1989; Su and Yu 2005b).

4.3.2.3 Aerobic-Granule-Based SBR

In the AQUASIM program, the volumes of the biofilm compartments have to be invariable (Reichert 1998). For the SBR used in our work, the influent was introduced from the reactor top. At the end of each cycle, the effluent was discharged from a port located at the middle of the reactor height. In order to simulate the fill and discharge processes for the SBR, the methods proposed by Beun et al. (2001) are adopted in this work, and three compartments are defined. The SBR is modeled with a biofilm compartment connected to a completely mixed compartment with an advective link, and a high recirculation flow rate ($Q_{recirculation}$) is incorporated from the biofilm compartment to the completely mixed one to ensure the same concentrations in the liquid of the biofilm compartment and in the completely mixed one (Beun et al. 2001). To simulate the effluent draw, another completely mixed compartment is introduced and connected to the first one with an advective link (Beun et al. 2001). Aeration is simulated by introducing a gas compartment (i.e., completely mixed compartment) connected with the biofilm one with a diffusive link. A schematic drawing of the aerobic-granule-based SBR configuration simulated with AQUASIM is illustrated in Fig. 4.1. The volume of the biofilm compartment accounts for the volumes of all the granules and bulk liquid, and is set at 1 L. The completely mixed compartment has the remaining liquid volume of the reactor, with a maximum volume of 1 L.

4.3.2.4 Model Simulation

The granules grown in the SBR have a size distribution of 0.41–3.5 mm with an average diameter of 1.2 mm. It is true that the granules with a wide range in size are formed in the reactor and that the size distribution keeps changing in time.

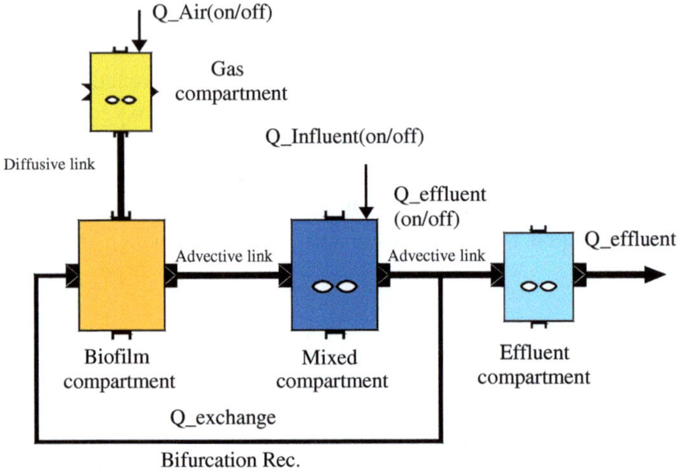

Fig. 4.1 Schematic diagram of the granule-based SBR as implemented in the AQUASIM: a biofilm compartment is connected to a completely mixed compartment with an advective link and a high recirculation flow rate, another completely mixed compartment is connected to the first one with an advective link to simulate the effluent draw, and aeration is simulated by a gas compartment connected with the biofilm one using a diffusive link. Biofilm compartment contains all soluble and particulate components of the biological model (Reprinted from Ni et al. (2008) with permission from Elsevier)

However, the granule size distribution is not taken into account in this model, as it would significantly increase the complexity of numerical computation and it is not expected to contribute to a better understanding of the system. Therefore, in the simulation the diameter is chosen to be 1.2 mm, the average diameter of the granules in this SBR. Simulation is performed under the same operational conditions as applied to the batch tests.

4.3.3 Model Calibration

AQUASIM (Reichert 1998) is used to perform the model calibration for the established model. The parameter values are estimated by minimizing the sum of squares of the deviations between the measured data and predicted values. The objective function to be minimized in the parameter estimation is as follows (Jubany et al. 2005):

$$F(p) = \left(\sum_{i=1}^{n} \left(y_{\exp,i} - y(p)_i \right)^2 \right)^{1/2} \tag{4.9}$$

where y_{exp} and y(p) are vectors of n measured value and model prediction at times t_i (i from 1 to n), and p is the vector of the model parameters.

The initial concentration of active heterotrophic biomass ($X_H(0)$) is estimated by using the baseline endogenous OUR level prior to substrate dose. In the OUR tests allylthiourea was added to inhibit nitrification:

$$OUR_{end-XH}(0) = (1 - f_I) \cdot b_H \cdot X_H(0) \tag{4.10}$$

where f_I is 0.2 g COD g^{-1} COD, as used in ASM3. Thus, for a given f_I and measured b_H, $X_H(0)$ can be calculated from the measured $OUR_{end\text{-}XH}(0)$ values.

Similarly, the initial concentration of active autotrophic biomass ($X_A(0)$) is determined by using the baseline endogenous OUR level for the autotrophs, which is the difference between the total endogenous OUR without allylthiourea dose and the $OUR_{end\text{-}XH}$. For a given f_I and b_A, $X_A(0)$ can be calculated from the $OUR_{end\text{-}XA}(0)$ values.

$$OUR_{end-XA}(0) = (1 - f_I) \cdot b_A \cdot X_A(0) \tag{4.11}$$

The initial X_{STO} concentration, $X_{STO}(0)$, is experimentally measured. Then, the initial value of X_I, $X_I(0)$, is set as the difference between the total VSS and the summation of $X_H(0)$, $X_A(0)$, and $X_{STO}(0)$.

To initiate the calibration procedure, an initial guess of the parameters involved is necessary. Such initial values are obtained on the basis of both measured results and data reported in literatures as shown in Table 4.3. To simplify the calibration process, the model calibration strategy is to change as few constants as possible (Xu and Hultman 1996), because of the limited variability of some parameters. In this study, the model is calibrated for the aerobic-granule-based SBR operated at an SRT of 20 d. Four parameters (i.e., k_{STO}, $\mu_{H,S}$, $\mu_{H,STO}$, and μ_A) are adjusted based on the OUR, X_{STO}, and COD profiles measured in one cycle. To make the calculated concentration profiles fit the measured data, the parameters, k_{STO}, $\mu_{H,S}$, $\mu_{H,STO}$, and μ_A, are changed based on the causality of the parameters to the model outputs.

The estimated parameter values are summarized in Table 4.3. There is a good agreement between the model predictions and experimental data. The maximum heterotrophic growth rate on S_S ($\mu_{H,S}$) is 0.07 h^{-1} after calibration, which is lower than the default value in ASM3, attributed to the partial growth of the heterotrophs on the storage polymers, rather than directly on the soluble substrates. The maximum heterotrophic growth rate on X_{STO} ($\mu_{H,STO}$) is calibrated as 0.15 h^{-1}, whereas the maximum storage rate of XH_H (k_{STO}) and maximum growth rate of X_A (μ_A) are determined to be 0.25 h^{-1} and 0.05 h^{-1}, respectively.

4.3.4 Model Verification

The model verification is based on the comparison between the experimental results and model predictions with the same input model parameters. Experimental data of the OUR, NH_4^+–N, COD, and VSS concentrations, which are not used for model calibration, are employed for model verification. The simulating results are

Table 4.3 Kinetic and stoichiometric coefficients for the optimum calibration (20 °C)

Parameter	Definition	Values	Unit	Source
Stoichiometry				
Y_{STO}	Yield coefficient for X_H storage	0.81	g COD g^{-1} COD	Sin et al. (2005)
Y_S	Yield for X_H growth on S_S	0.58	g COD g^{-1} COD	Sin et al. (2005)
Y_{STO}	Yield for X_H growth on X_{STO}	0.68	g COD g^{-1} COD	Sin et al. (2005)
Y_A	Yield coefficient for X_A growth	0.24	g COD g^{-1} COD	Gujer et al. (1999)
η_{NOx}	Anoxic reduction factor	0.55	–	Measured
f_I	Fraction of X_I in respiration	0.2	g COD g^{-1} COD	Gujer et al. (1999)
i_{NBM}	Nitrogen content of biomass	0.07	g N g^{-1} COD	Gujer et al. (1999)
i_{NXI}	Nitrogen content of X_I	0.02	g N g^{-1} COD	Gujer et al. (1999)
Kinetics				
k_{STO}	Maximum storage rate of X_H	0.25	h^{-1}	Estimated
μ_{STO}	Growth rate of X_H on X_{STO}	0.15	h^{-1}	Estimated
μ_S	Growth rate of X_H on S_S	0.07	h^{-1}	Estimated
K_S	Biomass affinity constant for S_S	11.38	g COD m^{-3}	Measured
K_{STO}	Affinity constant for X_{STO}	1.0	g COD g^{-1} COD	Gujer et al. (1999)
K_{O_2}	Dissolve oxygen affinity constant	0.20	$g_{O_2}m^{-3}$	Gujer et al. (1999)
b_{STO}	Respiration rate of X_{STO}	0.016	h^{-1}	Measured
b_H	Respiration rate coefficient of X_H	0.016	h^{-1}	Measured
μ_A	Maximum growth rate of X_A	0.05	h^{-1}	Estimated
b_A	Respiration rate coefficient of X_A	0.0063	h^{-1}	Gujer et al. (1999)
K_{A,O_2}	Oxygen affinity constant for X_A	0.50	g_{O_2} m^{-3}	Gujer et al. (1999)
K_{A,NH_4}	Biomass NH_4 affinity constant	0.01	g N m^{-3}	Gujer et al. (1999)
K_{A,NH_4}	NH_4 affinity constant for X_A	1.0	g N m^{-3}	Gujer et al. (1999)
K_{A,NO_3}	Biomass NO_3 affinity constant	0.50	g N m^{-3}	Gujer et al. (1999)

shown in Fig. 4.2. The model predicts the OUR, VSS, and NH_4^+–N profiles in one typical cycle and the COD data at steady-state well. Attributed to the initial rapid storage and consumption of substrate by the heterotrophs, the COD is significantly reduced at the initial stage of an operating cycle. The NH_4^+–N consumption exhibits a lag time, attributed to a relatively low oxidation rate of the autotrophs. The VSS profiles show a slight ascendant trend because of the growth and endogenous decay of both heterotrophs and autotrophs. The model is able to simulate all these variation trends well. The good agreement between the measured and predicted results suggests the validity of the model established in this work.

4.4 Discussion

4.4.1 Autotrophic and Heterotrophic Activities

The OUR fractions for the heterotrophs and autotrophs in one typical cycle of the aerobic-granule-based SBR are shown in Fig. 4.3a. The maximum total OUR of 350 mg O_2 L^{-1} h^{-1} is controlled by the combined oxygen uptake capacity of both

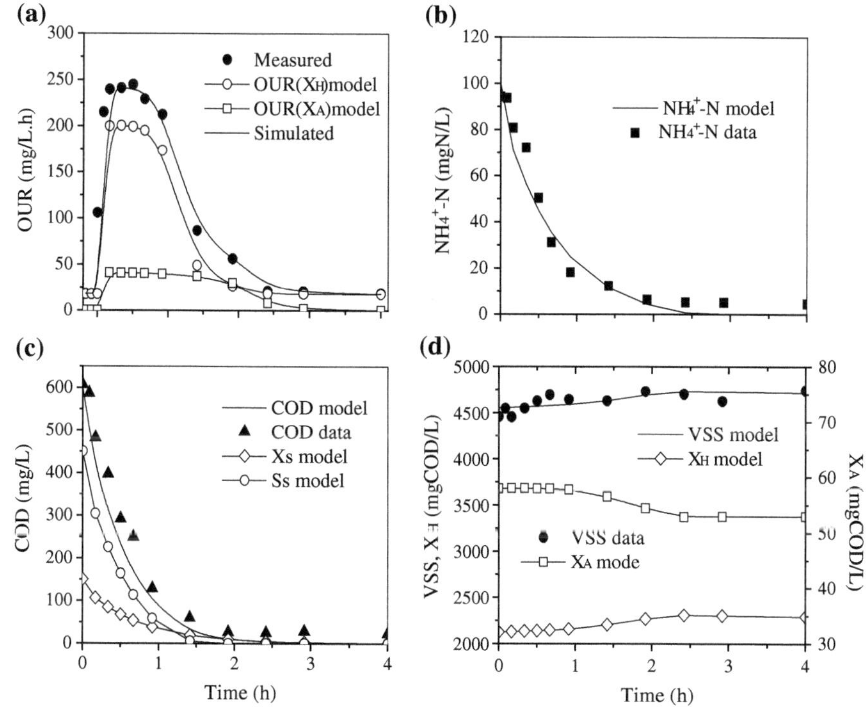

Fig. 4.2 Model validation: the measured OUR (**a**), COD (**b**), NH_4^+-N (**c**), and VSS (**d**) profiles in the steady-state operation of the granule-based SBR at an SRT of 20 d and those predicted by the established model, as well as the model simulations for the corresponding variation of OUR for X_H, OUR for X_A, and model components of X_S, S_S, X_H, and X_A (Reprinted from Ni et al. (2008) with permission from Elsevier)

heterotrophs and autotrophs. The maximum OURs for the autotrophs and heterotrophs are approximately 75 and 275 mg O_2 L^{-1} h^{-1}, respectively. However, these maximum rates are obtained in the different periods of the aeration phase. In addition, the heterotrophs account for 61 % of the total oxygen consumption of the reactor, while the autotrophs consume only 39 %.

Since denitrification can occur in the granule interior (Beun et al. 2001; Su and Yu 2006), special attention is paid to the denitrification processes in a granule-based SBR. The simulation is performed by setting adequate nitrate, instead of DO, in the input. The nitrate consumption profiles (NUR) of the different denitrification processes are calculated and shown in Fig. 4.3b. A relatively high amount of substrate is used for the primary anoxic growth, but a relatively low amount of substrate is utilized for anoxic storage. Similar to their aerobic counterparts, the two anoxic denitrification processes compete with each other for the electron donor (i.e., organic substrate) as well as electron acceptor (i.e., nitrate). The anoxic heterotrophic endogenous respiration process contributes to the main nitrate consumption at the famine phase.

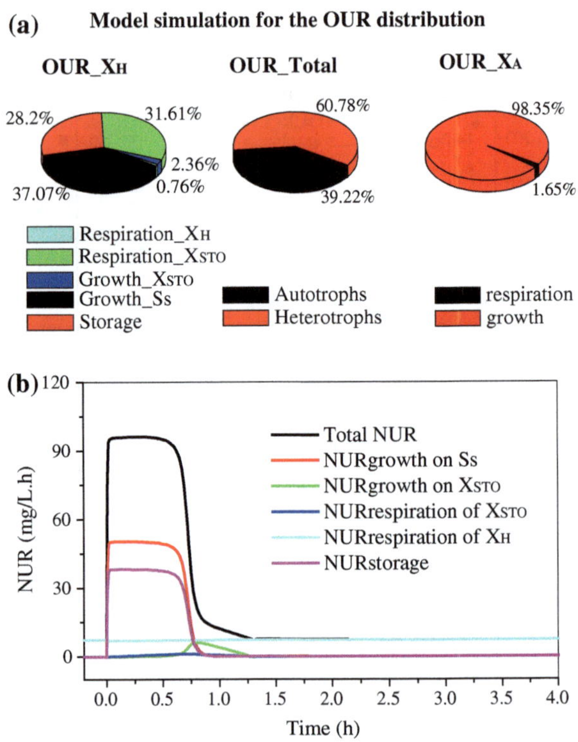

Fig. 4.3 Model simulation results for: **a** OUR fractions for the heterotrophs and autotrophs in one typical cycle of the granule-based SBR; and **b** nitrate consumption profiles (NUR) of different denitrification processes for the denitrifying heterotrophs (Reprinted from Ni et al. (2008) with permission from Elsevier)

Simulation results also show that storage polymers can be used as the carbon source in the denitrification when no external carbon source is available. Short- and long-term effects of decreased oxygen concentration on an aerobic granule reactor were experimentally studied by Mosquera-Corral et al. (2005). They found that the internal carbon source was degraded much slower than the soluble substrate and it could be used as the electron donor for denitrification when no external substrate is available. In particular, storage polymers stored in bacteria were located deeper in layers of granules where almost no oxygen existed, and then could readily serve as a carbon source for denitrification. This result also validates the findings of our work.

4.4.2 Autotrophic and Heterotrophic Biomass Fraction

The established model is also used to predict the effects of influent COD and NH_4^+–N concentration on the fraction of heterotrophs and autotrophs in aerobic granules at different SRTs under steady-state conditions (Fig. 4.4). The influent COD significantly influences the total heterotrophic and autotrophic biomass. The fraction of the heterotrophs increases considerably as a consequence of increasing

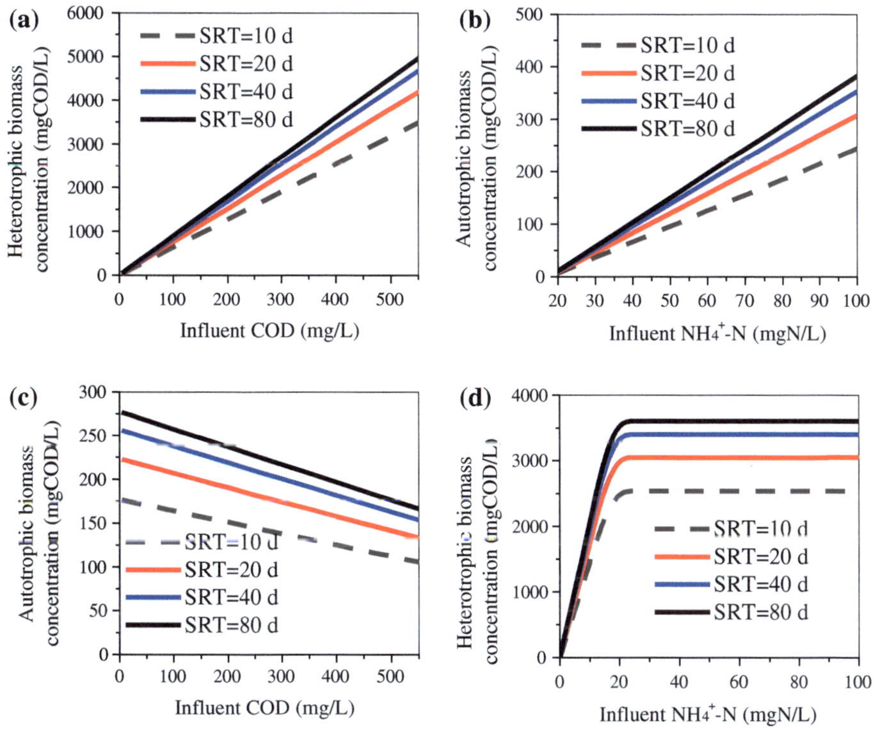

Fig. 4.4 Simulated effect of influent COD and NH_3-N concentrations on the content of heterotrophic and autotrophic biomass fraction in the granule-based SBR at different SRTs under steady-state conditions (Reprinted from Ni et al. (2008) with permission from Elsevier)

influent COD, but the fraction of the autotrophs decreases slowly (Figs. 4.4a and b). As the SRT is increased from 10 to 80 d, the heterotrophs and autotrophs increase by 42 % and 57 %, respectively (Fig. 4.4c). The influent NH_4^+–N concentration also has a significant effect on the composition of the heterotrophs and autotrophs in the granules (Fig. 4.4d).

The effects of influent NH_4^+–N concentration and SRT to the autotrophs are similar to those of influent COD and SRT to the heterotrophs. However, the effects of influent NH_4^+–N concentration and SRT to the heterotrophs are different. The heterotrophic biomass increases rapidly with the increasing influent NH_4^+–N concentration toward a constant level. At a fixed influent COD, the need of the heterotrophs for NH_4^+–N remains unchanged. These results demonstrate the key role of the input COD and NH_4^+–N in governing the composition of heterotrophic and autotrophic biomass in an aerobic-granule-based SBR. These simulation results are in accord with the experimental results of Liu et al. (2003), who observed that the relative abundance of nitrifying bacteria over heterotrophs in aerobic granules was closely related to the substrate N/COD ratio. The populations of both ammonia- and nitrite-oxidizers were significantly enriched with an

Table 4.4 Simulation results for the effect of the autotrophs on the active fraction of biomass in the SBR operated at an SRT of 20 and 50 d

Parameter	Unit	SRT = 20 d		SRT = 50 d	
		Absence	Presence	Absence	Presence
VSS	mg COD L^{-1}	5,491	5,637	10,365	10,686
X_H	mg COD L^{-1}	3,051	3,051	3,477	3,477
	%	55.56	54.12	33.55	32.54
X_A	mg COD L^{-1}	100	158	100	187
	%	1.82	2.81	0.96	1.75
X_I	mg COD L^{-1}	2,340	2428	6,788	7,022
	%	42.62	43.07	65.49	65.71
Active bacteria	%	55.56	56.93	33.55	34.29
Dead fraction	%	44.44	43.07	66.45	65.71

increase in the N/COD ratio, and a decreasing trend of heterotrophic population in the aerobic granules was observed. This observation confirms that a high N/COD ratio favors the selection of the autotrophs (i.e., ammonia- and nitrite-oxidation bacteria) in the aerobic granules.

4.4.3 Roles of Autotrophs in the Granule-Based SBR

The roles of the autotrophs in a granule-based SBR system are evaluated through simulation under similar operating conditions for the SBR, but all the autotrophic processes are switched off. The simulation results in the presence and absence of the autotrophs are shown in Table 4.4.

No significant changes in the total MLVSS and active biomass fraction are observed for the SBR in the presence or absence of the autotrophs. This could be attributed to the fact that the main fraction of the total biomass is particulate inert (e.g., 65.71 % at an SRT of 50 d) and heterotrophs (e.g., 32.54 % at an SRT of 50 d) in the aerobic granules. The autotrophs have two special growth characteristics. First, their maximum specific growth rate, 0.032 h^{-1}, is much smaller than that of the heterotrophs (0.083 h^{-1}). Secondly, their yield is small (0.24 g COD g^{-1} N). When 1 mg of NH_4^+–N is removed, only 0.129 mg of bacterial biomass is yielded. As a result, the autotrophs have no significant effect on the total biomass content (MLVSS), although they play an important role in nitrogen removal in an SBR. Xavier et al. (2007) established a multi-scale model to describe the complex dynamics of populations and microbial composition in the aerobic granules in an SBR. Their simulation results show that the fraction of total autotrophs was only about 1 % of the total biomass. This value is in agreement with the result of our work.

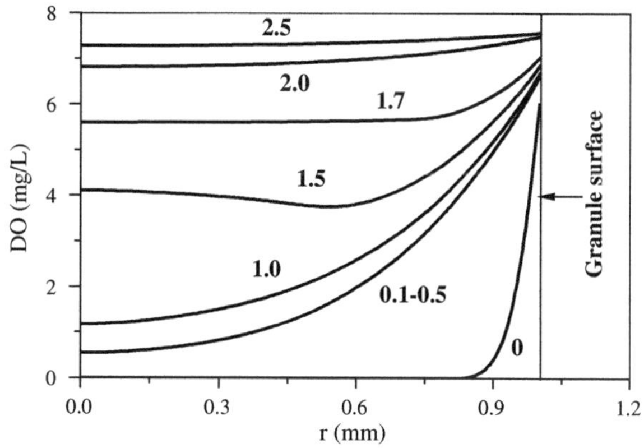

Fig. 4.5 Model simulation of O_2 penetration profiles as a function of the radius (*r*) of the aerobic granules in the SBR. Numbers in the figure indicate cycle time in hours (Reprinted from Ni et al. (2008) with permission from Elsevier)

4.4.4 Microbial Population Distribution in the Aerobic Granules

The DO has a substantial effect on the microbial population distribution in aerobic granules. Oxygen penetration depth in aerobic granules plays a crucial role in the conversion rates of different components and thus on overall nutrient removal efficiency (de Kreuk et al. 2007). Competition between the autotrophs and heterotrophs for DO and space in aerobic granules occurs when DO is not of a sufficient level. For the autotrophs, this is disadvantageous, because they have to grow under oxygen-diffusion-limitation conditions, which slows the nitrification process (van Benthum et al. 1997). The autotrophs can successfully compete for DO and space with the heterotrophs on the outer layers of granules only at a high S_O and NH_4^+–N concentrations because of their lower growth rate. In a discontinuously-fed reactor like aerobic-granule-based SBR, organic substrate is able to completely penetrate into granules because of the temporarily high substrate concentration in the liquid, but oxygen is present only on the outer layers of granules. The simulated oxygen penetration profiles in aerobic granules are shown in Fig. 4.5. At the feast phase (t = 0–0.5 h), the penetrated oxygen concentration is very small. As substrate is depleted, the calculated oxygen penetration concentration is 1.2 mg L^{-1}, as long as X_{STO} is present in the reactor. In this case, oxygen is used for the secondary growth. After X_{STO} becomes depleted, oxygen is able to completely penetrate into the granules.

Consequently, it is expected that the autotrophs are mainly located on the outer layer of aerobic granules to meet their basic requirement for DO to convert NH_4^+–N, and organic substrate is partially stored as X_{STO} by the heterotrophs inside granules. In contrast, the location of the heterotrophs in aerobic granules is less restricted. They can occupy the granule center, where they grow with organic substrate at the feast phase, using NO_3^-–N as the electron acceptor. They can also grow on the outer layers of granules where they use DO as an electron acceptor. These results are in accordance with the FISH analysis in our work (data not shown), in which the spatial distribution of the autotrophs (the ammonia- and nitrite-oxidation bacteria) in the aerobic granules is observed. The ammonia-oxidation bacteria were mainly located on the outer layer of granules, while the nitrite-oxidation bacteria were close to the ammonia-oxidation bacteria zone under the outer layer of granules.

These simulation results are somewhat opposite to those for a biofilm reactor reported by Nogueira et al. (2002). They found that the heterotrophs were mainly located on the outer layers, and that the autotrophs (nitrifying bacteria) grew in deeper inside the biofilm. However, after the reactor was operated in a shift from pure nitrification to combined nitrification and organic carbon removal, the growth of heterotrophs occurred on the layers where the autotrophs occupied. This finding is significantly different from our observation. The experimental and modeling work of Okabe et al. (1996) also confirmed such stratification in biofilms of a continuously fed reactor. They observed that the most of nitrifiers were located on the deeper layers of biofilm, whereas the heterotrophs dominated in the outermost layers. In fact, in the granule-based SBR the growth of the heterotrophs and autotrophs is simultaneous. Furthermore, because of the discontinuously-fed mode of the granule-based SBR, the organic substrate is able to completely penetrate into granules, but oxygen is available only on the outer layers of granules at feast phase. As a result, the autotrophs have to grow only on the outer layers of the aerobic granules to meet their essential requirements for DO. Beun et al. (2001) also reported similar results for microbial population distribution in aerobic granules. Their simulation shows that at a constant DO of 40 % the microbial distribution in the granules was significantly affected by DO. At a constant DO of 40 % the autotrophic biomass was located more in the outermost (i.e., aerobic) layer of the granules, compared with at a DO of 100 %. This is because the autotrophic biomass needed oxygen for the conversion of NH_4^+–N. The heterotrophic biomass could use both oxygen and NO_3^-–N and therefore were also present in the center of granules. More recently, through a simulation of the biomass distribution in aerobic granules and the oxygen penetration depths, de Kreuk et al. (2007) and Xavier et al. (2007) found that the increased oxygen penetration depth during the total aeration period (DO 10 mg L^{-1}) reduced the competition for oxygen between the autotrophs and heterotrophs in the outer layer, creating the possibility for the autotrophs to accumulate in the inner zones of granules.

4.5 Conclusions

The ASM3 is modified with a consideration of carbon removal, nitrification, and denitrification to describe the simultaneous autotrophic and heterotrophic dynamic growth in an aerobic-granule-based SBR. The model calibration and validation win a good agreement between the experimental and simulation results, demonstrating its capacity of elucidating the autotrophic and heterotrophic processes in the granule-based SBR. The heterotrophs account for major oxygen consumption than the autotrophs. More substrate is used for the primary anoxic growth, while a low amount of substrate is diverted to anoxic storage in the denitrifying processes. Biomass content increases with increasing SRT, but the active biomass ratio decreases. The autotrophs have no significant effect on the total biomass content, despite of their important role in nitrogen removal. The fraction of the heterotrophs considerably increases as a consequence of increasing influent COD, but the fraction of the autotrophs decreases slowly. The autotrophs are mainly located on the outer layers of granules for DO consumption, whereas the heterotrophs occupy the granule center or on the outer layers.

References

APHA.: Standard methods for the examination of water and wastewater. 19th ed. American Public Health Association, Washington (1995)

Arrojo, B., Mosquera-Corral, A., Garrido, J.M., Mendez, R.: Aerobic granulation with industrial wastewater in sequencing batch reactors. Water Res. **38**, 3389–3399 (2004)

Beun, J.J., Heijnen, J.J., van Loosdrecht, M.C.M.: N-removal in a granular sludge sequencing batch airlift reactor. Biotechnol. Bioeng. **75**, 82–92 (2001)

de Beer, D., Stoodley, P., Lewandowski, Z.: Liquid flow in heterogeneous biofilms. Biotechnol. Bioeng. **44**, 636–641 (1994)

de Kreuk, M.K., Heijnen, J.J., van Loosdrecht, M.C.M.: Simultaneous COD, nitrogen, and phosphate removal by aerobic granular sludge. Biotechnol. Bioeng. **90**, 761–769 (2005)

de Kreuk, M.K., Picioreanu, C., Hosseini, M., Xavier, J.B., van Loosdrecht, M.C.M.: Kinetic model of a granular sludge SBR—Influences on nutrient removal. Biotechnol. Bioeng. **97**, 801–815 (2007)

de Kreuk, M.K., van Loosdrecht, M.C.M.: Formation of aerobic granules with domestic sewage. J Environ Engin **132**, 694–697 (2006)

Grady, Jr, C.P.L., Glen, T.D., Henry, C.L.: Biological wastewater treatment, 2nd edn, Marcel Dekker, Inc (1999)

Gujer, W., Henze, M., Mino, T., van Loosdrecht, M.C.M.: Activated sludge model No. 3. Water Sci. Technol. **39**, 183–193 (1999)

Gujer, W., Larsen, T.A.: The implementation of biokinetics and conservation principles in ASIM. Water Sci. Technol. **31**, 257–266 (1995)

Henze, M., Grady, C.P.L., Gujer, W., Marais, G.V.R., Matsuo, T.: Activated Sludge Model No, 1, vol. **1**. IAWPRC, London (1987)

Insel, G., Celikyilmaz, G., Ucisik-Akkaya, E., Yesiladali, K., Cakar, Z.P., Tamerler, C., Orhon, D.: Respirometric evaluation and modeling of glucose utilization by *Escherichia coli* under aerobic and mesophilic cultivation conditions. Biotechnol. Bioeng. **96**, 94–105 (2007)

Jiang, H.L., Tay, J.H., Tay, S.T.L.: Aggregation of immobilized activated sludge cells into aerobically grown microbial granules for the aerobic biodegradation of phenol. Lett. Appl. Microbiol. **35**, 439–445 (2002)

Jubany, I., Baeza, J.A., Carrera, J., Lafuente, J.. Respirometric calibration and validation of a biological nitrite oxidation model including biomass growth and substrate inhibition. Water Res. **39**, 4574–4584 (2005)

Krishna, C., van Loosdrecht, M.C.M.: Substrate flux into storage and growth in relation to activated sludge modelling. Water Res. **33**, 3149–3161 (1999)

Liao, B.Q., Droppo, I.G., Leppard, G.G., Liss, S.N.: Effect of solids retention time on structure and characteristics of sludge flocs in sequencing batch reactors. Water Res. **40**, 2583–2591 (2006)

Liu, Y., Yang, S.F., Tay, J.H.: Elemental compositions and characteristics of aerobic granules cultivated at different substrate N/C ratios. Appl. Microbiol. Biotechnol. **61**, 556–561 (2003)

Moussa, M.S., Hooijmans, C.M., Lubberding, H.J., Gijzen, H.J., van Loosdrecht, M.C.M.: Modelling nitrification, heterotrophic growth and predation in activated sludge. Water Res. **39**, 5080–5098 (2005)

Mosquera-Corral, A., de Kreuk, M.K., Heijnen, J.J., van Loosdrecht, M.C.M.: Effects of oxygen concentration on N-removal in an aerobic granular sludge reactor. Water Res. **39**, 2676–2686 (2005)

Mu, Y., Yu, H.Q.: Biological hydrogen production in a UASB reactor with granules I: Physicochemical characteristics of hydrogen-producing granules. Biotechnol. Bioeng. **94**, 980–987 (2006)

Nogueira, R., Melo, L.F.: Competition between *Nitrospira* spp. and *Nitrobacter* spp. in nitrite-oxidizing bioreactors. Biotechnol. Bioengin. **95**, 169–175 (2006)

Nogueira, R., Melo, L.F., Pulrkhold, U., Wuertz, S., Wagner, M.: Nitrifying and heterotrophic population dynamics in biofilm reactors: effects of hydraulic retention time and the presence of organic carbon. Water Res. **36**, 469–481 (2002)

Ni, B.J., Yu, H.Q., Sun, Y.J.: Modeling simultaneous autotrophic and heterotrophic growth in aerobic granules. Water Res. **42**, 1583–1594 (2008)

Nicolella, C., van Loosdrecht, M.C.M., Heijnen, J.J.: Mass transfer and reaction in a biofilm airlift suspension reactor. Chem. Eng. Sci. **53**, 2743–2753 (1998)

Okabe, S., Hiratia, K., Ozawa, Y., Watanabe, Y.: Spatial microbial distributions of nitrifiers and heterotrophs in mixed-population biofilms. Biotechnol. Bioeng. **50**, 24–35 (1996)

Peng, D., Bernet, N., Delgenes, J.P., Moletta, R.: Aerobic granular sludge-a case study. Water Res. **33**, 890–893 (1999)

Pras, N.: Further kinetic characterization of alginate-entrapped cells of *Mucuna pruriens* L. Biotechnol. Bioeng. **33**, 1461–1468 (1989)

Pratt, S., Yuan, Z., Keller, J.: Modelling aerobic carbon oxidation and storage by integrating respirometric, titrimetric, and off-gas CO_2 measurements. Biotechnol. Bioeng. **88**, 135–147 (2004)

Reichert, P.: AQUASIM 2.0—User manual, computer program for the identification and simulation of aquatic systems. Swiss Federal Institute for Environmental Science and Technology (EAWAG) (1998)

Schwarzenbeck, N., Erley, R., Mc Swain, B.S., Wilderer, P.A., Irvine, R.L.: Treatment of malting wastewater in a granular sludge sequencing batch reactor (SBR). Acta Hydrochim. Hydrobiol. **32**, 16–24 (2004)

Sin, G., Guisasola A., De Pauw, D.J.W., Baeza, J.A., Carrera, J., Vanrolleghem, P.A.: A new approach for modelling simultaneous storage and growth processes for activated sludge systems under aerobic conditions. Biotechnol. Bioeng. **92**, 600–613 (2005)

Su, K.Z., Yu, H.Q.: Formation and characterization of aerobic granules in a sequencing batch reactor treating soybean-processing wastewater. Environ. Sci. Technol. **39**, 2818–2827 (2005a)

Su, K.Z., Yu, H.Q.: Gas holdup and oxygen transfer in an aerobic granule-based sequencing batch reactor. Biochem. Eng. J. **25**, 209–215 (2005b)

Su, K.Z., Yu, H.Q.: A generalized model for aerobic granule-based sequencing batch reactor. 1. Model development. Environ Sci Technol **40**, 4703–4708 (2006)

Tijhuis, L., Rekswinkel, H.G., van Loosdrecht, M.C.M., Heijnen, J.J.: Dynamics of population and biofilm structure in the biofilmairlift suspension reactor for carbon and nitrogen removal. Water Sci. Technol. **29**, 377–84 (1994)

van Benthum, W.A.J., van Loosdrecht, M.C.M., Heijnen, J.J.: Control of heterotrophic layer formation on nitrifying biofilms in a biofilm airlift suspension reactor. Biotechnol. Bioeng. **53**, 397–405 (1997)

Wanner, O., Reichert, P.: Mathematical modeling of mixed-culture biofilms. Biotechnol. Bioeng. **49**, 172–184 (1996)

Xavier, J.B., de Kreuk, M.K., van Loosdrecht, M.C.M.: Multi-scale individual-based model of microbial and bioconversion dynamics in aerobic granular sludge. Environ. Sci. Technol. **41**, 6410–6417 (2007). Picioreanu

Xu, S.L., Hultman, B.: Experiences in wastewater characterization and model calibration for the activated sludge process. Water Sci. Technol. **33**, 89–98 (1996)

Chapter 5
Thermodynamic Analysis of Wastewater Treatment by Aerobic Granules

A bioenergetic methodology was integrated with a modified activated sludge model No. 1 (ASM1) to analyze the activated sludge process, with the treatment of soybean-processing wastewater as an example. With the bioenergetic methodology established by McCarty and coworkers, the microbial yield was predicted and the overall stoichiometrics for biological reactions involving the key chemical and biological species in activated sludge were established. These obtained parameters were related to the ASM1 model, which was modified after coupling the biological reactions in activated sludge with electron balances. This approach was able to approximately describe the treatment of soybean wastewater by activated sludge in a sequencing batch reactor in terms of substrate utilization, biomass growth, and the electron acceptor consumption. Such an attempt provides useful information for accurate modeling of the complex activated sludge process.

5.1 Introduction

The activated sludge model No. 1 (ASM1), established by the International Water Association, was a basis for modeling of activated-sludge-based wastewater treatment systems (Henze et al. 2000). In this model, the wide variety of organic carbon and nitrogenous compounds are subdivided into a limited number of fractions based on biodegradability and solubility considerations. The ASM is being widely used to design and assess control strategies for various activated sludge systems (Keesman et al. 1998; Grady et al. 1999; Keesman and Spanjers 2000; Smets et al. 2003; Su and Yu 2006).

Accurate microbiological modeling requires evaluation of the effects of biological reactions on all important chemical and biological species in the activated sludge system (van Briesen and Rittmann 2000). The ASM requires a large number of empirical parameters, which are difficult to determine. Moreover, many

B.-J. Ni, *Formation, Characterization and Mathematical Modeling of the Aerobic Granular Sludge*, Springer Theses, DOI: 10.1007/978-3-642-31281-6_5,

important chemical and biological species are lumped into one assumed model component or one empirical process (Droste 1998; Henze et al. 2000). To sort out these problems, a thermodynamic analysis of the biological synthesis in activated sludge could be performed and the stoichiometrics could be more accurately estimated with the cell yield derived from thermodynamic considerations of the flows of energy and electrons in the catabolic and anabolic pathways. The microbial yield is fundamental to predicting the overall stoichiometrics of the growth/utilization reactions in microbial systems (van Briesen and Rittmann 2000). Since 1960s, McCarty had established a bioenergetic methodology to determine the energy and electron balances in microbial degradation systems such as activated sludge (McCarty 1965, 1971, 1972). This methodology has been found to be useful for designing and understanding the overall process in biological wastewater treatment systems (Droste 1998; van Briesen and Rittmann 2000).

In the present work, the bioenergetic methodology established by McCarty was integrated with a modified ASM1 to formulate a new approach to analyze the activated sludge process. The microbial yield was estimated and the overall stoichiometrics for the key chemical and biological species in the system were established by using the bioenergetic method. Then, they were incorporated into the ASM1 which was modified after coupling the biological reactions with electron balances. The performance of an SBR for the treatment of soybean wastewater was simulated with this established approach. The dynamic profiles of chemical species in the reactor were also measured and simulated.

5.2 Model Development

5.2.1 Bioenergetic Methodology

In the bioenergetic methodology established by McCarty (McCarty 1965, 1971, 1972), for mineralization reactions, all the electrons present in the electron-donor substrate were sent either to the acceptor (f_e) or to biomass synthesis (f_s) as shown in Fig. 5.1. The values of f_e and f_s were coupled through the electron balance, i.e., the sum of f_e and f_s must be equal to 1:

$$f_e + f_s = 1 \tag{5.1}$$

$$A = \frac{f_e}{f_s} = -\frac{\frac{\Delta G_p^{0'}}{K^m} + \Delta G_{\text{cells}}^{0'} + \frac{\Delta G_n^{0'}}{K}}{K \Delta G_R^{0'}} \tag{5.2}$$

$$f_s = \frac{1}{1+A} \quad \text{and} \quad f_e = \frac{A}{1+A} \tag{5.3}$$

in which $\Delta G_R^{0'}$ is the standard free energy of the energy-generating reaction; K is the efficiency of energy capture in the energy-generation reaction; f_e is the fraction of electron-donor electron equivalents sent to the acceptor to drive the energy-generating reaction; and f_s is the fraction of electron-donor electron equivalents invested in biomass via the synthesis reaction. The ratio of f_e/f_s could be defined as A. $\Delta G_p^{0'}$ is the energy consumed to transform the carbon source to pyruvate, $\Delta G_n^{0'}$ is the energy consumed to transform the nitrogen source to NH_4^+, and $\Delta G_{\text{cells}}^{0'}$ is the energy consumed to transform common cell intermediates (assumed to be at the energy level of pyruvate and ammonia) to cell biomass.

For the heterotrophic growth under optimum conditions, K varies from 0.4 to 0.8 with an average of 0.60 (McCarty 1972). The amount of energy required to transform ammonia and pyruvate (intermediate) into one equivalent of biomass $\left(\Delta G_{\text{cells}}^{0'}\right)$ is 7.5 kcal e-eq^{-1}. The exponent m is either +1 or −1, depending on whether energy is released (−1) or required (+1) for a reaction. The half-reactions for conversion of various forms of inorganic nitrogen into ammonia are used to determine$\Delta G_n^{0'}$. As sufficient ammonia nitrogen is present, $\Delta G_n^{0'} = 0$.

In the overall reaction stoichiometry half reactions are combined. Utilizing $C_5H_7O_2N$ for cells and assuming that the cells are formed from carbonate and ammonium, the case of soybean wastewater treatment was shown in this work, with $CH_{1.5}O_{0.31}N_{0.25}$ for proteins and CH_2O for carbohydrates (Grady et al. 1999).

$$\frac{8}{33}CO_2 + \frac{2}{33}NH_4^+ + \frac{31}{33}H^+ + e^- \rightarrow \frac{8}{33}CH_{1.5}O_{0.31}N_{0.25} + \frac{27}{66}H_2O \tag{5.4}$$

$$\frac{1}{4}CO_2 + H^+ + e^- \rightarrow \frac{1}{4}CH_2O + \frac{1}{4}H_2O \tag{5.5}$$

Reactions 4 and 5 indicate that full mineralization of proteins and carbohydrates yields 4.125 e-eq^{-1} mol of the former and 4 e-eq^{-1} mol of the latter, respectively. The oxidation of proteins and carbohydrates also yields free energy. Oxidation of proteins and carbohydrates as electron donors respectively produces 7.7 kcal ($\Delta G^{0'}$= −7.7 kcal e^{-1}-eq) and 10 kcal ($\Delta G^{0'}$= −10 kcal e^{-1}-eq) of energy per electron equivalent at pH 7.0.

5.2.2 Biological Reaction Model

In the present work, the ASM1 was modified through introducing the concept of endogenous decay regeneration (see Fig. 5.1). This modified ASM1 is given in matrix form in Table 5.1. In addition to the external substrate, the substrate released from endogenous decay was utilized for heterotrophic growth with a special biomass yield. The endogenous yield factor (Y_E) was not the same as the yield factor from the exogenous substrate (Y_H) as in the original ASM1 (Henze et al. 2000). This modified model had eight model components to describe the

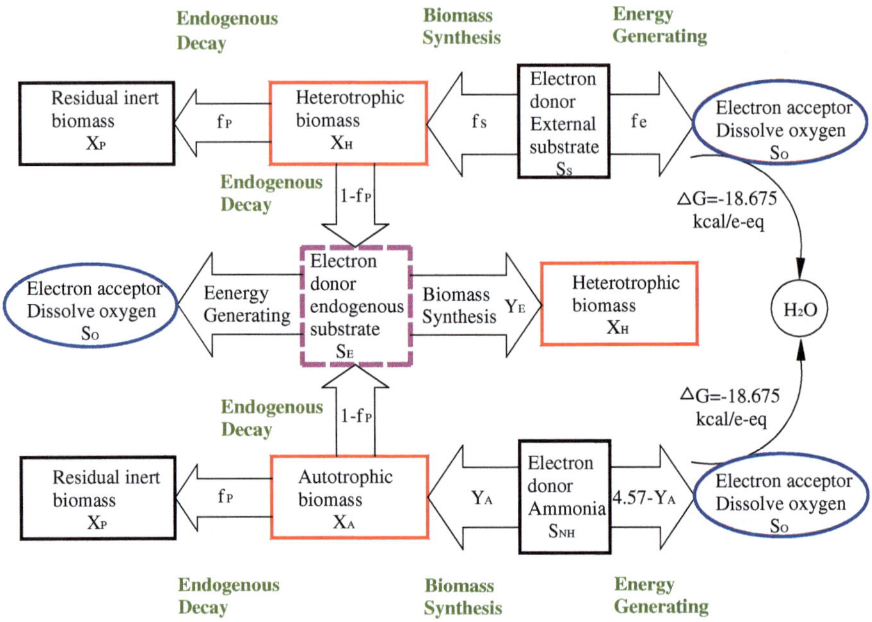

Fig. 5.1 Framework of the new approach to analyze activated sludge process. *fs* and *fe* are the fraction of electrons released from the donor substrate that go to cell synthesis and energy generation, respectively (From Ni and Yu (2009), reprinted with permission from Wiley-Blackwell)

biological reactions, i.e., heterotrophic microorganisms (X_H), autotrophic microorganisms (X_A), exogenous substrate (S_S), residual inert biomass (X_P), ammonia (S_{NH}), nitrate (S_{NO}), endogenous substrate (S_E), and dissolved oxygen (S_O). The units for all organic species were oxygen demand or oxygen (for DO), which is directly proportional to electron equivalents (i.e., 8 g O_2 per e^- equivalent).

An increase in volatile suspended solids (VSS) was the sum of biomass and residual inert biomass accumulations:

$$\frac{\mathrm{dVSS}}{\mathrm{d}t}=\frac{\mathrm{d}X_H}{\mathrm{d}t}+\frac{\mathrm{d}X_P}{\mathrm{d}t}+\frac{\mathrm{d}X_A}{\mathrm{d}t} \tag{5.6}$$

$$\frac{\mathrm{d}X_H}{\mathrm{d}t}=(\mu_1+\mu_2)X_H-b_HX_H \tag{5.7}$$

$$\frac{\mathrm{d}X_P}{\mathrm{d}t}=f_P(b_HX_H+b_AX_A) \tag{5.8}$$

$$\frac{\mathrm{d}X_A}{\mathrm{d}t}=\mu_3X_A-b_AX_A \tag{5.9}$$

Table 5.1 The modified ASM1

Component	S_C	S_{NH}	S_{NO}	S_E	S_S	X_P	X_A	X_H	Kinetics rate expression
Process	O_2	N	N	COD	COD	COD	COD	COD	
Heterotrophic									
Growth on exogenous substrate	$-\frac{1-Y_H}{Y_H}$	$-i_{XB}$			$-\frac{1}{Y_H}$			1	$\mu_H \frac{S_S}{K_S+S_S} \frac{S_O}{K_{O,H}+S_O} X_H$
Growth on endogenous substrate	$-\frac{1-Y_E}{Y_E}$	$-i_{XB}$		$-\frac{1}{Y_E}$				1	$\mu_H \frac{S_E}{K_S+S_E} \frac{S_O}{K_{O,H}+S_O} X_H$
Endogenous decay				$1\text{-}f_P$		f_P		-1	$b_H X_H$
Autotrophic									
Growth	$-\frac{4.57-Y_A}{Y_A}$	$-i_{XB}-\frac{1}{Y_A}$	$\frac{1}{Y_A}$				1		$\mu_A \frac{S_{NH}}{K_{NH}+S_{NH}} \frac{S_O}{K_{O,A}+S_O} X_A$
Endogenous decay				$1\text{-}f_P$		f_P	-1		$b_A X_A$

$$\mu_1 = \mu_{\mathrm{H}} \frac{S_{\mathrm{S}}}{K_{\mathrm{S}} + S_{\mathrm{S}}} \frac{S_{\mathrm{O}}}{K_{\mathrm{O,H}} + S_{\mathrm{O}}} \tag{5.10}$$

$$\mu_2 = \mu_{\mathrm{H}} \frac{S_{\mathrm{E}}}{K_{\mathrm{S}} + S_{\mathrm{E}}} \frac{S_{\mathrm{O}}}{K_{\mathrm{O,H}} + S_{\mathrm{O}}} \tag{5.11}$$

$$\mu_3 = \mu_{\mathrm{A}} \frac{S_{\mathrm{NH}}}{K_{\mathrm{NH}} + S_{\mathrm{NH}}} \frac{S_{\mathrm{O}}}{K_{\mathrm{O,A}} + S_{\mathrm{O}}} \tag{5.12}$$

where $\frac{\mathrm{dVSS}}{\mathrm{d}_t}$ is the rate of VSS accumulation, $\frac{\mathrm{d}x_H}{\mathrm{d}_t}$ is the rate of heterotrophic biomass accumulation, $\frac{\mathrm{d}x_A}{\mathrm{d}_t}$ is the rate of autotrophic biomass accumulation, $\frac{\mathrm{d}x_P}{\mathrm{d}_t}$ is the rate of residual inert biomass accumulation, μ_I is the biomass growth rate on S_{S}, μ_2 is the biomass growth rate on S_{E}, μ_{H} is heterotrophic the maximum biomass growth rate, b_{H} is heterotrophic endogenous decay rate, μ_{A} is the maximum autotrophic biomass growth rate, b_{A} is autotrophic endogenous decay rate, $K_{\mathrm{O,H}}$ is the dissolve oxygen affinity constant of X_{H}, $K_{\mathrm{O,A}}$ is the dissolve oxygen affinity constant of X_{A}, f_P is fraction of residual inert biomass COD formed from endogenous decay.

The substrate (SCOD) removal rate was the sum of the exogenous (S_{S}) and endogenous substrate (S_{E}) removal rates as follows:

$$\frac{\mathrm{dSCOD}}{\mathrm{d}t} = \frac{\mathrm{d}S_{\mathrm{S}}}{\mathrm{d}t} + \frac{\mathrm{d}S_{\mathrm{E}}}{\mathrm{d}t} = -\frac{I}{Y_{\mathrm{H}}} \mu_1 X_{\mathrm{H}} - \frac{1}{Y_{\mathrm{E}}} \mu_2 X_{\mathrm{H}} + (1 - f_P) b_{\mathrm{H}} X_{\mathrm{H}} \tag{5.13}$$

where $\frac{\mathrm{dSCOD}}{\mathrm{d}_t}$ is the rate of substrate removal, $\frac{\mathrm{d}s_{\mathrm{S}}}{\mathrm{d}_t}$ is the rate of exogenous substrate removal, $\frac{\mathrm{d}s_{\mathrm{E}}}{\mathrm{d}_t}$ is the rate of endogenous substrate removal.

The ammonia removal rate and nitrate production rate were described with the following equations:

$$\frac{\mathrm{d}S_{\mathrm{NH}}}{\mathrm{d}t} = -i_{\mathrm{XB}} \mu_1 X_{\mathrm{H}} - i_{\mathrm{XB}} \mu_2 X_{\mathrm{H}} - \left(i_{\mathrm{XB}} + \frac{1}{Y_{\mathrm{A}}} \right) \mu_3 X_{\mathrm{A}} \tag{5.14}$$

$$\frac{\mathrm{d}S_{\mathrm{NO}}}{\mathrm{d}t} = \frac{1}{Y_{\mathrm{A}}} \mu_3 X_{\mathrm{A}} \tag{5.15}$$

where $\frac{\mathrm{d}s_{\mathrm{NH}}}{\mathrm{d}_t}$ is the rate of ammonia removal, $\frac{\mathrm{d}s_{\mathrm{NO}}}{\mathrm{d}_t}$ is the rate of nitrate production, i_{XB} is nitrogen content of biomass, and Y_{A} is the yield factor of autotrophs.

The oxygen uptake rate (OUR or $\frac{\mathrm{d}s_{\mathrm{O}}}{\mathrm{d}_t}$) was the sum of the three respiration processes: utilization of S_{S}, S_{E} and S_{NH}.

$$\mathrm{OUR} = \frac{\mathrm{d}S_{\mathrm{O}}}{\mathrm{d}t} = -\frac{1 - Y_{\mathrm{H}}}{Y_{\mathrm{H}}} \mu_1 X_{\mathrm{H}} - \frac{1 - Y_{\mathrm{E}}}{Y_{\mathrm{E}}} \mu_2 X_{\mathrm{H}} - \frac{4.57 - Y_{\mathrm{A}}}{Y_{\mathrm{A}}} \mu_3 X_{\mathrm{A}} \tag{5.16}$$

5.3 Materials and Methods

5.3.1 Sludge, Wastewater, and Reactor

Sludge was cultivated in a laboratory-scale SBR, as described in Chap. 3 . The DO concentration in aerobic phase during SBR operation was above 4 mg L^{-1}. The desired SRT was set by controlling the amount of sludge wasted from the SBR in each cycle. In this work, the SRT for the SBR operation was designed at 20 days, and the initial MLVSS concentration of the SBR was approximately 2,000 mg L^{-1}. The SBR was fed with a soybean-processing wastewater (see Chap. 3) at a soluble COD of around 800 mg L^{-1}.

5.3.2 Experiments

A primary source of quantitative data used for a long-term reactor performance simulation was the routine measurements everyday for the SBR. The samples were analyzed for the COD, NH_4^+–N, and MLVSS concentrations.

In addition to the long-term SBR operation, batch experiments were also conducted to evaluate the established approach. Sludge was sampled from the SBR when no substrate was present in the medium. The 2-L batch reactors were inoculated with the activated sludge and continuously aerated until the DO concentration exceeded 4 mg L^{-1} for complete nitrification. Then, the external substrate, i.e., the soybean-processing wastewater, at a predetermined concentration after dilution was dosed and the OUR was monitored until the substrate became depleted and the endogenous activity was resumed. The initial external substrate in the batch reactor was approximately 220 mg L^{-1}. The samples were taken every 5.10 min for analysis. The tests were conducted at 20 °C and pH 7.0.

5.3.3 Analysis

The analysis of DO and OUR was described in Chap. 3. The carbohydrates were determined as glucose equivalent using enthrone sulfuric acid method (Dubois et al. 1956), whereas the proteins were measured as bovine albumin equivalent using the Lowry method (Lowry et al. 1951). Measurement of COD, NH_4^+–N, MLSS, and MLVSS was performed following the standard methods (APHA 1995).

5.4 Results and Discussion

5.4.1 Characterization of Soybean-Processing Wastewater

The raw wastewater contained proteins of 5.50 g L^{-1} and carbohydrates of 7.40 g L^{-1}. Efforts were initially made to estimate the contribution of each organic component to the total COD (TCOD) in this wastewater. For complete oxidation, 32 mol of proteins requires 33 mol of oxygen:

$$32CH_{1.5}O_{0.31}N_{0.25} + 33O_2 \rightarrow 12H_2O + 32CO_2 + 8NH_3 \qquad (5.17)$$

Further calculations give the amount of oxygen required per unit amount of proteins degraded:

$$\frac{33 * \text{mole} * O_2}{32 * \text{mole} * \text{protein}} \times \frac{1 * \text{mole} * \text{protein}}{22 * g * \text{protein}} \times \frac{32 * g * O_2}{1 * \text{mole} * O_2}$$

= 1.50 g O_2 g^{-1} proteins
= 1.50 g COD g^{-1} proteins
For 5.50 g L^{-1} proteins in the soybean wastewater,

$$\frac{1.5 * g * \text{COD}}{1 * g * \text{protein}} \times \frac{5.5 * g * \text{protein}}{1 * L * \text{wastewater}}$$

= 8.25 g COD L^{-1}

Therefore, the COD due to the presence of proteins in the wastewater was 8.25 g COD L^{-1}, which was 49.8 % of TCOD.

One mole of carbohydrates requires one mole of oxygen for complete oxidation.

$$CH_2O + O_2 \rightarrow H_2O + CO_2 \qquad (5.18)$$

Similarly, an estimate for the oxygen requirement per unit amount of carbohydrates is 1.07 g COD g^{-1} carbohydrates. The following calculations give the amount of oxygen required per unit amount of carbohydrates degraded:

$$\frac{1 * \text{mole} * O_2}{1 * \text{mole} * \text{carbohydrate}} \times \frac{1 * \text{mole} * \text{carbohydrate}}{30 * \text{g} * \text{carbohydrate}} \times \frac{32 * \text{g} * O_2}{1 * \text{mole} * O_2}$$

= 1.07 g O_2 g^{-1} carbohydrates
= 1.07 g COD g^{-1} carbohydrates
For 7.40 g L^{-1} carbohydrates in the wastewater,

$$\frac{1.07 * \text{g} * \text{COD}}{1 * g * \text{carbohydrate}} \times \frac{7.4 * g * \text{carbohydrate}}{1 * L * \text{wastewater}}$$

= 7.92 g COD L^{-1}

In a similar way, the COD exerted by carbohydrates in this wastewater was calculated to be 7.92 g COD L^{-1}, which was 47.8 % of TCOD.

Based on analyzes of organics in the soybean wastewater, it was concluded that the proteins and carbohydrates were the major organic components contributing the TCOD in this wastewater. A total of 97.6 % of the TCOD in the wastewater was accounted for by proteins and carbohydrates.

5.4.2 Overall Stoichiometry

For the full biodegradation stoichiometry (R), the reactions coupling the electron donor and acceptor were combined with the reactions for cell synthesis using the f_e and f_s values (McCarty 1972).

For proteins, the released energy (ΔG_{e-R}) could be calculated as follows:

$$\Delta G_R^{0'} = \Delta G_a^{0'} - \Delta G_d^{0'} = -7.7 - 18.675 = -26.375 \text{ kcal/e} - \text{eq.} \tag{5.19}$$

The energy consumed in cell synthesis (ΔG_{e-s}) was calculated as:

$$\Delta G_{\text{syn}}^{0'} = \frac{\Delta G_p^{0'}}{K^m} + \Delta G_{\text{cells}}^{0'} + \frac{\Delta G_n^{0'}}{K} = \frac{8.545 - 7.7}{0.6} + 7.5 + 0 = 8.908 \text{ kcal/e} - \text{eq}, \tag{5.20}$$

$$A = -\frac{\Delta G_{\text{syn}}^{0'}}{K\Delta G_R^{0'}} = \frac{8.908}{0.6 * 26.375} = 0.563e - \text{eq/e} - \text{eq.} \tag{5.21}$$

According to the bioenergetic methodology, $f_s = 0.640$ e-eq e-eq^{-1}, and $f_e = 0.360$ e-eq e-eq^{-1}. Normalizing the result to 1 mol of the substrate, the full stoichiometry for the mineralization of proteins was as below:

$$\begin{aligned} CH_{1.5}O_{0.31}N_{0.25} + 0.1193H^+ + 0.3701O_2 &\rightarrow 0.132C_5H_7O_2N \\ &+0.111H_2O + 0.3372CO_2 + 0.1176NH_4^+ \end{aligned} \tag{5.22}$$

Similarly, the full stoichiometry for the mineralization of carbohydrates was as follows:

$$\begin{aligned} CH_2O + 0.1444HCO_3^- + 0.1444NH_4^+ + 0.278O_2 &\rightarrow 0.1444C_5H_7O_2N \\ &+0.8556H_2O + 0.4224CO_2 \end{aligned} \tag{5.23}$$

Table 5.2 summarizes the mole-based stoichiometric coefficients for the two reactions, where a negative value indicates a reactant and a positive sense indicates a product.

Table 5.2 Stoichiometrics for the mineralization of proteins and carbohydrates

$CH_{1.5}O_{0.31}N_{0.25}$	CH_2O	H^+	H_2O	CO_2	NH_4^+	HCO_3^-	$C_5H_7O_2N$
–1.0		–0.1193	0.111	0.3372	0.1176		0.132
	–1.0		0.8556	0.4224	–0.1444	–0.1444	0.1444

5.4.3 Exogenous Heterotrophic Biomass Yield

The theoretical biomass yield is a stoichiometric constant which relates only to the growth mechanisms and is not affected by the growth reaction rate (Rittmann and McCarty 2001). It could be evaluated through electron equivalent without the interference of endogenous metabolism (McCarty 1972). The evaluation process involves the stoichiometry of the electron donor and acceptor, the free energy released per unit electron transferred, different biochemical pathways responsible for the electron transfer, energy utilization, and the corresponding energy transfer efficiencies. Therefore, the microbial yield could be predicted with a thermodynamic method (Heijnen et al. 1992; Rittmann and McCarty 2001; Xiao and van Briesen 2006; McCarty 2007). The f_s value determined using the electron and energy balances could be used to predict the theoretical biomass yield (Y) in the following pathway (van Briesen and Rittmann 2000):

$$\frac{\text{Ymol} * \text{cells}}{\text{mol} * \text{substrate}} = f_s \times \frac{\text{mol} * \text{cells}}{\text{EQC} * \text{e}^-\text{eq}} \times \frac{\text{EQS} * \text{e}^-\text{eq}}{\text{mol} * \text{substrate}} \tag{5.24}$$

where EQS is the electron equivalents available per mole of substrate and EQC is the electron equivalents required to synthesize cells.

To calculate the values of EQC and EQS, the stoichiometry was predicted based on the structured overall reaction, assuming that ammonia was the nitrogen source for cell synthesis:

$$CH_mO_lN_k + aNH_3 + bO_2 \rightarrow y_cCH_pO_nN_g + cH_2O + dCO_2 \tag{5.25}$$

where $CH_mO_lN_k$ is the carbon source electron-donor formula, $CH_pO_n\underline{N}_g$ is the cell formula, y_c is the fraction of the substrate carbon converted to biomass (the carbon yield of cells), and a, b, c, d are the stoichiometric coefficients for the other compounds in the reaction.

The degree of reductance of carbon in the substrate (γ_s) and the degree of reductance of carbon in cells (γ_c) could be computed from the following equations. Then, EQS and EQC were equivalent to γ_s and γ_c when the cells and carbon source are written in terms of 1 C-mol.

$$\gamma_s = 4 + m - 2l - 3k, \tag{5.26}$$

$$\gamma_c = 4 + p - 2n - 3g. \tag{5.27}$$

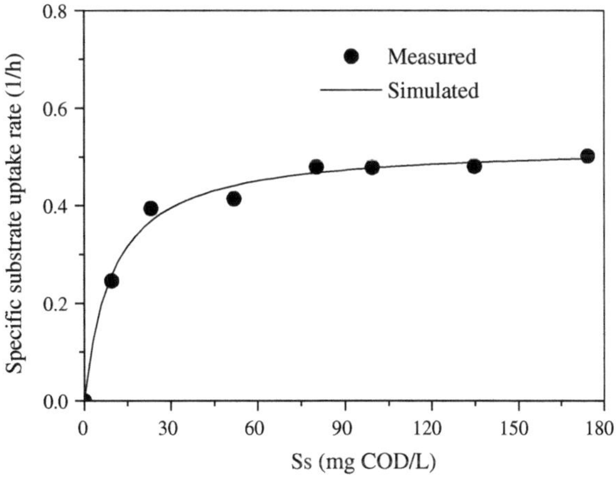

Fig. 5.2 Plot of substrate uptake rate versus substrate concentration to determine the specific maximum growth rate on substrate (μ_H) and the substrate half-saturation constant (From Ni and Yu (2009), reprinted with permission from Wiley-Blackwell)

In the present work, EQC was 20 when $C_5H_7O_2N$ is selected as a cell formulation. The EQS was 4.125 for proteins and 4 for carbohydrates. With Eq. 5.24, the theoretical biomass yields for growth on proteins (Y_I) and carbohydrates (Y_2) were 0.132 mol cells mol^{-1} substrate (i.e., 0.667 mg COD_X mg^{-1} COD_S) and 0.1444 mol cells mol^{-1} substrate (i.e., 0.544 mg COD_X mg^{-1} COD_S), respectively. Thus, from the characterization results of the soybean wastewater for the fractionation of proteins and carbohydrates, the biomass yields for growth on this wastewater (Y_H) was estimated to be 0.607 mg COD_X mg^{-1} COD_S.

5.4.4 Biological Reaction Parameters

The specific maximum growth rate on substrate (μ_H) and the substrate half-saturation coefficient (K_S) were experimentally determined following Henze et al. (2000). The increase in the substrate uptake rate with the increasing S_S concentration is depicted in Fig. 5.2. The experiments were carried out with dose of substrate at various levels to the endogenous sludge, resulting in various substrate uptake rates, i.e., exogenous respiration rates, up to a maximum rate. The parameters μ_H and K_S could be estimated by fitting the curve in Fig. 5.2. The changing pattern was well described with a high correlation coefficient of 0.9914. The μ_H and K_S were determined to be 0.35 h^{-1} and 9.8 g COD m^{-3}, respectively.

Y_E is the net biomass yield from growth on the endogenously released substrate. Biomass is the substrate and it would be partially oxidized for energy, while the remaining portion is resynthesized. The yield in the endogenous substrate resynthesis depends on the state of reduction of nitrogen in the system. If ammonia is in abundance, the approach outlined by Droste could be adopted Droste (1998). The determination of Y_E followed the procedure for determining Y_H with the pyruvate as electron donor for the half reactions (see Eq. 5.28). The utilization of 1 g COD

Table 5.3 Parameters used in the modified ASM1

Parameter	Definition	Values	Unit	Source
Y_H	Yield for X_H growth on S_S	0.607	g COD g^{-1} COD	(1)
Y_E	Yield for X_H growth on S_E	0.648	g COD g^{-1} COD	(1)
Y_A	Yield coefficient for X_A growth	0.24	g COD g^{-1} COD	(2)
f_P	Fraction of X_P in respiration	0.08	g COD g^{-1} COD	(2)
i_{XB}	Nitrogen content of biomass	0.086	g N g^{-1} COD	(2)
μ_H	Growth rate of X_H on S_S	8.4	d^{-1}	(1)
K_S	Biomass affinity constant for S_S	9.8	g COD m^{-3}	(1)
$K_{O,H}$	DO affinity constant of X_H	0.20	g O_2 m^{-3}	(2)
b_H	Respiration rate of X_H	0.62	d^{-1}	(2)
μ_A	Maximum growth rate of X_A	0.8	d^{-1}	(2)
b_A	Respiration rate of X_A	0.15	d^{-1}	(2)
$K_{O,A}$	Oxygen affinity constant for X_A	0.4	g O_2 m^{-3}	(2)
K_{NH}	biomass S_{NH} affinity constant	1.0	g N m^{-3}	(2)
f_e	Electron-donor substrate is sent to the electron acceptor	0.36* 0.278**	e-eq e^{-1}-eq	(1)
f_s	Electron-donor substrate is sent to the biomass synthesis	0.64* 0.722**	e-eq e^{-1}-eq	(1)

Notes (1) This work, (2) Henze et al. (1987), * for proteins, ** for carbohydrates

of the endogenously released substrate in the biomass decay process would yield 0.648 g COD of resynthesized biomass, i.e., $Y_E = 0.648$ mg COD_X mg^{-1} COD_S. The ultimate accumulation of endogenous residual inert biomass is usually appropriately 20 % of the newly produced biomass (Christensen and McCarty 1975). Typical values for f_P and b_H are 0.08 and 0.62 d^{-1}, respectively (Henze et al. 2000). Table 5.3 shows all the parameters used in our modified model, their symbols, and their units.

$$\frac{1}{10}CH_3COCOO^- + \frac{2}{5}H_2O \rightarrow \frac{1}{5}CO_2 + \frac{1}{10}HCO_3^- + H^+ + e^- \tag{5.28}$$

5.4.5 Modeling the Treatment of Soybean Wastewater

Figure 5.3 compares the simulated and measured results for the OUR, MLVSS, NH_4^+–N, and soluble COD in the batch tests. The initial concentration of active biomass was estimated using the baseline endogenous OUR prior to the substrate addition. The maximum difference between the measured and calculated values was 20 and 65 % of the results had a difference of less than 5 %. Furthermore, the model shows no systematic deviations. Model results matched the experimental

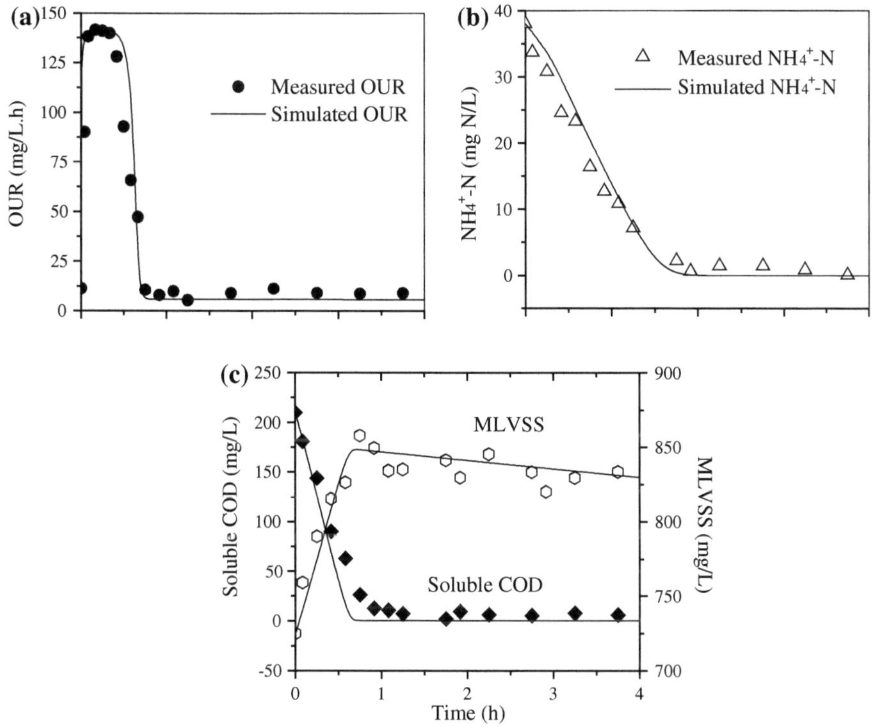

Fig. 5.3 Model evaluation of the experimentally measured OUR, NH_4^+–N, soluble COD, data and MLVSS profiles in batch tests (Line for model prediction, and symbols for experimental measurement) (From Ni and Yu (2009), reprinted with permission from Wiley–Blackwell)

data for the four parameters. After the dose of substrate, the OUR increased because of the rapid oxidation of the external substrate, while part of the electron flow went directly to respiration (Fig. 5.3a). The sharp bending point in the OUR curve corresponded with the complete removal of the exogenous substrate dosed (Fig. 5.3b). The soluble COD concentration decreased rapidly and continuously in the initial half hour (Fig. 5.3b), as the external substrate was consumed. The NH_4^+-N consumption exhibited a lag time, attributed to a relatively low oxidation rate of the autotrophs (Fig 5.3c). The MLVSS initially increased from 720 to 850 mg L^{-1} in the initial half hour, but later decreased gradually to 830 mg L^{-1} at 4 h (Fig. 5.3b).

The model was also evaluated by simulating the long-term SBR operation results of 200 days. Figure 5.4 compares the simulation and experimental results for the MLVSS and effluent COD and NH_4^+–N concentrations. Similarly, the average difference between the measured and calculated values was 12 % and the model shows no systematic deviations. Thus, the model was able to simulate all these variations well. Such a good agreement between the measured and predicted results suggests the validity of the model established in this work. The modeling

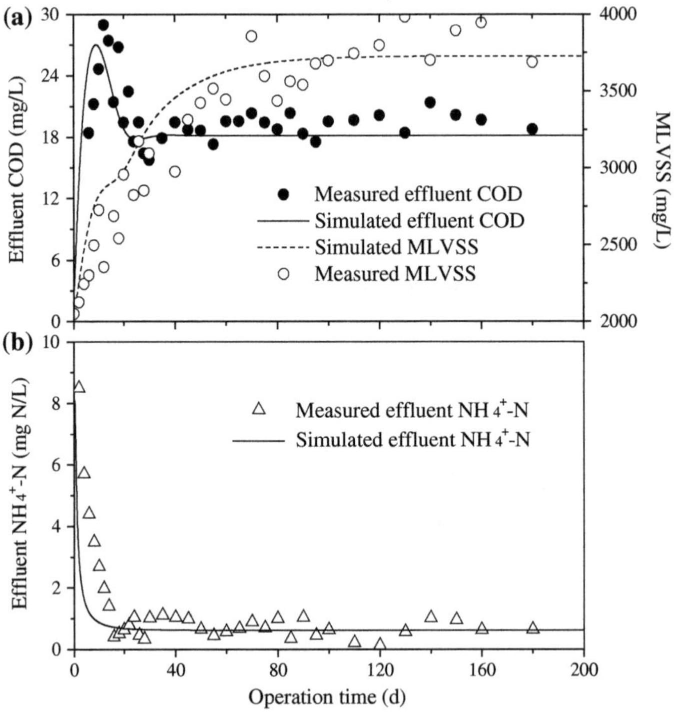

Fig. 5.4 Model evaluation of the long-term operational performance of the SBR in terms of effluent COD and NH_4^+–N concentrations as well as MLVSS profiles (From Ni and Yu (2009), reprinted with permission from Wiley-Blackwell)

results show that the effluent COD and NH_4^+–N take about 30 days to approach their plateaus. On the other hand, the MLVSS needed a long run (at least 100 days) to reach its plateau. At steady state, the active microorganisms were about 30 % of MLVSS. The remaining biomass was comprised of inert biomass.

5.4.6 Analysis of the Degradation Process

Biological degradation of wastewater by activated sludge is always achieved through multiple-substrate utilization for complete mineralization. These multiple substrates utilization might proceed at different rates, and thus prediction of the microbial growth and the elimination of degradable substrates in such systems require an accurate method for modeling. In our approach, determination of the rate of substrate utilization was coupled to the stoichiometry, which was linked to our modified ASM1 model. Therefore, the kinetic biodegradation routine could predict rates for all species in the degradation reaction and returned changes to the total concentration of each affected species.

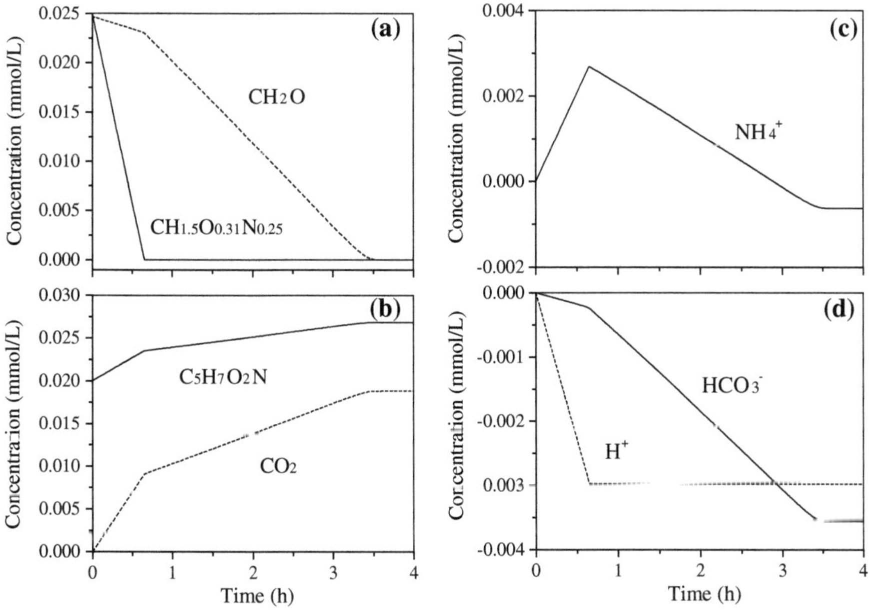

Fig. 5.5 Model simulation for the profiles of proteins ($CH_{1.5}O_{0.31}N_{0.25}$), carbohydrates (CH_2O), cells ($C_5H_7O_2N$), released CO_2, NH_4^+, HCO_3^-, and H^+ for the heterotrophic growth in soybean wastewater treatment (From Ni and Yu (2009), reprinted with permission from Wiley-Blackwell)

The substrate utilization rate for each electron-donor substrate in the soybean wastewater was coupled with the stoichiometry presented in Table 5.2 to predict the concentration changing rate of all reactants and products. Figure 5.5 shows the model predictions for the profiles of proteins ($CH_{1.5}O_{0.31}N_{0.25}$), carbohydrates (CH_2O), cells ($C_5H_7O_2N$), H^+, NH_4^+, HCO_3^-, and released CO_2 from the heterotrophic growth in the degradation process of soybean wastewater. It should be noted that the species profiles here were the relative dynamic variations attributed to the degradation process, rather than their absolute concentrations. The initial values for H^+, NH_4^+, HCO_3^- and CO_2 were set at zero, and the negative values indicate their consumption. Due to the rapid consumption of substrate by the active cells for new synthesis, the concentration of $CH_{1.5}O_{0.31}N_{0.25}$ decreased sharply and down to zero within half hour (Fig. 5.5a). The concentration of CH_2O exhibited a decreasing trend (Fig. 5.5a) attributed to the variation of the concentration of cells ($C_5H_7O_2N$). In addition, the carbohydrates in the soybean wastewater of this work were a polymer, rather than a simple sugar. Therefore, their consumption rate was assumed to be slower than that of proteins. In initial half hour, the active cells utilized the external substrate for growth. The content of active cells increased because of the rapid consumption of the external substrate (Fig. 5.5b). The increasing rate of cells in this phase was greater than the subsequent phase, in which CH_2O was the only external substrate (Fig. 5.5a).

Both reactions 22 and 23 released CO_2 when $CH_{1.5}O_{0.31}N_{0.25}$ and CH_2O were consumed by cells (Fig. 5.5b). Formation of NH_4^+ occurs only when $CH_{1.5}O_{0.31}N_{0.25}$ was being utilized. The NH_4^+ concentration increases initially in the rapid microbial $CH_{1.5}O_{0.31}N_{0.25}$ consumption period. After the depletion of $CH_{1.5}O_{0.31}N_{0.25}$, NH_4^+ is utilized by cells for growth on CH_2O, resulting in a decline of NH_4^+ concentration (Fig. 5.5c). A linear decrease in H^+ concentration occurs in the $CH_{1.5}O_{0.31}N_{0.25}$-consumption phase because of the microbial assimilation of H^+ (Reaction 5-22). In the CH_2O-consumption phase, the H^+ concentration did not change further (Fig. 5.5d). The HCO_3^- concentration decreased continually when CH_2O was consumed at an increasing rate. Thus, the variation of HCO_3^- concentration was similar to that of CH_2O concentration (Fig. 5.5d). These results clearly illuminate the overall bioreaction processes in the degradation of soybean wastewater.

In summary, the approach developed in this work through integrating the thermodynamic methodology and a modified ASM1 was able to precisely describe the treatment of soybean wastewater by activated sludge in an SBR, in terms of substrate utilization, biomass growth, and elector acceptor consumption. The effects of biological reactions on all important chemical and biological species in the system, i.e., the dynamic profiles of cells, H^+, NH_4^+, HCO_3^- and release of CO_2, could be simulated well with this new approach. Hopefully, information provided in this chapter might be useful for more accurate modeling of wastewater treatment processes.

5.5 Conclusions

A bioenergetic model for wastewater treatment was developed using the thermodynamic methodology combined with a modified ASM1. A systematic method for developing reaction stoichiometry for waste biodegradation reactions was used, and the thermodynamic methodology was applied to evaluate the coefficients for the true biomass yield from substrate. The obtained parameters were related to the ASM1 model, which coupled biological reactions with bioenergetics. The established model was applied to the treatment of a soybean-processing wastewater in an SBR. The model was able to accurately describe the consumption of substrate, the accumulation of MLVSS, and the OUR profiles in the wastewater treatment process and the long-term operation results of the SBR. Model simulation results clearly show the variations of $C_{16}H_{24}O_5N_4$, CH_2O, cells ($C_5H_7O_2N$), H^+, NH_4^+, HCO_3^- concentration, and the release of CO_2.

References

APHA: Standard methods for the examination of water and wastewater, 19th edn. American Public Health Association, Washington, DC (1995)

Christensen, D.R., McCarty, P.L.: Multi-process biological treatment model. J. Water Pollut. Contr. Fed. **47**, 2652–2664 (1975)

Droste, R.L.: Endogenous decay and bioenergetics theory for aerobic wastewater treatment. Water Res. **32**, 410–418 (1998)

Grady, C.P.L. Jr, Glen, T.D., Henry, C.L.: Biological wastewater treatment. 2nd edn. Marcel Dekker, Inc (1999)

Henze, M., Gujer, W., Mino, T., van Loosdrecht, M.C.M.: Activated sludge models ASM1, ASM2, ASM2d, and ASM3. IWA Scientific and Technical Report No. 9. IWA Publishing, London, UK (2000)

Heijnen, J.J., van Loosdrecht, M.C.M., Tijhuis, L.: A black box mathematical model to calculate auto- and heterotrophic biomass yields based on Gibbs energy dissipation. Biotechnol. Bioeng. **40**, 1139–1154 (1992)

Keesman, K.J., Spanjers, H.: Endogenous model state and parameter estimation from an extensive batch experiment. Biotechnol. Bioeng. **68**, 422–429 (2000)

Keesman, K.J., Spanjers, H., van Straten, G.: Analysis of endogenous process behavior in activated sludge. Biotechnol. Bioeng. **57**, 155–163 (1998)

McCarty, P.L.: Energetics and bacterial growth. In: Faust, S.D., Hunter, J.V. (eds.) Organic compounds in aquatic environments. Marcel Dekker, New York (1971)

McCarty, P.L.: Stoichiometry of biological reactions. Paper presented at the International Conference toward a Unified Concept of Biological Waste Treatment Design, Atlanta, 6 Oct 1972.

McCarty, P.L.: Thermodynamics of biological synthesis and growth. In: Baers, J. (ed.) Advances in Water Pollution Research: Proceedings of the 2nd International Conference on Water Pollution Research, pp. 169–199. Pergamon Press, Oxford (1965)

McCarty, P.L.: Thermodynamic electron equivalents model for bacterial yield prediction: Modifications and comparative evaluations. Biotechnol. Bioeng. **97**, 377–388 (2007)

Ni, B.J., Yu, H.Q.: A new approach to analyze the activated sludge process: Application to the soybean wastewater treatment in a sequencing batch reactor. AIChE J. **55**, 2737–2745 (2009)

Rittmann, B.E., McCarty, P.L.: Environmental biotechnology: Principles and applications, p. 754. McGraw-Hill, New York (2001)

Smets, I.Y., Haegebaert, J.V., Carrette, R., van Impe, J.F.: Linearization of the activated sludge model ASM1 for fast and reliable predictions. Water Res. **37**, 1831–1851 (2003)

Su, K.Z., Yu, H.Q.: A generalized model for aerobic granule-based sequencing batch reactor. 1. Model development. Environ. Sci. Technol. **40**, 4703–4708 (2006)

van Briesen, J.M., Rittmann, B.E.: Mathematical description of microbiological reactions involving intermediates. Biotechnol. Bioeng. **67**, 35–52 (2000)

Xiao, J.H., van Briesen, J.M.: Expanded thermodynamic model for microbial true yield prediction. Biotechnol. Bioeng. **93**, 110–121 (2006)

Chapter 6
Storage and Growth Processes in Aerobic Granular Sludge

The internal storage mechanisms and electron flow from the external substrate occurring in the aerobic granule sludge under both aerobic and anoxic conditions were explored with extensive storage experiments. The simultaneous growth and storage processes in aerobic granules were accurately modeled. It was found that the aerobic granules in an SBR, subjected to alternative feast and famine conditions, were able to rapidly take up carbon substrate in wastewater and to store it as intracellular storage products when the substrate was in excess.

6.1 Growth and Storage Processes Under Aerobic Conditions

6.1.1 Introduction

Recently, aerobic granules in SBR systems have attracted increasing interests, because the granules have good settling properties and high biomass activity (Schwarzenbeck et al. 2004; Su and Yu 2005). Such a reactor has been proven to be applicable to the treatment of various industrial wastewaters, such as dairy (Arrojo et al. 2004), malting (Schwarzenbeck et al. 2004), and soybean processing wastewaters (Su and Yu 2005), as well as municipal wastewater (de Kreuk and van Loosdrecht 2006) and many special contaminations (Tay et al. 2005; Yi et al. 2006; Jiang et al. 2006).

Modeling is always a useful tool for understanding and optimizing substrate utilization in the activated sludge wastewater treatment system. Model simulation and prediction can provide a solid foundation for design and operation of biological treatment systems. Thus, a mathematical model which can describe the bioreaction processes occurring in an SBR is essential. For the conventional floc-based activated sludge systems, modeling has evolved from ASM1 (Henze et al. 1987) to a more complicated model involving the description of storage

B.-J. Ni, *Formation, Characterization and Mathematical Modeling of the Aerobic Granular Sludge*, Springer Theses, DOI: 10.1007/978-3-642-31281-6_6,

phenomena, i.e., ASM3 (Gujer et al. 1999). The main reason behind such a shift is the increasing understanding on the important roles of storage polymers as an essential intermediate in the overall substrate removal process by activated sludge, in particular when sludge is subjected to feast and famine conditions (Dircks et al. 1999; Majone et al. 1999; Krishna and van Loosdrecht 1999; Carucci et al. 2001; Pratt et al. 2004; Sin et al. 2005).

However, when ASM3 is used to interpret the data obtained in short-term respirometric batch experiments, numerical inconsistencies often arise (Krishna and van Loosdrecht 1999). In ASM3, it is assumed that all readily biodegradable substrate are initially stored as internal storage products before it is used for growth in the famine phase (Gujer et al. 1999). However, it is not true for the case of an aerobic-granule-based SBR. Such a system is repeatedly subjected to feast and famine conditions (Arrojo et al. 2004; Su and Yu 2005). As a result, simultaneous growth and storage processes always occur (Beun et al. 2001; Su and Yu 2006). Thus, ASM3 should be extended to account for the simultaneous growth and storage occurring in activated sludge. In an aerobic-granule-based reactor, numerous internal interactions among process variables (growth, storage, maintenance, and endogenous respiration, etc.) are present, and sludge characteristics (e.g., diffusion) are influential to the overall reactor performance (Su and Yu 2006). However, a mathematic model to describe an aerobic-granule-based SBR with considerations of these factors is not available yet.

Therefore, in this work a generalized model is established in order to describe the simultaneous growth and storage processes in an aerobic-granule-based SBR. In addition, this new model also takes adsorption, microbial maintenance, and substrate diffusion into account. The established model is calibrated and validated by using experimental results of an SBR fed with soybean-processing wastewater. In addition, its comparison with the original ASM3 is also carried out to explore the simultaneous growth and storage processes in an aerobic-granule-based SBR.

6.1.2 Materials and Methods

6.1.2.1 Reactor Setup and Operation

Aerobic granules were cultivated in a lab-scale SBR with the soybean-processing wastewater, as described in Chap. 3.

6.1.2.2 Batch Experiments

The batch experiments were conducted at 20 °C. The granules were sampled from the SBR when no substrate was present in the reactor. The granules were washed twice with distilled water and aerated them continuously to remove the external organic material. Then, the granules were transferred to beakers with a working

Table 6.1 Stoichiometric evaluation of the experimental results

Parameter	Unit	Set 1	Set 2	Set 3	Set 4	Set 5
Substrate feed (COD)	mg COD L^{-1}	195	256	378	509	792
Initial granule (VSS)	mg L^{-1}	1,215	1,205	1,511	1,794	2, 040
Overall net oxygen consumption (ΔO_{tot})	$mgO_2\ L^{-1}$	35.44	40.54	51.29	63.08	81.09
Oxygen consumption for storage $(\Delta O_{stoASM3})$	$mgO_2\ L^{-1}$	16.11	18.43	23.31	27.35	36.86
Oxygen consumption for growth $(\Delta O_{groASM3})$	$mgO_2\ L^{-1}$	19.33	22.11	27.98	35.73	44.23
Feast phase length according to OUR	Min	15	17	20	22	25

volume of 1 L. The soybean wastewater was characterized into particulate COD (X_S) and soluble COD (S_S) after filtration through 0.45-μm pore sized membrane. Five sets of batch tests were conducted in this work. The initial concentrations of the granules (VSS: 1,200, 1,200, 1,500, 1,800, 2,000 mg L) and substrate (COD: 190, 260, 380, 510, 800 mg L^{-1}) were variable in these tests (Table 6.1), while the DO was kept at a saturation level. Samples were analyzed for COD, storage polymers (polyhydroxyalkanoate, PHA), OUR, and volatile solids for replications.

6.1.2.3 Parameter Determination

Kinetic and stoichiometric parameters of the aerobic granules were determined at 20 °C and pH 7.0. The heterotrophic storage yield coefficient was measured using the method proposed by Karahan et al. (2002), whereas the heterotrophic yield, maximum growth rate on substrate, substrate half-saturation coefficient, and decay coefficient were determined as described by Henze et al. (1987). The measured data were used for parameter estimation with the methods suggested by Avcioglu et al. (Avcioglu 2003).

6.1.2.4 Analytical Procedures

The analysis of DO, OUR, and sludge image was described in Chap. 3. Measurements of COD, MLSS, and MLVSS followed the standard methods (APHA 1995). PHA was measured according to Pratt et al. (2004).

6.1.3 Model Development

6.1.3.1 Conceptual Basis

Granules with a wide range in size are modeled with a consideration of component diffusion limitation within the granules. For the components involved in biological reactions, the first step is their diffusion into granules where the reactions take place. Then, the slowly biodegradable substrate included in wastewater is rapidly adsorbed and hydrolyzed to readily biodegradable substrate by the heterotrophs. The biomass can use the readily biodegradable substrate for simultaneous storage and growth. When the readily biodegradable substrate is depleted (as low as the half-saturation concentration for the primary growth), the degradation (secondary growth) of the storage polymers takes place. In addition to growth, microbial maintenance on the external and internal substrates is also an important compartment in an integrated model for describing an aerobic-granule-based reactor. Finally, the biomass is subjected to decay, yielding an inert organic carbon.

6.1.3.2 SubModel for Biological Reactions

The submodel for describing the biological reactions in aerobic granules reflects an appropriate modification of ASM3, with a consideration of simultaneous growth and substrate storage, together with the addition of adsorption process, and microbial maintenance. The submodel has eight model components, i.e., DO, S_O; soluble inert COD, S_I; readily biodegradable COD, S_S; slowly biodegradable COD, X_S; adsorbed COD, X_{ads}; heterotrophic biomass, X_H; particulate inert COD, X_I , and storage products, X_{STO}. It involves eight microbial processes: adsorption, hydrolysis, growth on S_S, aerobic storage, growth on X_{STO}, maintenance on S_S, maintenance on X_{STO}, and decay.

Related process kinetics and stoichiometry are presented in a matrix format in Table 6.2 to highlight the interactions among the model components and processes. In this model the adsorption process, in which X_S is adsorbed as X_{ads}, has been described with first-order kinetics, as given in Eq. 6.1, because the substrate disappearance is rapid in all the batch experiments of this study.

$$\frac{dX_S}{dt} = -k_{ads}X_S \tag{6.1}$$

The adsorption substrate is hydrolyzed with a rate governed by surface reaction kinetics (Eq. 6.2). Hydrolysis process is presumably a rate-limiting step for the further utilization of the readily biodegradable substrate generated.

$$\frac{dX_{ads}}{dt} = -k_H \frac{X_{ads}/X_H}{K_X + X_{ads}/X_H} X_H \tag{6.2}$$

Table 6.2 Stoichiometric matrix for simultaneous storage and growth processes

	Component Process	S_O O_2	S_I COD	S_S COD	X_I COD	X_{ads} COD	X_S COD	X_H COD	X_{STO} COD	Kinetics rates expressions
1	Adsorption					1	-1			$k_{ads} X_S$
2	Hydrolysis		f_{SI}	$1 - f_{SI}$		-1				$k_H \frac{X_{ads}/X_H}{K_X + X_{ads}/X_H} X_H$
3	Aerobic storage	$-\frac{1-Y_{STO}}{Y_{STO}}$		$-\frac{1}{Y_{STO}}$					1	$k_{STO} \frac{S_S}{K_S+S_S} \frac{S_O}{K_O+S_O} X_H$
4	Growth on S_S	$-\frac{1-Y_{H,S}}{Y_{H,S}}$		$-\frac{1}{Y_{H,S}}$				1		$\mu_{H,S} \frac{S_S}{K_S+S_S} \frac{S_O}{K_O+S_O} X_H$
5	Growth on X_{STO}	$-\frac{1-Y_{H,STO}}{Y_{H,STO}}$						1	$-\frac{1}{Y_{H,STO}}$	$\mu_{H,STO} \frac{K_S}{K_S+S_S} \frac{S_O}{K_O+S_O} \frac{X_{STO}/X_H}{K_{STO}+X_{STO}/X_H} X_H$
6	Maintenance on S_S	-1		-1						$m_{H,S} \frac{S_S}{K_S+S_S} \frac{S_O}{K_O+S_O} X_H$
7	Maintenance on X_{STO}	-1							-1	$m_{H,STO} \frac{K_S}{K_S+S_S} \frac{S_O}{K_O+S_O} \frac{X_{STO}/X_H}{K_{STO}+X_{STO}/X_H} X_H$
8	Decay				f_I		$1 - f_I$	-1		$b_H \frac{K_S}{K_S+S_S} \frac{K_{STO}}{K_{STO}+X_{STO}/X_H} X_H$

In the feast phase, because of a pulse addition of the readily biodegradable substrate, the biomass is likely to primarily utilize the available substrate via storage and growth. The kinetic expression for the primary growth ($r_{\text{grow Ss}}$) is given in Eq. 6.3, while the rate of storage, r_{STO}, is defined in Eq. 6.4.

$$r_{\text{grow Ss}} = \mu_{\text{H,S}} \frac{S_S}{K_S + S_S} \frac{S_O}{K_O + S_O} X_H \tag{6.3}$$

$$r_{\text{STO}} = k_{\text{STO}} \frac{S_S}{K_S + S_S} \frac{S_O}{K_O + S_O} X_H \tag{6.4}$$

In the famine phase, the secondary growth process can take place only if S_S is as low as the half-saturation concentration for the primary growth, K_S, ,i.e., the microbial growth rate on X_{STO} is assumed to occur under strictly famine conditions. Thus, a Monod inhibition function for the external substrate is added to the kinetic description ($r_{\text{grow STO}}$) to yield the following equation:

$$r_{\text{grow STO}} = \mu_{\text{H,STO}} \frac{K_S}{K_S + S_S} \frac{S_O}{K_O + S_O} \frac{X_{\text{STO}}/X_H}{K_{\text{STO}} + X_{\text{STO}}/X_H} X_H \tag{6.5}$$

However, in addition to the energy for anabolism, the heterotrophs also require maintenance energy for their activity and chemical processes. Maintenance rates on the external substrate (S_S, $r_{\text{maintenance Ss}}$) and the internal substrate (X_{STO}, $r_{\text{maintenance STO}}$) are given in Eqs. 6.6 and 6.7, respectively, based on a previous work for aerobic granular sludge modeling (Beun et al. 2001).

$$r_{\text{maintenance Ss}} = m_{\text{H,S}} \frac{S_S}{K_S + S_S} \frac{S_O}{K_O + S_O} X_H \tag{6.6}$$

$$r_{\text{maintenance STO}} = m_{\text{H,STO}} \frac{K_S}{K_S + S_S} \frac{S_O}{K_O + S_O} \frac{X_{\text{STO}}/X_H}{K_{\text{STO}} + X_{\text{STO}}/X_H} X_H \tag{6.7}$$

Finally, Eq. 6.8 is used to describe the accumulation of the particulate inert COD (X_I) as a result of decay of active biomass.

$$\frac{dX_I}{dt} = f_I b_H \frac{K_S}{K_S + S_S} \frac{K_{\text{STO}}}{K_{\text{STO}} + X_{\text{STO}}/X_H} X_H \tag{6.8}$$

6.1.3.3 SubModel for Aerobic Granules

In modeling, aerobic granules are assumed to be spherical in shape and their size distribution to be constant in one cycle of operation. A biofilm model (Wanner and Reichert 1996) implemented in AQUASIM (Reichert 1998) is used to model the aerobic granules in this work.

One-dimensional conservation laws (Eq. 6.9) are formulated as a balance between the mass conserved and utilized in this granule model (Reichert 1998).

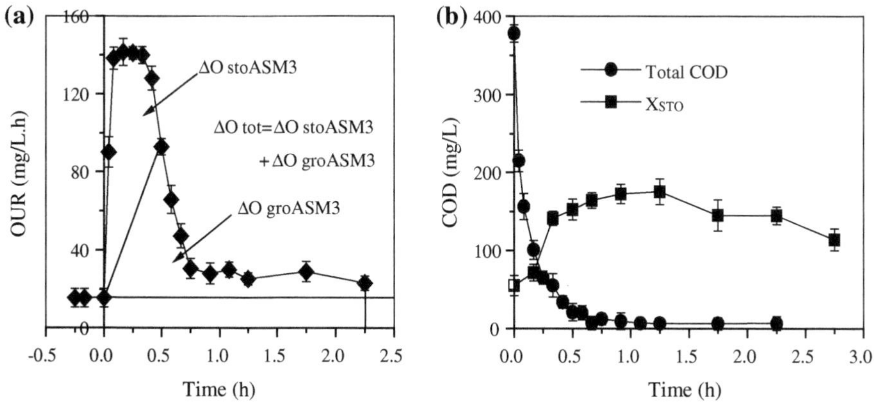

Fig. 6.1 Typical batch experimental results of Set 3: **a** OUR response; and **b** COD removal and the X_{STO} profiles (From Ni and Yu (2008), reprinted with permission from Wiley-Blackwell)

$$\frac{\partial \hat{\rho}}{\partial t} + \frac{\partial \hat{j}}{\partial z} = \hat{r} \tag{6.9}$$

where $\hat{\rho}$ is the one-dimensional density (amount of conserved quantity per unit compartment length), $\hat{j}$ is the one-dimensional flux (amount of the conserved quantity transported per unit time), $\hat{r}$ is the one-dimensional source term (amount produced per unit compartment length and per unit time), t is time, and z is the space coordinate.

6.1.3.4 Model Simulation

Modeling and simulation are performed using the software package AQUASIM (Reichert 1998). The granules grown in the SBR have a size distribution of 0.41–3.5 mm with an average diameter of 1.2 mm, when they are physiologically stabilized. It is true that granules with a wide range in size are formed in the reactor and that their size distribution also changes in time. However, the granule size distribution is not taken into account in this model, as it would significantly increase the complexity of numerical computation and it is not expected to contribute to a better understanding of the system. Therefore, in the simulation the diameter is chosen to be 1.2 mm, which is the most representative for the aerobic granules in this SBR.

6.1.4 Results and Discussion

6.1.4.1 Experimental Data and General Observations

The experimental results are summarized in Table 6.1. The evaluation of the reactions and the amounts of consumed oxygen calculated for different phases are based on ASM3. A typical data set (Set 3) is illustrated in Fig. 6.1 to provide explanations on the stoichiometric considerations given in Table 6.1. As shown in Fig. 6.1a, the area under the entire OUR curve and above the endogenous OUR level, determined at the beginning of the cycle prior to substrate dose, represents the amount of oxygen utilized for the consumption of substrate fed to the SBR (ΔO_{tot}). The amount of oxygen utilized for the substrate storage ($\Delta O_{stoASM3}$) is determined using the area under the OUR curve and above the straight line drawn from the endogenous decay level to the inflection point in the OUR curve, as suggested by Karahan et al. (2006a). The remaining amount of oxygen consumed in the SBR ($\Delta O_{groASM3}$) is devoted to the heterotrophic growth (Karahan et al. 2006a). According to ASM3, the amount of oxygen consumption for growth is higher than that for storage, when substrate is fed to the reactor (Table 6.1).

The substrate disappearance in the bulk liquid is rapid and exhibits a similar pattern in all cases, i.e., an initial rapid uptake, followed by a gradual decrease until nearly complete depletion. Such a rapid COD removal seems not to correspond with the relatively low overall oxygen consumption for substrate (ΔO_{tot}, Table 6.1). This can be attributed to the rapid adsorption of X_S and result in the model approach used in this work (process 1 in Table 6.2). Prior to the substrate addition, OURs are constant and low, which is caused by the endogenous respiration processes adopted in ASM3 or the maintenance processes introduced in this model. As depicted in Fig. 6.1a, after the substrate addition the OURs increase, attributed to the exogenous oxidation of carbon source. Although the content of X_S in the wastewater is relatively high and the hydrolysis process is the limiting step, the high OUR response at the beginning of experiment can be associated with the relatively high hydrolysis rate. This is in accordance with the experimental results of Karahan et al. (2006b). The sharp bending points in the OURs curve correspond with the nearly complete removal of the substrate added. However, the OURs do not immediately return to the endogenous level. This is likely to be attributed to the oxidation of storage polymers formed in the substrate removal and maintenance processes.

The storage polymer production after a pulse addition of substrate and later growth on the internally accumulated storage polymers is also explored. In the pulse experiment two distinct phases are observed (Fig. 6.1b). The production of the storage polymer, i.e., PHA, is the first phase: a rapid uptake of substrate which is used for growth and storage polymer production. The next is the storage polymer consumption phase: after the depletion of the substrate, growth on the accumulated PHA occurs. The PHA concentration initially increases in the rapid microbial growth period when the external substrate is consumed at an increasing rate. Some of the consumed external substrate are converted into PHA. In the second phase,

Table 6.3 Kinetic and stoichiometric coefficients for the optimum calibration of the experimental data

Parameter	Definition	Method	Values	Unit
Stoichiometry				
Y_{STO}	Yield coefficient for storage	Measured	0.79	g COD g^{-1} COD
$Y_{H,S}$	Yield for growth on S_S	Measured	0.61	g COD g^{-1} COD
$Y_{H,STO}$	Yield for growth on X_{STO}	Estimated	0.67	g COD g^{-1} COD
f_I	Fraction of X_I in decay	Assumed	0.2	g COD g^{-1} COD
f_{SI}	Fraction of S_I in hydrolysis	Assumed	0.05	g COD g^{-1} COD
Kinetics				
k_{STO}	Maximum storage rate	Estimated	0.18	h^{-1}
$\mu_{H,STO}$	Maximum growth rate on X_{STO}	Estimated	0.043	
$\mu_{H,S}$	Maximum growth rate on S_S	Measured	0.083	h^{-1}
K_S	Substrate affinity constant	Measured	9.5	g COD m^{-1}
K_{STO}	Affinity constant for X_{STO}	Assumed	1.0	g COD g^{-1} COD
K_O	DO affinity constant	Assumed	0.2	g COD m^{-1}
$m_{H,S}$	Maintenance rate on S_S	Estimated	0.001	h^{-1}
$m_{H,STO}$	Maintenance rate on X_{STO}	Estimated	0.11	h^{-1}
b_H	Decay rate coefficient	Measured	0.011	h^{-1}
k_H	Maximum hydrolysis rate	Estimated	1.03	h^{-1}
K_X	Hydrolysis affinity constant	Assumed	0.15	g COD g^{-1} COD
k_{ads}	Maximum adsorption rate	Assumed	15	h^{-1}

after the depletion of the external substrate, growth on the accumulated storage polymer occurs. In this way, PHA is utilized when it is the sole carbon and energy source for microbial growth, resulting in a decline in the PHA concentration.

6.1.4.2 Calibration of the Model: Parameter Estimation

Model calibration procedure is a process of adjusting model parameters values, so that the results simulated by the model with these parameters closely agree with the measured data. To initiate the calibration procedure, an initial guess of the parameters is necessary. Such initial values are obtained with the experimental results and data in the literature. As shown in Table 6.3, several parameters, including Y_{STO}, $Y_{H,S}$, b_H, $u_{H,S}$, and K_S are determined experimentally and independently, and are later used in the model. The initial concentration of active biomass is estimated using the baseline endogenous OUR level (Fig. 6.2) prior to substrate addition (Sin et al. 2005):

$$OUR_{end}(0) = (1 - f_I) \cdot b_H \cdot X_H(0) \quad (6.10)$$

In this equation, f_I is fixed at 0.2 mg COD mg^{-1}COD, as used in ASM3. Hence, for a given f_I and measured b_H, $X_H(0)$ can be calculated from the $OUR_{end}(0)$ data.

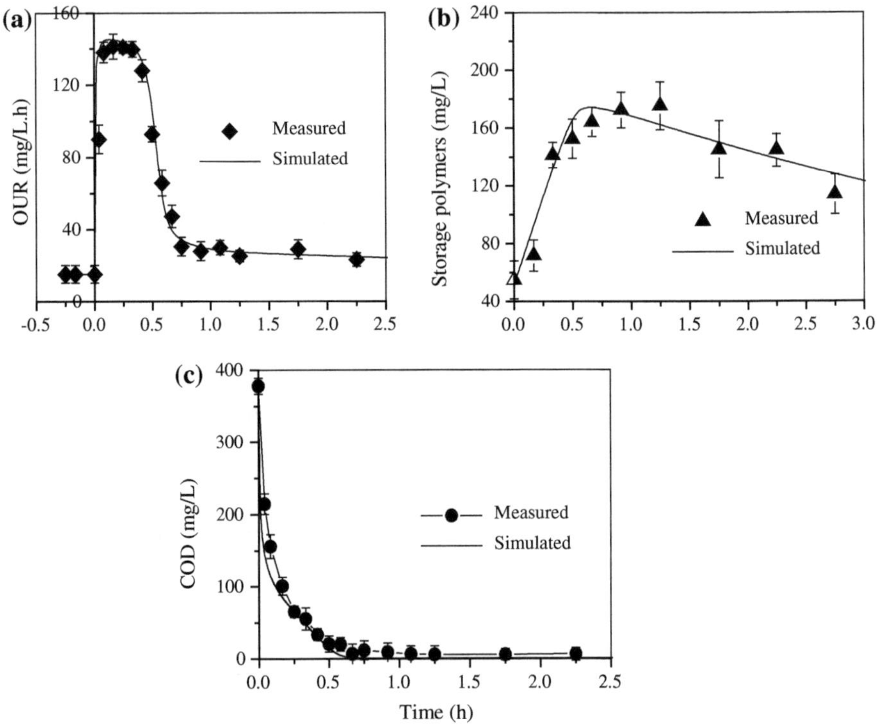

Fig. 6.2 Model calibration for the experimental data of Set 3: **a** OUR profiles, average absolute error is 11.84 mg L^{-1}; **b** COD profiles, average absolute error is 4.15 mg $L^{-1}h^{-1}$; and **c** storage polymer profiles, average absolute error is 24.81 mg L^{-1} (From Ni and Yu (2008), reprinted with permission from Wiley-Blackwell)

The model is calibrated for the COD, OUR, and X_{STO} concentrations in the batch experiments (Set 3). To simplify the calibration process and to consider the limited variability of some parameters, the calibration strategy is to change as few constants as possible (Xu and Hultman 1996). Six parameters (e.g., $Y_{H,STO}$, k_{STO}, $\mu_{H,STO}$, $m_{H,S}$, $m_{H,STO}$, and k_H) are changed based on the causality of the parameters on the COD, OUR, and X_{STO} concentrations. The standard deviation for parameter estimation defined globally for all data points is set to 50 % to ensure the validity of the parameter values obtained.

The estimated parameter values are summarized in Table 6.3, whereas the simulation results are shown in Fig. 6.2. There is a good agreement between the model predictions and experimental data. The average absolute errors obtained are 11.84 mg L^{-1}, 4.15 mg $L^{-1}h^{-1}$, and 24.81 mg L^{-1} for COD, OUR, and PHA concentrations, respectively.

The heterotrophic growth rate is lower than the typical range of values reported in the literature, partially attributed to the heterotrophic growth on PHA, rather than directly on the soluble substrate. The calibrated heterotrophic growth rate on

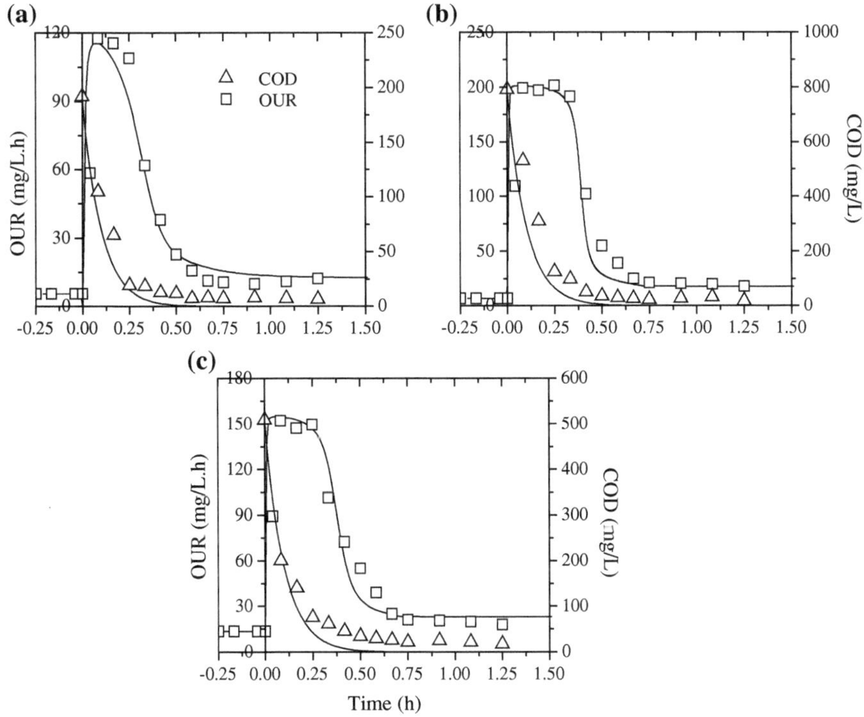

Fig. 6.3 Model evaluation of the measured OUR profiles (□ measured, — simulated) and COD (△ measured, — simulated) removals for: **a** Set 1; **b** Set 4; and **c** Set 5 (From Ni and Yu (2008), reprinted with permission from Wiley-Blackwell)

PHA is 0.043 h^{-1}, whereas the substrate half-saturation coefficient is determined to be 9.5 g m^{-3}. The substrate affinity constants, K_S, are found to be of the same order of magnitude as the values obtained from other batch experiments (Vanrolleghem et al. 2004). The maximum storage rate of biomass, k_{STO}, is higher than that on substrate. This result is in agreement with the experimental findings of Pratt et al. (2004). The biomass decay rate, b_H, is experimentally determined to be 0.011 h, while the hydrolysis rate k_H is calibrated to be 1.03 h^{-1}, which is much higher than that given in ASM3.

In our two-step modeling approach, X_S is assumed to be initially adsorbed as X_{ads} to better describe the experimental data. Then, X_{ads} is hydrolyzed to S_S. However, in ASM3, X_S is hydrolyzed to S_S in one step, resulting in a lower hydrolysis rate of 0.125. The yield coefficients (Y_{STO} and $Y_{H,STO}$) and kinetic rates (k_{STO} and $\mu_{H,STO}$) for storage and growth in our work are different from those in ASM3 (Table 6.3) because of the discrepancy between two model structures for modeling the storage mechanisms. In our approach, substrate electrons are diverted to both microbial storage products and active biomass simultaneously, and the heterotrophs grow on PHA rather than directly on the soluble substrates. Different wastewater characteristics, reactor operating conditions, and microbial

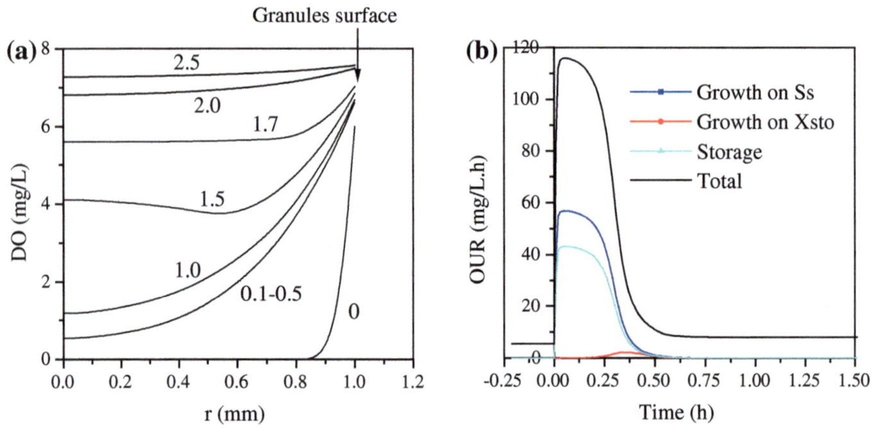

Fig. 6.4 **a** Model simulation of the oxygen penetration profiles as a function of granule radius (r), numbers in the figure indicate cycle time in hours; and **b** OUR fractions for storage and growth calculated from the established model (From Ni and Yu (2008), reprinted with permission from Wiley-Blackwell)

communities might be responsible for the different K_S and b_H values shown in Table 6.3.

6.1.4.3 Model Evaluation

A tailing-off of the DO-consumption curve is observed in the respiration tests (Fig. 6.3). This phenomenon can be described by the established model with simultaneous considerations of storage and growth and maintenance-decay concept in this work. The model evaluation results are illustrated in Fig. 6.3 for the selected sets, i.e., batch tests 1, 4, and 5 as representative examples, which are not previously used for parameter estimation. The model results match the experimental data.

Evaluation of the experimental data with the established model indicates that a high hydrolysis rate is applicable in this study. The high OUR response at the beginning of the experiment can be simulated only using a high amount of readily biodegradable substrate, and therefore a high hydrolysis rate is required (Karahan et al. 2006a). The initial high amount of readily biodegradable substrate results in a relatively rapid storage process with a maximum rate, k_{STO}, of 0.18 h^{-1}. The growth rates on PHA in famine phase ($\mu_{H,STO}$) are estimated to be 0.043 h^{-1} with a K_{STO} value of 1.0 g COD g^{-1}COD. These model results support the idea that the heterotrophs utilize substrate storage as a tool for maximizing substrate uptake when retaining an optimum growth rate (Karahan et al. 2006a). Furthermore, the validity of the model is verified with the OUR, COD, and X_{STO} data for the two experimental sets. This confirms that the consumption of dissolved oxygen, removal and utilization of substrate, and production and consumption of storage polymers by the aerobic granules can be described well by our model.

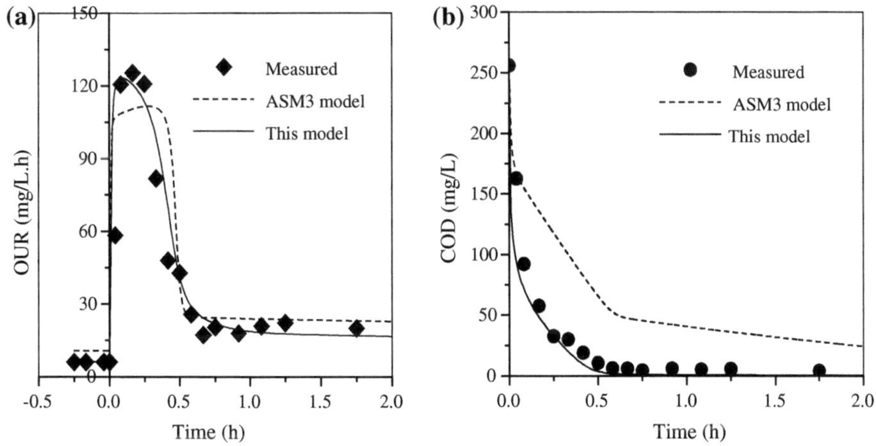

Fig. 6.5 Model simulation for the experimental results of Set 2 with ASM3 and the established model: **a** OUR profiles; and **b** COD removal (From Ni and Yu (2008), reprinted with permission from Wiley-Blackwell)

6.1.4.4 Simulation for Oxygen Consumption

In this model proposed, five oxygen-consumption processes, storage of S_S, growth on S_S, growth on X_{STO}, maintenance on S_S, and maintenance on X_{STO}, are included. Thus, an in-depth interpretation of oxygen consumption should be performed. Simulation makes it possible to perform steady-state calculation of the contributions of each process in the system. This information is useful for understanding the effect of each process on the reactor performance.

The simulation is performed under the same conditions for model evaluation of the batch test 1. The initial COD is 195 mg L^{-1} and initial VSS in the reactor is 1215 mg L^{-1}. The oxygen penetration profiles in the granules are shown in Fig. 6.4a. In the feast period, the oxygen penetration concentration is very low. As the substrate is depleted, the oxygen penetration concentration is about 1.2 mg L^{-1} in the simulation, as long as X_{STO} is present in the reactor. This oxygen is used for second growth. When X_{STO} is depleted, oxygen penetrates into the granules completely.

Oxygen utilization due to primary substrate consumption, oxygen utilization attributed to the second substrate (X_{STO}) consumption, and oxygen utilization due to substrate (S_S) storage are defined by the following process kinetic relationships:

$$OUR_{\mathrm{grow},S_S} = \mu_{H,S} X_H \frac{1 - Y_{H,S}}{Y_{H,S}} \frac{S_O}{K_O + S_O} \frac{S_S}{K_S + S_S} \quad (6.11)$$

$$OUR_{grow\mathrm{STO}} = \mu_{\mathrm{H,STO}} X_{\mathrm{H}} \frac{1 - Y_{\mathrm{H,STO}}}{Y_{\mathrm{H,STO}}} \frac{K_{\mathrm{S}}}{K_{\mathrm{S}} + S_{\mathrm{S}}} \frac{S_{\mathrm{O}}}{K_{\mathrm{O}} + S_{\mathrm{O}}} \frac{X_{\mathrm{STO}}/X_{\mathrm{H}}}{K_{\mathrm{STO}} + X_{\mathrm{STO}}/X_{\mathrm{H}}} \quad (6.12)$$

$$OUR_{\mathrm{STO}} = k_{\mathrm{STO}} X_{\mathrm{H}} \frac{1 - Y_{\mathrm{STO}}}{Y_{\mathrm{STO}}} \frac{S_{\mathrm{O}}}{K_{\mathrm{O}} + S_{\mathrm{O}}} \frac{S_{\mathrm{S}}}{K_{\mathrm{S}} + S_{\mathrm{S}}} \tag{6.13}$$

The model interpretation of the OUR profiles according to the equations above (Fig. 6.4b) indicates that a relatively high amount of substrate is used for primary growth, but a relatively low amount of substrate is utilized for storage. The primary growth on substrate may proceed at a low level in some cases, depending on the type and structure of substrate and operating conditions (Karahan et al. 2006a). However, the presence of excessive substrate in the reactor will favor the process in terms of the utilization of substrate and electron acceptor (Karahan et al. 2006a). Thus, the models for an aerobic-granule-based reactor should take simultaneous storage and growth processes into account.

6.1.4.5 Comparison Between ASM3 and the Proposed Model

The experimental measured data (Set 2) are also simulated with the original ASM3 model for comparison with the proposed model. For the ASM3, all biodegradable soluble substrate are initially stored as the internal storage products before it is used for growth, and the adsorption and maintenance processes are not considered (Gujer et al. 1999). The initial stoichiometry and kinetics and simulation conditions used for the calibration of ASM3 are adopted for the calibration of the proposed model. As shown in Fig. 6.5, ASM3 is also capable of simulating the high OUR response at the initial stage. However, a sharp transition toward the second OUR level is simulated. This does not agree with the experimental data. Furthermore, the rapid COD removal process cannot be precisely simulated by the ASM3.

To sort out this problem, the proposed model, with considerations of rapid adsorption process, substrate diffusion, storage process, and heterotrophic growth on storage polymers, is used to simulate the experimental results of Set 2. In contrast, the proposed model is able to predict both the initial OUR response and the smooth transition to the famine phase (Fig. 6.5). The COD profile is also well predicted. In the proposed model, the initial OUR response is determined dominantly by the rapid storage process and primary growth process. These two processes compete with each other for substrate, especially at a low substrate concentration, near the substrate affinity coefficient (K_{S}) value (Karahan et al. 2006a). In this way, the smooth transition to the famine phase can be modeled well. Thus, the process rates and substrate affinity constants are crucial for the simulation of the OUR response to substrate storage and primary growth. In addition, with the introduction of the adsorption process, the model is able to describe the experimental results well. Model comparison results also demonstrate that the proposed model is able to properly simulate the biological reactions occurring in the granule-based reactor under the alternative feast and famine conditions.

6.1.5 Conclusions

This study establishes a mathematic model for describing the simultaneous storage and growth processes in an aerobic-granule-based SBR treating soybean wastewater, with considerations of microbial maintenance and substrate diffusion. Aerobic granules subjected to alternative feast and famine environments in an SBR are able to take up the carbon substrate in the soybean wastewater rapidly and to store it as PHA when the substrate is in excess. The model is successfully calibrated by using experimental results. The validity of the model is verified with both OUR and COD data for the three experimental sets. This model is able to predict the fate of all major model components, i.e., PHA, COD, and OUR. The comparison with ASM3 clearly underlines the merit of incorporating the simultaneous storage and growth process in the model structure. The established model demonstrates its capacity of elucidating the biological processes, such as microbial maintenance, substrate diffusion, simultaneous storage, and growth process. It is a helpful tool to get insight into the biochemical processes occurring in an aerobic-granule-based SBR.

6.2 Storage and Growth of Denitrifiers in Aerobic Granules—Part I: Model Development

6.2.1 Introduction

Denitrification is an important process in biological nutrient removal (BNR) systems. Knowledge about denitrification kinetics is essential towards the optimum design and operation of BNR plants. Denitrification is based on the reduction of nitrate which, in the absence of oxygen, serves as the terminal electron acceptor in the electron transport chain (Sozen and Orhon 1999). This process, also called anoxic respiration as an alternative to aerobic respiration, is accomplished by heterotrophic bacteria utilizing the same metabolic pathway both for anoxic and aerobic respirations, with the only difference being the nitrate reductase generated and the terminal electron acceptor under anoxic conditions. This difference is inevitably related with the kinetic characteristics of the denitrifiers.

Respirometric methods relying upon the measurement of electron acceptor uptake rate have been used as one of the main methods for the characterization of wastewater composition and activated sludge biokinetics (Vanrolleghem et al. 1999; Petersen et al. 2002; Avcioglu 2003). The kinetic behavior of the heterotrophs is determined by means of the OUR under aerobic conditions. Similarly, the denitrification kinetics can be evaluated using the nitrogen uptake rate (NUR) (Naidoo et al. 1998; Kujawa and Klapwijk 1999; Petersen et al. 2002; Avcioglu 2003). It has also been used for assessing anoxic rate reduction factors through comparing OUR and NUR values (Kristensen et al. 1992; Orhon et al. 1996; Soen et al. 1998).

Simultaneous microbial storage and growth have been demonstrated to occur through interpretation of substrate removal mechanisms in activated sludge systems under both aerobic (van Loosdrecht et al. 1997; Krishna and van Loosdrecht 1999; Beun et al. 2000a; Dircks et al. 2001; Carucci et al. 2001) and anoxic conditions (Beun et al. 2000b; Dionisi et al. 2001). From a microbial survival point of view, it is clear that the microbes capable of storing substrate as internal storage polymers when external substrate is available, will gain a selective advantage over the microbes with the sole growth strategy under famine conditions (van Loosdrecht and Heijnen 2002). Moreover, experimental observations also confirm the existence of storage phenomenon (Beun et al. 2000b; Dionisi et al. 2001). For instance, a higher growth yield, e.g. 0.75, is usually required for adequately modeling the OUR profiles of activated sludge from batch experiments (Gernaey et al. 2002). This means that part of the substrate is used for other purposes, in addition to the formation of new bacterial cells. Consequently, several mathematical models are proposed to describe such aerobic simultaneous storage and growth processes. These models are mainly based on the metabolic pathways in ASM approaches (Krishna and van Loosdrecht 1999; Carucci et al. 2001; Karahan et al. 2003; Sin et al. 2005; Karahan et al. 2006a). They are usually evaluated using OUR measurements obtained from respirometers, which are well-established tools for monitoring biomass activity under aerobic conditions.

However, so far information on the storage phenomenon under anoxic conditions is still sparse, compared with its aerobic counterpart. A simulation model was established to describe the nitrification and denitrification processes in aerobic granules based on ASM3 (Beun et al. 2001), and was used to evaluate the role of decreased DO levels in reactors and to predict the effect of several operating parameters on N removal. Another approach was proposed by Sin et al. (2005) to model simultaneous storage and growth processes in activated sludge systems under aerobic conditions, and a second-order model was developed for describing the degradation of storage products at famine phase. This model was successfully calibrated using OUR data obtained from batch tests. However, in an aerobic-granule-based reactor simultaneous storage and growth under anoxic conditions is a different complex case, with numerous internal interactions among process variables and sludge characteristics.

Therefore, the main objective of this work is to establish a new mathematic model to describe the simultaneous storage and growth processes in aerobic granules under anoxic conditions. In this model, mass transfer, hydrolysis, simultaneous storage and growth, maintenance, and endogenous decay are all taken into account. Parameter estimation and model verification are presented in an accompanying paper (Ni et al. 2008).

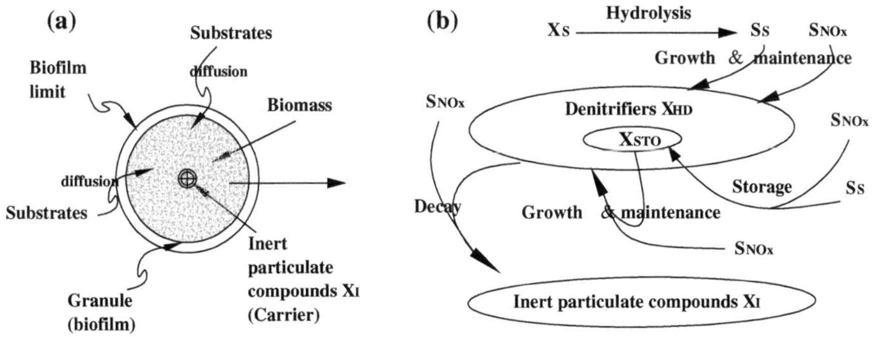

Fig. 6.6 Modeling storage and growth of the granules under anoxic conditions: **a** Characteristics of granule and substrate diffusion; and **b** the proposed food web illustrating the flow of the external and internal substrates, and the interactions among the involved components (From Ni and Yu (2008), reprinted with permission from Wiley-Blackwell)

6.2.2 Materials and Methods

6.2.2.1 Reactor Setup and Operation

Granules were cultivated under aerobic conditions in a laboratory-scale SBR fed with a fatty-acids-rich wastewater, as described in Chap. 3.

6.2.2.2 Determination of Anoxic Reduction Factor

With the process of experiment, the number and average diameter of the granules kept increasing. After 1 month operation, the granules became physiologically stabilized. They were yellow in color and sphere or ellipse in shape. Then, batch experiments for OUR and NUR measurements with external carbon sources were conducted under both anoxic and aerobic conditions to determine the anoxic reduction factor.

The respirometric tests were carried out with two identical 1 L batch reactors with the same wastewater and aerobic granules. One reactor was for OUR measurement through monitoring the OUR, and another was for the NUR determination by measuring the nitrate utilization rate. The two reactors were inoculated with the granular sludge, which was withdrawn from the SBR at the end of the aeration phase and kept aeration for 10 h to ensure the depletion of any external substrate. Later, the fatty-acids-rich wastewater was added as the external substrate for the two reactors. The test lasted for 4–6 h at 20 °C and pH of 7.0–8.0. Samples were withdrawn from the two reactors every 6–10 min, and were immediately centrifuged and then stored at 4 °C until analysis was performed.

6.2.2.3 Analytical Procedures

The analysis of DO, OUR, and sludge image was described in Chap. 3. Measurements of COD, nitrite (NO_2-N), nitrate (NO_3-N), MLSS, and MLVSS followed the standard methods (APHA 1995). The concentration of internal storage polymers, i.e., PHA, was measured according to Pratt et al. (2004).

6.2.3 Model Development

6.2.3.1 Conceptual Basis

The model developed in this work has six components, i.e., nitrate and nitrite, S_{NOx}; particulate inert COD, X_I; readily biodegradable COD, S_S; slowly biodegradable COD, X_S; denitrifier biomass, X_{HD}; and storage products, X_{STO}. It involves eight processes, including mass transfer, hydrolysis, anoxic growth on S_S, anoxic storage, anoxic growth on X_{STO}, anoxic maintenance on S_S, anoxic maintenance on X_{STO}, and biomass decay. A schematic diagram is given in Fig. 6.6 to describe the relationships among the different groups of components within an aerobic granule under anoxic conditions. Granules with a wide range in size are modeled with a consideration of component diffusion limitation. For the components involved in the biological reactions, the first step is their diffusion into the granules where the reactions take place. Then, the influent X_s is hydrolyzed to S_s by X_{HD}. These bacteria can use S_s for anoxic simultaneous storage and growth. As S_s is depleted, the anoxic degradation of X_{STO} occurs. In addition to anoxic growth and maintenance, microorganisms in the granules are subjected to endogenous decay, generating X_I.

In modeling, simultaneous storage and growth of aerobic granules under anoxic conditions, two distinct but complementary phases: feast and famine, should be considered like its aerobic counterpart. Under feast conditions, based on a conventional ASM-type model structure, three distinctive yield coefficients independent from each other are used for storage, direct growth on external substrate, and growth on internal storage products, respectively: (Krishna and van Loosdrecht 1999; Carucci et al. 2001; Karahan et al. 2003; Pratt et al. 2004; Sin et al. 2005; Karahan et al. 2006a). Under famine conditions, degradation of the storage polymers is the rate-limiting step and governs the growth rate (Sin et al. 2005).

6.2.3.2 Electron Acceptor for Denitrification

In activated sludge models, denitrification is often modeled using Monod kinetics for growth and a reduced yield coefficient, by assuming complete denitrification in one step (from NO_3^- to nitrogen gas). However, numerous studies have demonstrated that denitrification occurs following a sequential reduction path for oxidized nitrogen species (Thomsen et al. 1994; Almeida et al. 1995) as follows:

$$NO_3^- \Rightarrow NO_2^- \Rightarrow NO \Rightarrow 0.5N_2O \Rightarrow 0.5N_2 \quad (6.14)$$

Moreover, partial denitrification may or may not occur, depending on several factors such as COD/N, substrate type, biomass type, pH, and temperature, and so on. (Thomsen et al. 1994; Oh and Silverstein 1999). Since there is still no consensus—rather conflicting result, on the mechanisms and occurrence of partial denitrification (e.g., nitrite build up or emission of N_2O in denitrification), in the present study denitrification is modeled in a modified single step (Soen et al. 1998).

However, the accumulation of nitrite in denitrification is considered for modeling in the present study. Thus, the accurate amount of electron acceptor in the NUR tests should be known in such a modeling. For all calculations associated with substrate utilization, the NO_3^--N measurement does not generate acceptable results when coupled with appreciable NO_2^--N accumulation in the experimental course. In such a case, the electron equivalence of the readily biodegradable COD consumption could be calculated with the following equation (Kujawa and Klapwijk 1999; Soen et al. 1998):

$$N = NO_3^- - N + 0.6NO_2^- - N \quad (6.15)$$

This calculation is based on the fact that the reduction of 1 g NO_2^-–N to 1 g N_2 requires the same amount of electrons as the reduction of 0.6 g NO_3^-–N to 0.6 g N_2. The sum of nitrate and nitrite is expressed as the amount of nitrate that needs the same amount of electrons by reducing it to nitrogen (Cokgor et al. 1998).

6.2.3.3 Mass Transfer

A mass transfer model (Wanner and Reichert 1996) implemented in AQUASIM (Reichert 1998) is used to model the substrate diffusion within granules in this work. Similar to biofilm, aerobic granules consist of a solid matrix with pore water that contains dissolved substances and suspended solids (Reichert 1998). The growth or decay of microorganisms forming the solid matrix leads to the expansion or shrinking of the granules (Reichert 1998). The substrate consumption by a high concentration of microorganisms in the granule solid matrix results in a growth limitation by the diffusive mass transfer into the granule depth (Reichert 1998).

One-dimensional conservation law (Eq. 6.16) is formulated as a balance between the mass conserved and utilized in this model (Wanner and Reichert 1996).

$$\frac{\partial \hat{\rho}}{\partial t} + \frac{\partial \hat{j}}{\partial z} = \hat{r} \quad (6.16)$$

where $\hat{\rho}$ is the one-dimensional density (amount of conserved quantity per unit compartment length), $\hat{j}$ is the one-dimensional flux (amount of the conserved quantity transported per unit time), $\hat{r}$ is the one-dimensional source term (amount

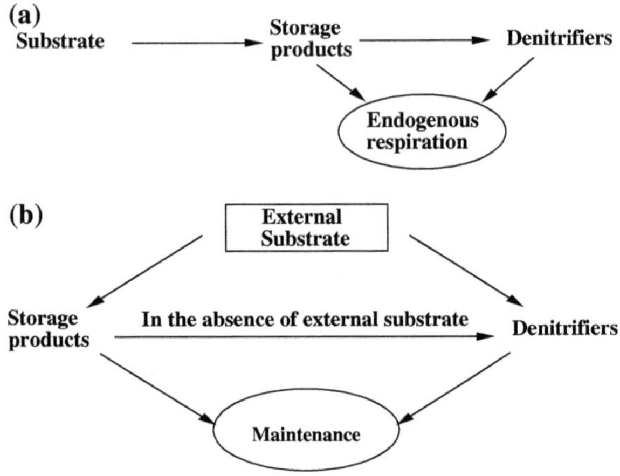

Fig. 6.7 **a** Mechanisms of the anoxic substrate utilization in ASM3; and **b** the proposed mechanisms for the anoxic substrate utilization (From Ni and Yu (2008), reprinted with permission from Wiley-Blackwell)

produced per unit compartment length and per unit time), t is time, and z is the space coordinate.

6.2.3.4 Simultaneous Storage and Growth Under Anoxic Conditions

In the aerobic storage process in ASM3 (Gujer et al. 1999), it is assume that all the readily biodegradable organic substrates are first taken up by the heterotrophs and converted to stored materials, which are subsequently assimilated to biomass (Fig. 6.7a) However, in this work the ASM3 model is modified by assuming the utilization of substrate for simultaneous growth and storage under anoxic conditions. It is considered that the substrate is used for the production of storage polymers and biomass, and that the hydrolysis process is present for the slowly biodegradable substrate. After substrate depletion, microorganisms utilize the storage polymers accumulated for anoxic growth (Fig. 6.7b).

Slowly biodegradable substrate, X_S, is hydrolyzed at a rate determined by surface reaction kinetics as in ASM models (Eq. 6.17). Hydrolysis process is presumably rate-limiting stage for the anoxic utilization of the partial readily biodegradable substrate.

$$\frac{dX_S}{dt} = -k_{HD}\frac{X_S/X_{HD}}{K_X + X_S/X_{HD}}X_{HD} \tag{6.17}$$

where k_{HD} is the maximum hydrolysis rate, and K_X is the hydrolysis affinity constant.

At feast phase, because of a pulse addition of S_S, the microbes primarily utilize the substrate available through storage and growth. The kinetic expression for the primary growth is given in Eq. 6.18, and the rate of storage, $r_{anoxic\text{-}STO}$, is defined below:

$$r_{\text{anoxic-growS}} = \mu_{\text{HD,S}}\eta_{\text{NOx}}\frac{S_S}{K_s + S_s}\frac{S_{\text{NOx}}}{K_{\text{NOx}} + S_{\text{NOx}}}X_{\text{HD}} \tag{6.18}$$

$$r_{\text{anoxic-STO}} = k_{\text{STO,D}}\eta_{\text{NOx}}\frac{S_S}{K_S + S_S}\frac{S_{\text{NOx}}}{K_{\text{NOx}} + S_{\text{NOx}}}X_{\text{HD}} \tag{6.19}$$

where $\mu_{\text{HD,S}}$ is the maximum growth rate on S_S, K_S is the external substrate affinity constant, K_{NOx} is the biomass affinity constant for S_{NOx}, $k_{\text{STO,D}}$ is the maximum storage rate of biomass, and η_{NOx} is the anoxic reduction factor.

The nitrate utilization kinetics associated with the primary substrate consumption is described by Eq. 6.20, while the nitrate utilization kinetics associated with the anoxic substrate storage (X_{STO}) is expressed as Eq. 6.21.

$$NUR_{\text{growS}} = \mu_{\text{HD,S}}\eta_{\text{NOx}}X_{\text{HD}}\frac{1 - Y_{\text{HD,S}}}{2.86Y_{\text{HD,S}}}\frac{S_S}{K_S + S_S}\frac{S_{\text{NOx}}}{K_{\text{NOx}} + S_{\text{NOx}}} \tag{6.20}$$

$$NUR_{\text{STO}} = k_{\text{STO,D}}\eta_{\text{NOx}}X_{\text{HD}}\frac{1 - Y_{\text{STO,D}}}{2.86Y_{\text{STO,D}}}\frac{S_S}{K_S + S_S}\frac{S_{\text{NOx}}}{K_{\text{NOx}} + S_{\text{NOx}}} \tag{6.21}$$

where $Y_{\text{HD,S}}$ is the anoxic yield coefficient for growth on S_S, and $Y_{\text{STO,D}}$ is the anoxic yield coefficient for storage.

6.2.3.5 Anoxic Degradation of Storage Products

After the consumption of the primary readily biodegradable substrate, the secondary anoxic growth occurs on the stored X_{STO} at famine phase (Fig. 6.7b). The secondary anoxic growth process can only take place if the readily biodegradable substrate concentration S_S is as low as the half saturation concentration for the primary growth, K_S. In this study, the anoxic growth rate on X_{STO} is modeled as:

$$r_{\text{anoxic-growSTO}} = \mu_{\text{HD,STO}}\eta_{\text{NOx}}\frac{K_S}{K_S + S_S}\frac{S_{\text{NOx}}}{K_{\text{NOx}} + S_{\text{NOx}}}\frac{X_{\text{STO}}/X_{\text{HD}}}{K_{\text{STO}} + X_{\text{STO}}/X_{\text{HD}}}X_{\text{HD}} \tag{6.22}$$

where $\mu_{\text{HD,STO}}$ is the maximum growth rate on X_{STO}, and K_{STO} is the biomass affinity constant for X_{STO}. The item $\frac{X_{\text{STO}}/X_{\text{HD}}}{K_{\text{STO}}+X_{\text{STO}}/X_{\text{HD}}}$ describes the surface-saturation-type degradation kinetics of X_{STO}. The anoxic microbial growth rate on X_{STO} is assumed to occur under strict famine conditions (van Loosdrecht and Heijnen 2002). Thus, a Monod inhibition function for the external substrate $\frac{K_S}{K_S+S_S}$ is added to the kinetic description of $r_{\text{anoxic-growSTO}}$.

The nitrate utilization associated with the second anoxic substrate (X_{STO}) consumption is defined below:

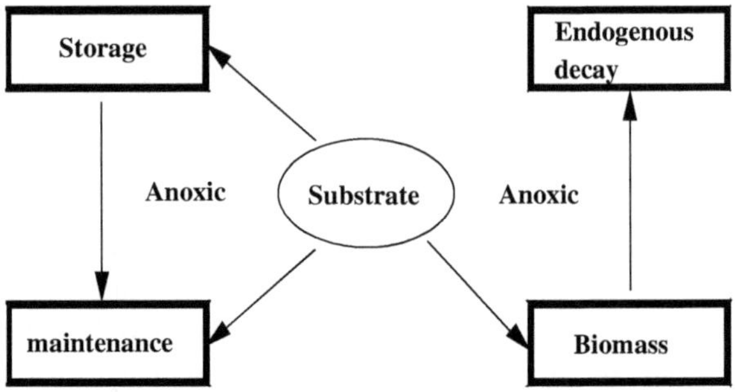

Fig. 6.8 The anoxic maintenance and endogenous decay concept proposed (From Ni and Yu (2008), reprinted with permission from Wiley-Blackwell)

$$NUR_{\text{growSTO}} = \mu_{\text{HD,STO}}\eta_{\text{NOx}}X_{\text{HD}}\frac{1-Y_{\text{HD,STO}}}{2.86Y_{\text{HD,STO}}}\frac{K_S}{K_S+S_S}\frac{X_{\text{STO}}/X_{\text{HD}}}{K_{\text{STO}}+X_{\text{STO}}/X_{\text{HD}}}\frac{S_{\text{NOx}}}{K_{\text{NOx}}+S_{\text{NOx}}} \tag{6.23}$$

where $Y_{\text{HD,STO}}$ is the anoxic yield coefficient for growth on X_{STO}.

However, when X_{STO} decreases and reaches a minimum level, the degradation rate of X_{STO} becomes limited. The anoxic degradation rate of X_{STO} strongly depends on X_{STO} content in biomass (Dircks et al. 2001). The biomass always has a minimum X_{STO} content, so that a minimum level of storage products can be used for the maintenance by the denitrifiers at famine phase, similar to the maintenance on S_s under feast conditions.

6.2.3.6 Anoxic Maintenance

Maintenance is usually defined based on the influent substrate (primary substrate), if the secondary substrates like storage polymers are taken into account, the maintenance can then be related with these materials as well (van Loosdrecht and Henze 1999). In this study, the anoxic maintenance of the denitrifiers is separately considered, and is not included in the endogenous respiration coefficient, in order to approach the actual situations (Fig. 6.8). The denitrifiers also need energy for maintenance and other activities, in addition to for anabolism.

When the external substrate (S_s) is present at feast phase, the denitrifiers (X_{HD}) consume S_s for their anoxic maintenance at a rate given in Eq. 6.24. As S_s is depleted at famine phase, the denitrifiers consume X_{STO} for their anoxic maintenance at a rate shown in Eq. 6.25.

$$r_{\text{anoxic-mainS}} = m_{\text{HD,S}}X_{\text{HD}}\frac{S_S}{K_S+S_S}\frac{S_{\text{NOx}}}{K_{\text{NOx}}+S_{\text{NOx}}} \tag{6.24}$$

$$r_{\text{anoxic-mainSTO}} = m_{\text{HD,STO}}X_{\text{HD}}\frac{K_S}{K_S+S_S}\frac{S_{\text{NO}x}}{K_{\text{NO}x}+S_{\text{NO}x}}\frac{X_{\text{STO}}/X_{\text{HD}}}{K_{\text{STO}}+X_{\text{STO}}/X_{\text{HD}}} \tag{6.25}$$

Table 6.4 Kinetic and stoichiometric coefficients used in the established model

Parameter	Definition	Unit	Value	Source
$Y_{STO,D}$	Anoxic yield for storage	g COD g^{-1} COD	0.55	(1)
f_I	Fraction of X_I in decay	g COD g^{-1} COD	0.20	(2)
η_{NOx}	Anoxic reduction factor	–	0.60	(2)
$Y_{HD,S}$	Yield for growth on S_S	g COD g^{-1} COD	0.40	(1)
$Y_{HD,STO}$	Yield for growth on X_{STO}	g COD g^{-1} COD	0.67	(1)
$k_{STO,D}$	Maximum storage rate	h^{-1}	0.21	(2)
$\mu_{HD,STO}$	Maximum growth rate on X_{STO}	h^{-1}	0.083	(2)
$\mu_{HD,S}$	Maximum growth rate on S_S	h^{-1}	0.083	(2)
K_S	Substrate affinity constant	g COD m^{-3}	2.0	(2)
K_{NOx}	Affinity constant for S_{NOx}	g N m^{-3}	0.50	(2)
$b_{HD,NOx}$	Anoxic decay rate coefficient	h^{-1}	0.0042	(2)
K_{STO}	Affinity constant for X_{STO}	g COD g^{-1} COD	1.0	(2)
K_X	Hydrolysis affinity constant	g COD g^{-1} COD	1.0	(2)
k_{HD}	Maximum hydrolysis rate	h^{-1}	0.13	(2)
$m_{HD,S}$	Anoxic maintenance rate on *SS*	h^{-1}	0.0066	(3)
$m_{HD,STO}$	Anoxic maintenance rate on X_{STO}	h^{-1}	0.0044	(3)

Notes (1) Sin (2004), (2) Gujer et al. (1999), (3) Beun et al. (2001)

Table 6.5 Diffusion parameters of granules

Parameter	Description	Value	Reference
Granules			
R_{min}	Minimum radius, mm	0.41	Measured
R_{max}	Maximum radius, mm	3.50	Measured
R_{mean}	Mean radius, mm	1.20	Measured
Mass transport			
DSs	Effective diffusivity of Ss, dm2 h^{-1}	0.000576	Beun et al. (2001)
D^{NO_2}	Effective diffusivity of NO_2, dm^2 h^{-1}	0.000576	Beun et al. (2001)
D^{NO_3}	Effective diffusivity of NO_3, dm^2 h^{-1}	0.000576	Beun et al. (2001)

where $m_{HD,S}$ is the anoxic maintenance coefficient on S_S, and $m_{HD,STO}$ is the anoxic maintenance coefficient on X_{STO}.

The nitrate utilization associated with anoxic maintenance on the external substrate consumption is defined by Eq. 6.26, whereas the nitrate utilization associated with anoxic maintenance on the substrate storage (X_{STO}) is described as Eq. 6.27.

$$NUR_{mainS} = \frac{1}{2.86} m_{HD,S} X_{HD} \frac{S_S}{K_S + S_S} \frac{S_{NOx}}{K_{NOx} + S_{NOx}} \tag{6.26}$$

$$NUR_{mainSTO} = \frac{1}{2.86} m_{HD,STO} X_{HD} \frac{K_S}{K_S + S_S} \frac{S_{NOx}}{K_{NOx} + S_{NOx}} \frac{X_{STO}/X_{HD}}{K_{STO} + X_{STO}/X_{HD}} \tag{6.27}$$

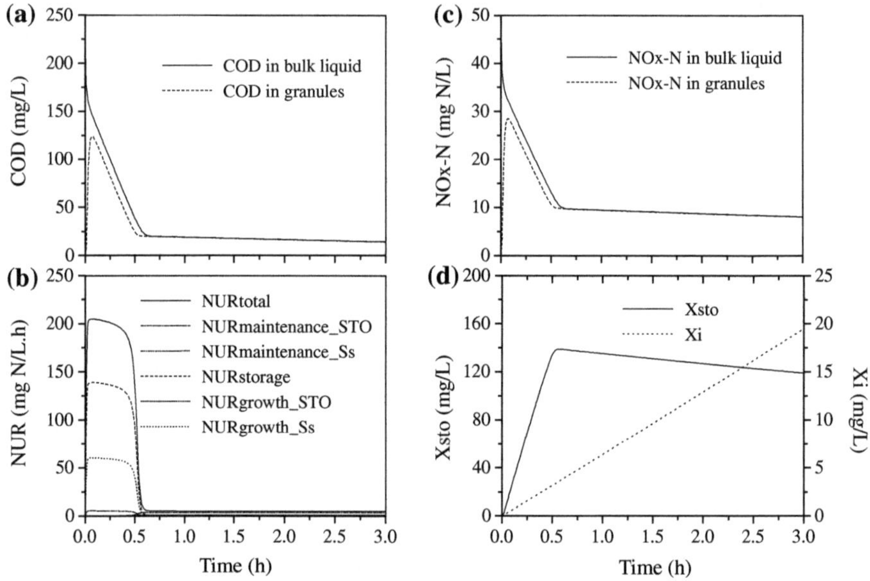

Fig. 6.9 The model simulation results: **a** Variations of COD concentration with time; **b** Distribution of NUR; **c** Variations of S_{NOx} concentration with time; and **d** Variations of storage product, X_{STO}, and inert product, X_{I}, concentration with time (From Ni and Yu (2008), reprinted with permission from Wiley-Blackwell)

6.2.3.7 Biomass Decay

Biomass endogenous decay caused by predation and cell lysis leads to a significant loss in biomass mass and its activity. The assumption of solids formation in biomass endogenous decay in the ASM models is based on the observation that the model fits the data better if the generation of inert COD is assumed (van Loosdrecht and Henze, 1999). Thus, the inert production in respiration is introduced in this model. Equation 6.28 describes the accumulation of particulate inert COD (X_{I}) as a result of decay of active biomass.

$$\frac{\mathrm{d}X_{\mathrm{I}}}{\mathrm{d}t} = f_{\mathrm{I}} b_{\mathrm{HD,NOx}} \frac{K_{\mathrm{S}}}{K_{\mathrm{S}} + S_{\mathrm{S}}} \frac{K_{\mathrm{STO}}}{K_{\mathrm{STO}} + X_{\mathrm{STO}}/X_{\mathrm{HD}}} X_{\mathrm{HD}} \tag{6.28}$$

where f_{I} is the fraction of X_{I} in decay, and $b_{\mathrm{HD,NOx}}$ is the anoxic decay rate coefficient.

6.2.3.8 Model Simulation

The model simulation is carried out with the AQUASIM computation program (Reichert 1998). The model developed in this work is implemented in the well-established simulation software. A combination of completely mixed reactor and biofilm reactor compartments (Beun et al. 2001) provided by AQUASIM is used to simulate the mass transport and conversion processes occurring in the bulk liquid and granules. The granules cultivated in this study have a size distribution between 0.41 and 3.5 mm with an average of 1.2 mm. It is true that granules with a wide range in size are formed in the reactor and that the size distribution also changes in time. However, the granule size distribution is not taken into account in this model, as it would significantly increase the complexity of numerical computation and it is not expected to contribute to a better understanding of the system. Therefore, in the simulation the diameter is chosen to be 1.2 mm, which is the most representative for the aerobic granules in this SBR. Simulation is performed under the same operational conditions as applied to the batch tests. The parameters of the model developed to generate outputs are respectively shown in Tables 6.4 and 6.5.

6.2.4 Results and discussion

6.2.4.1 Simulation results

With the model setup and parameters listed in Tables 6.4 and 6.5, model simulation is performed. The simulation results of the fates of X_{STO}, X_I, S_{NOx}, and total COD concentration are illustrated in Fig. 6.9. Prediction of the contributions of each process to the total NUR is also given.

It is clear from the Fig. 6.9a that the COD concentration in bulk liquid decreases sharply. The COD concentration in granules increases initially, attributed to the diffusion process, but gradually decreases in the anoxic utilization process. The modeling results explicitly show that the organic substrate is immediately utilized for the storage and growth at feast phase.

As shown in Fig. 6.9b, at feast phase the NUR increases initially, and peaks within 2–3 min. This peak level keeps almost unchanged until reaching the famine phase. At famine phase, the NUR decreases to a baseline level beyond the endogenous NUR level. This elevated level of NUR can be attributed to the additional uptake of nitrate for oxidation of the polymers stored previously. In this model, the consumption of nitrate at feast phase is associated with the growth and storage. The storage process consumes more nitrate, compared to the biomass production (Fig. 6.9b). At famine phase, nitrate is consumed by a secondary growth process on the internal substrate, X_{STO}, and anoxic maintenance of biomass. In this way, this model is able to model the feast-famine phase more mechanistically, compared to the ASM3.

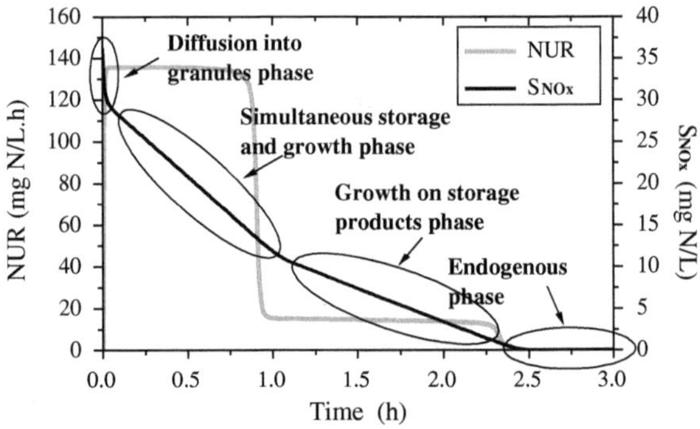

Fig. 6.10 Kinetic structure analysis on NUR (From Ni and Yu (2008), reprinted with permission from Wiley-Blackwell)

Figure 6.9c shows the simulation results for the nitrate concentration, S_{NOx}. Several linear phases (slopes) of nitrate reduction in denitrification are observed. In the substrate utilization process the concentration of the residual particulate inert COD, X_I, increases linearly (Fig. 6.9d). The storage product concentration (X_{STO}) increases at feast phase, but decreases gradually at the subsequent operation.

6.2.4.2 Kinetic Structure of Nitrogen Uptake Rate (NUR)

For the NUR tests, four linear phases (slopes) of nitrate reduction rate in denitrification are observed. The first slope reflects the diffusion of nitrate into the granules. The second slope of nitrate utilization rate is related to the simultaneous denitrification and storage on S_S and maintenance on *Ss*. This is followed by a lower denitrification rate resulting from the anoxic utilization and maintenance of the storage products X_{STO}. The lowest denitrification rate is reached after the depletion of S_s and X_{STO}.

A typical NUR profile of the granules can be evaluated according to the proposed model as shown in Fig. 6.10. The first phase is associated with the diffusion of nitrate and nitrite within the granules. The simultaneous storage and growth phase emerges after the addition of the exogenous substrate. The nitrate utilization rate at this phase is governed by the three anoxic processes, i.e., storage, growth, and maintenance:

$$\begin{aligned}\mathrm{NUR}_{\mathrm{Sto-Grow,phase}} &= \left(\frac{\mathrm{d}S_{\mathrm{NOx}}}{\mathrm{d}t}\right)_{\mathrm{Sto-Grow,phase}} \\ &= \left(\frac{\mathrm{d}S_{\mathrm{NOx}}}{\mathrm{d}t}\right)_{\mathrm{Sto}} + \left(\frac{\mathrm{d}S_{\mathrm{NOx}}}{\mathrm{d}t}\right)_{\mathrm{Grow},Ss} + \left(\frac{\mathrm{d}S_{\mathrm{NOx}}}{\mathrm{d}t}\right)_{\mathrm{Main},Ss}\end{aligned} \tag{6.29}$$

The third phase in Fig. 6.10 is the growth of the denitrifiers on the storage products X_{STO}, and the overall rate is attributed to the processes of growth on X_{STO}, and anoxic maintenance on X_{STO}:

$$\mathrm{NUR}_{\mathrm{Grow}X_{\mathrm{STO}},\mathrm{phase}} = \left(\frac{\mathrm{d}S_{\mathrm{NOx}}}{\mathrm{d}t}\right)_{\mathrm{Grow}X_{\mathrm{STO}},\mathrm{phase}} = \left(\frac{\mathrm{d}S_{\mathrm{NOx}}}{\mathrm{d}t}\right)_{\mathrm{Grow},X_{\mathrm{STO}}} + \left(\frac{\mathrm{d}S_{\mathrm{NOx}}}{\mathrm{d}t}\right)_{\mathrm{Main},X_{\mathrm{STO}}} \tag{6.30}$$

The last phase is associated with the anoxic maintenance on X_{STO}:

$$\mathrm{NUR}_{\mathrm{Endo,phase}} = \left(\frac{\mathrm{d}S_{\mathrm{NOx}}}{\mathrm{d}t}\right)_{\mathrm{Endo,phase}} = \left(\frac{\mathrm{d}S_{\mathrm{NOx}}}{\mathrm{d}t}\right)_{\mathrm{Main},X_{\mathrm{STO}}} \tag{6.31}$$

6.2.4.3 Anoxic Reduction Factor: ηNOx

Heterotrophs have the capability of switching the mode of their metabolic activity, and readily adapting themselves to consume nitrate as the terminal electron acceptor, instead of DO (Soen et al. 1998). However, the electron acceptor utilization rate under anoxic conditions is always lower than that under aerobic conditions. This is because only a portion of heterotrophs can use nitrate as the terminal electron acceptor. Another explanation is that the maximum specific growth rate of heterotrophs may be reduced under anoxic conditions (Soen et al. 1998). Since no experimental procedure is available for separately quantifying the contribution ratios of the two cases, an approach is to adopt the aerobic rate expression with the addition of an empirical coefficient (η_{NOx}) acting as an overall correction factor for the anoxic case. The value of $\eta_{\mathrm{NO}x}$ should always be lower than 1.0 ($\eta_{\mathrm{NOx}} < 1.0$) as the aerobic respiration is a more efficient process than anoxic respiration from a thermodynamic point of view (Orhon et al. 1996).

Experimental assessment of η_{NOx} relies on respirometric methods through measuring the electron acceptor utilization rates in parallel under both aerobic and anoxic conditions (Soen et al. 1998; Orhon et al. 1996). Electron acceptor is consumed by both growth and maintenance processes. In this work, the oxygen utilization rate, OUR, for an aerobic reactor may be expressed as:

$$\mathrm{OUR} = \left(\mu_{\mathrm{H,S}}\frac{1 - Y_{\mathrm{H,S}}}{Y_{\mathrm{H,S}}} + k_{\mathrm{STO}}\frac{1 - Y_{\mathrm{STO}}}{Y_{\mathrm{STO}}}\right)\frac{S_{\mathrm{S}}}{K_{\mathrm{S}} + S_{\mathrm{S}}}X_{\mathrm{H}} + m_{\mathrm{H}}\frac{S_{\mathrm{S}}}{K_{\mathrm{S}} + S_{\mathrm{S}}}X_{\mathrm{H}} \tag{6.32}$$

For the counterpart, NUR, the nitrate utilization rate is expressed as follows:

$$\mathrm{NUR} = \left(\mu_{\mathrm{HD,S}}\frac{1 - Y_{\mathrm{HD,S}}}{2.86Y_{\mathrm{HD,S}}} + k_{\mathrm{STO,D}}\frac{1 - Y_{\mathrm{STO,D}}}{2.86Y_{\mathrm{STO,D}}}\right)\eta_{\mathrm{NOx}}\frac{S_{\mathrm{S}}}{K_{\mathrm{S}} + S_{\mathrm{S}}}X_{\mathrm{HD}} + m_{\mathrm{HD}}\frac{S_{\mathrm{S}}}{K_{\mathrm{S}} + S_{\mathrm{S}}}X_{\mathrm{HD}} \tag{6.33}$$

A correction factor is experimentally determined using the following equation:

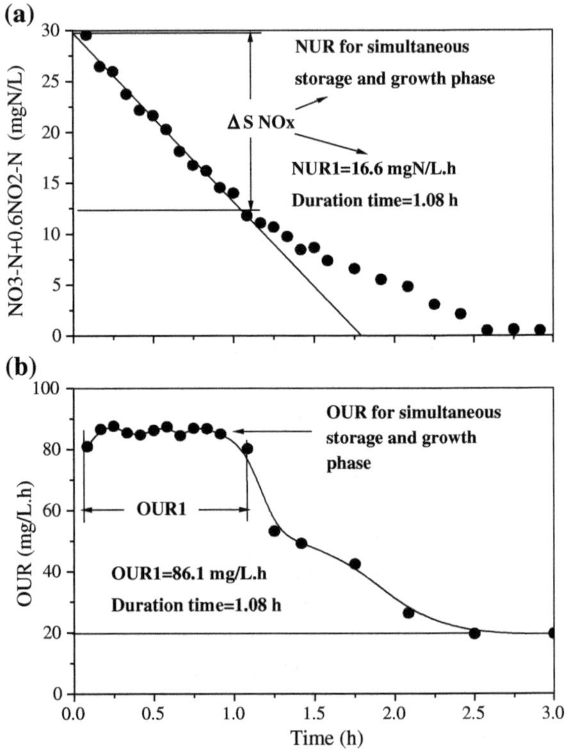

Fig. 6.11 Respirometric tests for the assessment of anoxic reduction factor η_{NOx} of the fatty-acids-rich wastewater: **a** NUR test; and **b** OUR test (From Ni and Yu (2008), reprinted with permission from Wiley-Blackwell)

$$\frac{2.86\mathrm{NUR}}{\mathrm{OUR}} = \frac{\left(\mu_{HD,S}\frac{1-Y_{HD,S}}{Y_{HD,S}} + k_{STO,D}\frac{1-Y_{STO,D}}{Y_{STO,D}}\right)\eta_{NOx}\frac{S_S}{K_S+S_S}X_{HD} + m_{HD}\frac{S_S}{K_S+S_S}X_{HD}}{\left(\mu_{H,S}\frac{1-Y_{H,S}}{Y_{H,S}} + k_{STO}\frac{1-Y_{STO}}{Y_{STO}}\right)\frac{S_S}{K_S+S_S}X_H + m_H\frac{S_S}{K_S+S_S}X_H} \tag{6.34}$$

As the experimental setup is adjusted to characterize the maximum growth conditions, the equation above may be rearranged with the assumption that the electron acceptor utilization rate attributed to maintenance is very low and may be neglected (Orhon et al. 1996), thus:

$$\frac{2.86 \times \mathrm{NUR}_1}{\mathrm{OUR}_1} = \frac{\left(\mu_{HD,S}\frac{1-Y_{HD,S}}{Y_{HD,S}} + k_{STO,D}\frac{1-Y_{STO,D}}{Y_{STO,D}}\right)\eta_{NOx}\frac{S_S}{K_S+S_S}X_{HD}}{\left(\mu_{H,S}\frac{1-Y_{H,S}}{Y_{H,S}} + k_{STO}\frac{1-Y_{STO}}{Y_{STO}}\right)\frac{S_S}{K_S+S_S}X_H} \tag{6.35}$$

Equation 6.35 shows that, if the yield coefficient value remains unchanged under both aerobic and anoxic conditions, the correction factor η_{NOx} could be estimated from the respirometric data. When $Y_{HD,S}$ is different from $Y_{H,S}$, the item (2.86 NUR/OUR ratio) should be related to η_{NOx}, as defined below:

$$\eta'_{\mathrm{NOx}} = \frac{2.86 * \mathrm{NUR}_1}{\mathrm{OUR}_1} = \frac{\left(\mu_{\mathrm{HD,S}} \frac{1-Y_{\mathrm{HD,S}}}{Y_{\mathrm{HD,Ss}}} + k_{\mathrm{STO,D}} \frac{1-Y_{\mathrm{STO,D}}}{Y_{\mathrm{STO,D}}}\right)}{\left(\mu_{\mathrm{H,S}} \frac{1-Y_{\mathrm{H,S}}}{Y_{\mathrm{H,S}}} + k_{\mathrm{STO}} \frac{1-Y_{\mathrm{STO}}}{Y_{\mathrm{STO}}}\right)} \eta_{\mathrm{NOx}} \tag{6.36}$$

In the present study, the correction factor for the anoxic growth has been experimentally assessed for the granules grown on the fatty-acids-rich wastewater. The anoxic correction factor η_{NOx} is calculated with Eq. 6.35, using the initial OUR in the aerobic batch tests and the initial NUR in the anoxic batch tests. The OUR and NUR tests calculated are illustrated in Fig. 6.11. The anoxic correction factor value is estimated to be $\eta_{\mathrm{NOx}} = 0.55$.

However, it should be noticed that the experimental assessment of the correction factor for denitrification is likely to have an error if the yield coefficient values are significantly different under aerobic and anoxic conditions:

$$\frac{\mu_{\mathrm{HD,S}} \frac{1-Y_{\mathrm{HD,S}}}{Y_{\mathrm{HD,S}}} + k_{\mathrm{STO,D}} \frac{1-Y_{\mathrm{STO,D}}}{Y_{\mathrm{STO,D}}}}{\mu_{\mathrm{H,S}} \frac{1-Y_{\mathrm{H,S}}}{Y_{\mathrm{H,S}}} + k_{\mathrm{STO}} \frac{1-Y_{\mathrm{STO}}}{Y_{\mathrm{STO}}}} \tag{6.37}$$

In this study, the difference between the yield coefficient values under aerobic and anoxic conditions is not significant. Hence, the calibrated experimental value of the anoxic correction factor $\eta_{\mathrm{NO}x}$ is used for the model simulation.

6.2.5 Conclusions

In this study a mathematical model is developed to describe the simultaneous storage and growth activities of denitrifiers in aerobic granules under anoxic conditions. The respirometric tests are carried out with two identical SBRs for OUR and NUR measurements. A mass transfer model is used to characterize the mass diffusion within the aerobic granules. With a modified ASM3, biochemical reactions under anoxic conditions, including hydrolysis, anoxic simultaneous storage and growth, anoxic maintenance and biomass decay are taken into account. The model simulation indicates that the NUR of granules-based denitrification process includes four linear phases of nitrate reduction. The most important parameter, i.e., the anoxic correction factor η_{NOx}, is experimentally determined to be 0.55. With the parameters initialized, model simulation is performed to elucidate the anoxic denitrification process in aerobic-granule-based reactors.

6.3 Storage and Growth of Denitrifiers in Aerobic Granules—Part II: Model Calibration and Verification

6.3.1 Introduction

A new mathematical model to describe the simultaneous storage and growth activities of denitrifiers in aerobic granular sludge under anoxic conditions has been developed in the accompanying paper (Ni and Yu 2008). ASM3 is modified to establish this new model. In the present model, the assessment of the accurate amount of electron acceptor is modified in the NUR test, as an accumulation of nitrite is not always negligible. A combination of completely mixed reactor and biofilm reactor compartments provided by AQUASIM, a well-established simulation software (Reichert 1998), is used to simulate the mass transport and conversion processes occurring in the bulk liquid and granules. In this model the anoxic biochemical reactions described with the modified ASM3 through taking hydrolysis, anoxic simultaneous storage and growth, anoxic maintenance, and endogenous decay into account.

The modeling results explicitly show that the external substrate is immediately utilized for storage and growth at feast phase. More nitrate is consumed in the storage process, compared to the production of biomass. The simulated NUR curve of the established model includes four linear phases (slopes) of nitrate reduction. The methodology for determining the most important parameter, i.e., anoxic reduction factor (η_{NOx}), in this model is also developed. The proposed model gives a good prediction of the anoxic growth yield and the maximum anoxic growth rate of the denitrifiers in aerobic granular sludge, since it addresses the anoxic storage phenomenon. However, in the model established a significant number of parameters require estimation with adequate methods.

A model can be used effectively only if the parameters are known. Most of the model parameters can be directly obtained from experimental measurements or experiential default values. However, some other parameters should be determined through a mathematical calibration procedure, which needs appropriate experimental data (Flora et al. 1999). The determination of changeable parameters in the calibration process is dependent on the sensitivity of these parameters to the model output (Koch et al. 2000; Hao et al. 2002). The calibration procedure requires an initial guess and a logical domain of each coefficient. These values are either selected from the literature or measured through experiments (Petersen 2000). In a calibration procedure, the parameter values are varied within their feasible domain, until a minimum of the selected objective function is achieved (Shahalam et al. 1996; Petersen 2000). After all of the parameters are determined with model calibration, the model should be verified using sets of experimental values different from those used for calibration.

In this part of study, the sensitivity analysis of the model parameters is performed to gain better insight into the model structure in order to improve the parameter estimation procedure. Thus, a calibration procedure using the batch experimental data is developed and applied successfully to calibrate the model. In addition, model verification and comparison with the ASM1 and ASM3 are also carried out. Information provided in this chapter might be useful for modeling the anoxic denitrification in aerobic granules as well as for designing and controlling of the aerobic-granule-based process.

6.3.2 Materials and Methods

6.3.2.1 Batch Tests for Model Calibration and Verification

Batch experiments with various initial conditions were conducted under anoxic conditions for the model calibration and verification. One-Liter anoxic batch reactor was set up to monitor the NUR. This reactor was equipped with a magnetic stirrer and rubber stopper with piping for sampling, nitrogen supply, and outflow. Argon was purged into the mixed liquor for 1 min to ensure anaerobic conditions before the reactor was sealed. The reactor was inoculated with the granular sludge withdrawn from the SBR at the end of the aeration phase, and the aeration was kept for 10 h to ensure the depletion of any external substrate. Nitrate was added externally in a predetermined concentration. The final volume in the reactors was adjusted to 1 L using tap water. A fatty-acids-rich wastewater was added as the external substrate. A pulse dose of this anaerobic mixture of nitrate and carbonaceous components was added at the beginning of the experiments. The test lasted for 4–6 h at 20 °C and pH of 7.0–8.0. Samples were withdrawn from the reactor every 5–10 min, and were immediately centrifuged and then stored at 4 °C until analysis was performed.

6.3.2.2 Parameter Estimation

The parameter values were estimated by minimizing the sum of squares of the deviations between the experimental and simulating results. AQUASIM was used to perform the simulation and parameter estimation. The objective function to be minimized in the parameter estimation was:

$$F(p) = \left(\sum_{i=1}^{n} \left(y_{\mathrm{exp},i} - y(p)_i \right)^2 \right)^{1/2} \tag{6.38}$$

in which y_{exp} and $y(p)$ are vectors of n measured values and model predictions at times t_i (i from 1 to n), and p is the vector of the parameters of the model.

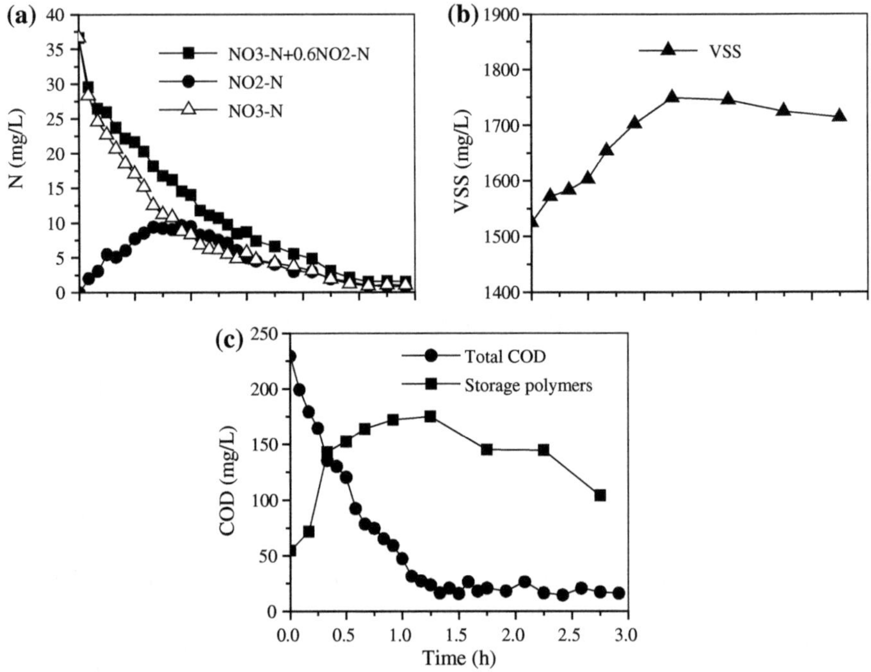

Fig. 6.12 Typical anoxic measurements obtained using aerobic granular biomass: **a** Nitrite and nitrate profiles; **b** COD and storage polymer profiles; and **c** VSS profiles (From Ni et al. (2008), reprinted with permission from Wiley-Blackwell)

6.3.2.3 Model Simulation

A computer grogram, AQUASIM 2.0 (Reichert 1998), was used for modeling the biological process carried out by aerobic granules. AQUASIM 2.0 was a powerful program designed mainly for estimation of the coefficients and parameters involved in a bioreaction model. This program offered flexible definition of the kinetic model, flow scheme, and process control strategies; it also provided support for graphic display of the support of the simulation results, corresponding experimental data, and communication with spreadsheet programs (Siegrist et al. 2002). In consideration of the mass transfer feature of aerobic sludge granules in a bioreactor, the biofilm compartment of the AQUASIM was specifically employed in the modeling work. The simulation cases were performed under the same operational conditions as applied during the operation of the experiments.

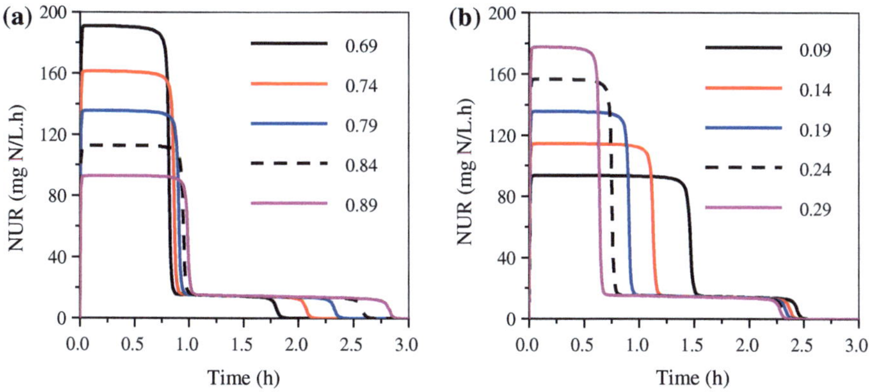

Fig. 6.13 Impact of varying anoxic storage parameters on the reactor performance (NUR) for. **a** $0.69 \leq Y_{STO,D} \leq 0.89$ (base value is 0.79); and **b** $0.09 \leq k_{STO,D} \leq 0.29$ (base value is 0.19) (From Ni et al. (2008), reprinted with permission from Wiley-Blackwell)

6.3.3 Results and Discussion

6.3.3.1 Monitoring Denitrification Processes

Monitoring the denitrification process with nitrate and nitrite measurements obtained from anoxic respirometric experiments is shown in Fig. 6.12. Nitrite build up is monitored regularly in the tests and it is found not to be negligible with a maximum NO_2^- concentration of 10.04 mg N L^{-1}. This is not surprising as in several studies nitrite build up is observed in denitrification (Almeida et al. 1995; Wild et al. 1996; Sozen and Orhon 1999; Beun et al. 2000). Moreover, the results are expected to be different when using a denitrifying sludge previously stabilized under anoxic conditions. This stabilized denitrifying bacteria is adapted to the anoxic conditions, i.e., in the presence of nitrate, and may completely reduce nitrate to nitrogen without nitrite accumulation (Oh and Silverstein 1999; Malisse 2002; Sin 2004). There is no consensus on factors leading to nitrite accumulation in denitrification process in literatures. The substrate type (Wilderer et al. 1987; Sozen and Orhon 1999) and many parameters are suggested to be responsible for this mechanism, but without clear and conclusive evidences. In the present study, nitrite accumulation is observed in all the anoxic tests conducted under various operational and environmental conditions.

The removal of substrate exhibits a similar pattern in all cases, i.e., a relatively rapid initial uptake, followed by a gradual decrease until complete depletion (Fig. 6.12b). Upon pulse addition of external substrate, S_{NOx} in bulk liquid starts immediately to decrease. This follows the first and the maximum S_{NOx} reduction rate (Fig. 6.12a), or fatty-acids-rich wastewater anoxic degradation process, which takes about 60 min (Fig. 6.12b). After the external substrate is completely consumed by biomass, the S_{NOx} decreases with a lower reduction rate, i.e., the second

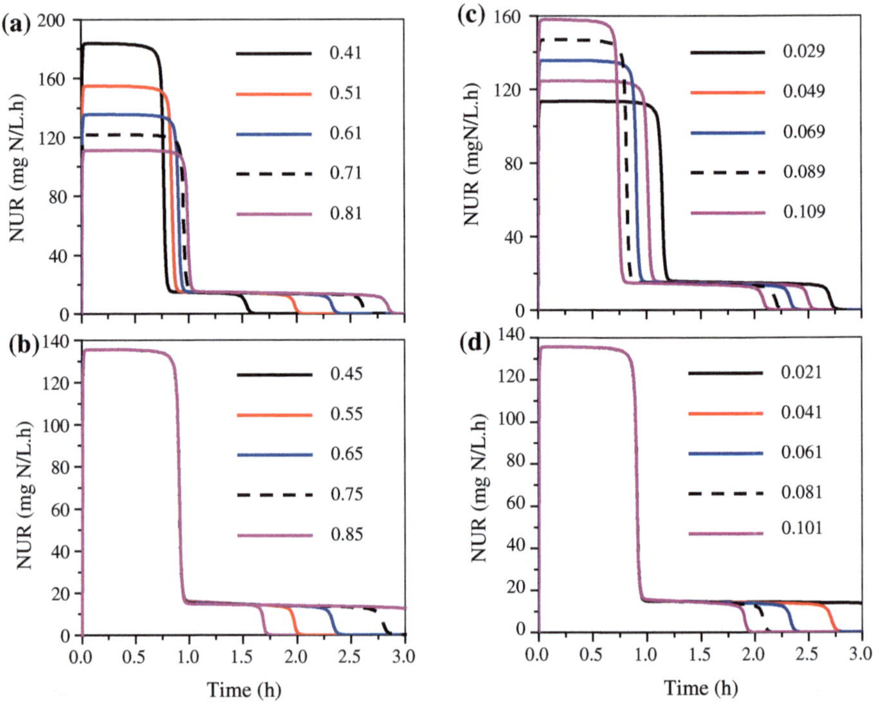

Fig. 6.14 Impact of varying anoxic growth parameters on the reactor performance (NUR) for: **a** $0.41 \leq Y_{HD,S} \leq 0.81$ (base value is 0.61); **b** $0.45 \leq Y_{HD,STO} \leq 0.85$ (base value is 0.65); **c** $0.029 \leq \mu_{HD,S} \leq 0.109$ (base value is 0.069); and **d** $0.021 \leq \mu_{HD,STO} \leq 0.101$ (base value is 0.061) (From Ni et al. (2008), reprinted with permission from Wiley-Blackwell)

S_{NOx} reduction process, which is clearly higher than the endogenous S_{NOx} reduction rate level observed (Fig. 6.12a). This elevated level of NUR or S_{NOx} reduction process could be attributed to the additional uptake of S_{NOx} for oxidation of carbon polymers previously stored, i.e., X_{STO} (Fig. 6.12b). The anoxic storage polymer production after a pulse addition of substrate and later growth on the internally accumulated storage polymers are also explored. In the pulse experiment two distinct phases are observed (Fig. 6.12b). The anoxic storage polymer production is the first phase: a relatively rapid uptake of substrate which is used for anoxic growth and storage polymer production. The next is the anoxic storage polymer consumption phase: after the depletion of the substrate, anoxic growth on the accumulated storage polymer occurs. Since the production of storage polymers is measured, the net microbial growth can be determined from the difference between MLVSS and X_{STO}. MLVSS profiles in the test are shown in Fig. 6.12c. The MLVSS initially increases from 1,500 to 1,750 mg L^{-1} at ~75 min, and later gradually decreases to 1,700 mg L^{-1} at 3 h.

6.3.3.2 Sensitivity Analysis

Sensitivity analysis is a useful tool to evaluate whether the values of theoretical identifiable parameters can be reliably obtained from experimental data (Reichert 1998; Pratt et al. 2004). For a parameter to be reliably identifiable, the measured data should be highly sensitive to changes in this parameter. A low sensitivity indicates that it is difficult to assign a unique value to the given parameter. Furthermore, the similarity in the shapes of the sensitivity profiles of a measured variable to different parameters indicates that the estimates of these parameters are likely to be correlated (Vanrolleghem et al. 1995).

The first step of this study is devoted to an extensive testing of the sensitivity of the model to most of the parameters involved in the different biochemical processes, in order to reduce the selection down to a limited number of significant parameters for obtaining a reasonable model fit. The sensitivity analysis is performed with the adjusted parameters to evaluate their effects on the NUR profiles using a one-variable-at-a-time approach as mentioned above. The NUR measurement provides ample information on the microbial activity, process kinetics, and COD balance over the system (Insel et al. 2003). Model coefficients are changed in the simulation one by one. The sensitivities of the measured NUR to the changes in these parameters are then determined, and are shown in Figs. 6.13 and 6.14.

Effect of anoxic storage parameters. Anoxic storage of easily biodegradable substrates by the denitrifiers plays an important role in the substrate removal. Also, growth of the denitrifier on the storage polymers has a considerable influence on the biomass growth. Anoxic storage parameters show significantly influential to the NUR output. As shown in Fig. 6.13, an increment of $k_{STO,D}$ (maximum anoxic storage rate of denitrifiers) results in an increase in the maximum nitrate uptake rate and a decrease in the reaction time. Such an influence is similar to that of the anoxic reduction factor η_{NOx}, attributed to the same mechanisms. However, the yield coefficient of storage $Y_{STO,D}$ has an influence on the model output in an opposite way. A lower $Y_{STO,D}$ value results in a less lasting time for the maximum NUR value. This is associated with the indirect effect of this yield on the denitrifier growth rate by affecting the storage polymers content.

Effect of anoxic growth parameters. Anoxic growth parameters are evaluated, and some of them are proven to have little influence on the overall SBR performance. However, the maximum anoxic specific growth rates and the anoxic yield coefficients of the denitrifiers on S_S and on X_{STO} show a great sensitivity to the calculation results.

The results for the effect of the maximum anoxic specific growth rate ($\mu_{HD,S}$) on the reactor performance are presented in Fig. 6.14a for $0.029 \leq \mu_{HD,S} < 0.109$ with a base value of 0.069. The reactor has a significantly different behaviors as the value of $\mu_{HD,Ss}$ is changed. It is quite obvious that the maximum NUR value decreases with the decrease of $\mu_{HD,S}$. For instance, a 40 % decrease in $\mu_{HD,S}$ results in 12 % decrease in the maximum NUR value. The effect of changes of $Y_{HD,S}$, a yield coefficient relating the amount of cells yielded to the amount of

Table 6.6 Parameters providing optimum calibration for the experimental data

Parameter	Unit	Definition	Value
$k_{STO,D}$	h^{-1}	Maximum storage rate of biomass	0.19
$\mu_{HD,STO}$	h^{-1}	Maximum growth rate on X_{STO}	0.061
$\mu_{HD,S}$	h^{-1}	Maximum growth rate on S_S	0.069
K_S	g COD m^{-3}	Substrate affinity constant	0.45
K_{NOx}	g N m^{-3}	Biomass affinity constant for *SNOx*	0.050
K_{STO}	g COD g^{-1} COD	Biomass affinity constant for X_{STO}	3.5

substrate consumed, on the system performance is shown in Fig. 6.14b for $0.41 \leq Y_{HD,S} \leq 0.81$ with a base value of 0.61. Figure 6.14b clearly shows that both NUR and S_{NOx} concentration are significantly affected by changes of the $Y_{HD,S}$ value. An increase in $Y_{HD,S}$ reduced the depletion rate of the soluble substrate S_s, and hence decreases the formation rate of active biomass (X_{HD}), which results in a lower maximum NUR value. A 20 % increase in $Y_{HD,S}$ leads to a 10 % decrease in the maximum NUR value.

The changes in the system performance for $Y_{HD,STO}$, a yield coefficient relating the amount of intermediate product formed to the amount of cells formed, are illustrated in Fig. 6.14c. The sensitivity analysis is performed in a range $0.45 \leq Y_{HD,STO} \leq 0.85$ with a base value of 0.65. As shown in Fig. 6.14c, the intermediate COD concentration X_{STO} is greatly affected by the change in $Y_{HD,STO}$, which directly affects the NUR in the third phase (i.e., growth on X_{STO} phase). The effect of decreasing $\mu_{HD,STO}$ on the system performance is similar to that of increasing $Y_{HD,STO}$ due to the same mechanisms (Fig. 6.14d).

6.3.3.3 Model Calibration Procedures

The model calibration procedure is the process of adjusting coefficient values of the model so that the results produced by the model with these coefficients agree well, in the context of a selected objective function, with a set of measured data. The observed NUR, COD and storage polymer responses, obtained from the experiments conducted in this work, are used to calibrate the model. To initiate the calibration procedure, an initial guess of the parameters involved is necessary. Such initial values are obtained on the basis of both experimental results and the data reported in other literature. The most important parameter, η_{NOx}, determined experimentally and independently in the accompanying paper, and is used in the model. In order to simplify the calibration process and due to the limited variability of some parameters, the basic idea in the model calibration strategy is to change as few constants as possible (Xu and Hultman 1996). The software package AQUASIM (Reichert 1998) is used to estimate the values of the model parameters K_{STO}, K_S, K_{NOx}, $\mu_{HD,STO}$, $\mu_{HD,S}$ and $k_{STO,D}$.

For the model calibration, the initial concentration of active denitrifiers, $X_{HD}(0)$, is estimated using the baseline endogenous NUR level prior to substrate addition,

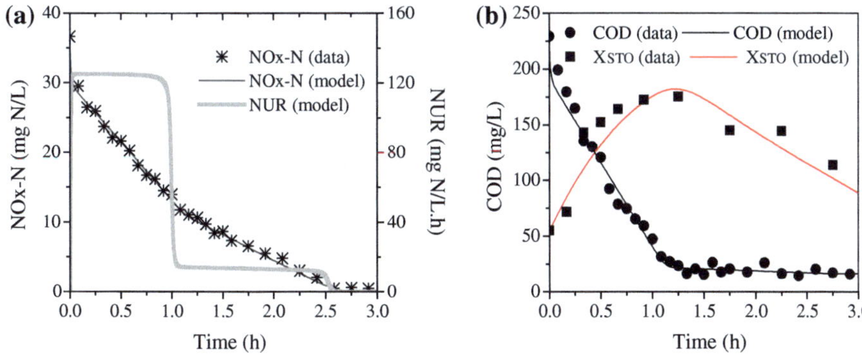

Fig. 6.15 Model calibration using NUR, COD and X_{STO} measurements: **a** NOx-N; and **b** COD and X_{STO} (From Ni et al. (2008), reprinted with permission from Wiley-Blackwell)

when the anoxic endogenous decay rate coefficient, $b_{HD,NOx}$ is fixed to the value mentioned in the ASM3 model (Gujer et al. 1999), i.e., 0.1 day^{-1}. According to ASM3, the endogenous NUR prior to substrate addition is calculated as follows:

$$\mathrm{NUR_{end}}(0) = \frac{1}{2.86}(1 - f_\mathrm{I})b_{\mathrm{HD,NOx}}X_{\mathrm{HD}}(0) \tag{6.39}$$

where $\mathrm{NUR_{end}}(0)$ is the endogenous NUR prior to substrate addition, f_I is the fraction of X_I in decay, and $b_{HD,NOx}$ is the anoxic decay rate coefficient.

In the present work, f_I is fixed to the value mentioned in ASM3, 0.2 mg COD mg^{-1} COD. It is not possible to obtain unique values of both $b_{HD,NOx}$ and $X_{HD}(0)$ using short-term (e.g., 15–30 min) endogenous NUR measurements. This is because the anoxic endogenous respiration of the denitrifiers is practically negligible within such a short period. Long-term (e.g., 20 h) measurement of the endogenous NUR is needed for unique estimation of $b_{HD,NOx}$ (Keesman et al. 1997). Hence, for given f_I and $b_{HD,NOx}$ values, the $X_{HD}(0)$ can be calculated from the NUR(0) data.

6.3.3.4 Model Calibration Results

A set of kinetic parameters for the anoxic storage and growth of the denitrifiers X_{HD} can be estimated on the basis of batch anoxic experiments with the aerobic granular sludge and fatty-acid-rich wastewater as the organic carbon substrate. The final parameter results estimated using the NUR, COD, and storage polymer measurements are summarized in Table 6.6, while the best fits of the model to the experimental data are shown in Fig. 6.15. There is an excellent agreement between the model predictions and experimental data.

Parameter estimation results in Table 6.6 show that the rate of substrate diverted for storage is rapid in the experiment with $k_{STO,D} = 0.19$ h^{-1}. This favors

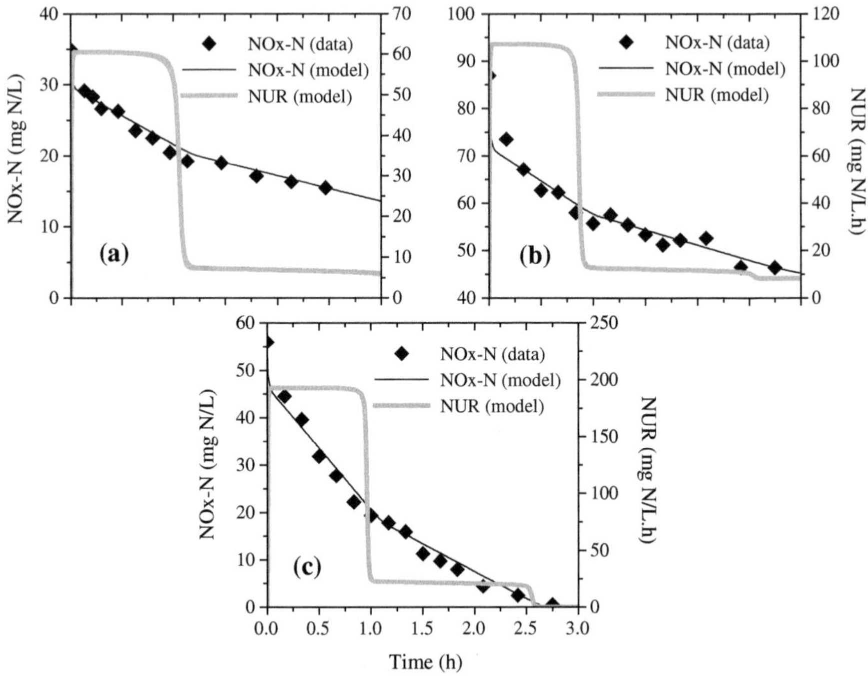

Fig. 6.16 Model verification results of the aerobic granule-based denitrification for three sets of experimental data (From Ni et al. (2008), reprinted with permission from Wiley-Blackwell)

a rapid anoxic substrate storage accompanied by an optimized primary anoxic growth on the available substrate in the system. In this case, the denitrifiers utilize substrate storage as a tool for maximizing substrate uptake while retaining an optimum growth rate. Moreover, the anoxic growth rate on substrate of the denitrifiers estimated with this model is a little lower than the growth rate with the ASM3 model (Table 6.6). The growth rate ($\mu_{HD,S}$) of the denitrifiers is reduced from 0.083 h^{-1} to 0.069 h^{-1} after calibration, attributed to part of denitrifier growth on the storage polymers instead of directly on the soluble substrates. Thus, this model is able to provide more realistic estimates for the maximum anoxic growth rate of the denitrifiers (0.069) than the corresponding value predicted by the ASM3 (0.083). Such a difference is attributed to the simultaneous anoxic storage and growth phenomenon, which is confirmed by the batch experiments.

The affinity constant for nitrate, K_{NOx}, could be estimated for the model with the experimental data when nitrate becomes limited at the end of the experiment. The estimated value for K_{NOx} is significantly lower than the value, 0.5, reported for the ASM3 models (Gujer et al. 1999). Different model structure might be the reason for the difference. Additionally, reactor conditions and microbial communities may partially contribute to such a difference. The affinity constants for the storage products and substrates predicted by this model and ASM3 are in not close agreement. The K_{STO} value is estimated as high as 3.5 mg COD mg^{-1} COD,

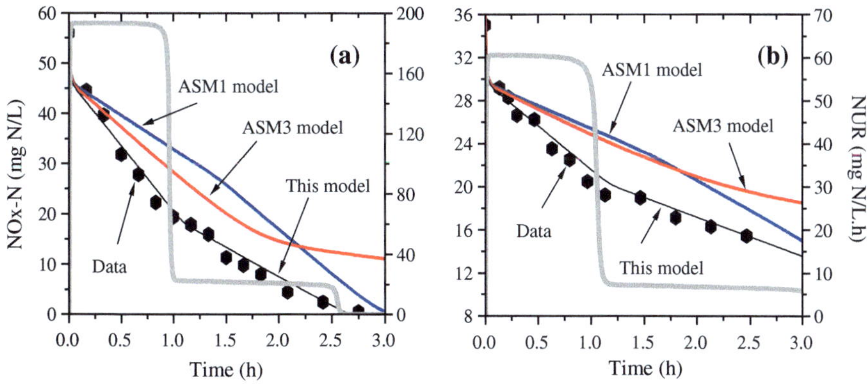

Fig. 6.17 Model simulation of the NUR profiles for two examples with ASM1, ASM3 and the model established in this study (From Ni et al. (2008), reprinted with permission from Wiley-Blackwell)

significantly higher than the default values reported for the ASM3 model (1.0 mg COD mg^{-1} COD). The K_S value is estimated to be as low as 0.45 mg COD L^{-1}, much lower than the default values reported for the ASM3 model (2.0 mg COD L^{-1}). Such differences are likely to be associated with the different conditions for parameter estimation, different wastewater characteristics and different reactor operation.

After the addition of substrate the NUR increase, attributed to the exogenous anoxic consumption of external substrate for biomass growth. However, the NUR do not immediately return to the endogenous level, caused by the oxidation of X_{STO} which are formed under feast phase. The second-order model adopted for describing anoxic degradation of the storage products in this established model is found to adequately fit this smooth NUR transition process. The anoxic growth rate on the storage products, $\mu_{HD,STO}$, is calibrated to be 0.061 h^{-1} (Table 6.6).

6.3.3.5 Model Verification

The model verification is based on a comparison between the experimental results and the calibrated model predictions with the same input model parameters. The experimental results, different from those used for model calibration, are employed for model verification. The simulating results for the NURs profiles are shown in Fig. 6.16 and a good fit is observed. All of the four linear phases (slopes) of nitrate reduction included in the kinetic structure of NUR in the aerobic granular sludge are predicted satisfactorily.

The simultaneous rapid storage and consumption of substrates by the denitrifiers results in a rapid increase in NUR. As shown in Fig. 6.16, the high S_{NOx} reduction response at the initial stage of the experiments can be modeled by the established model (the maximum nitrate reduction rate). The S_{NOx} reduction process before the endogenous phase due to additional uptake of S_{NOx} for oxidation of previously anoxic stored carbon polymers can also be simulated well. The good agreement between the experimental and model predicted results suggests the validity of the model to elucidate the denitrification process in aerobic granules.

6.3.3.6 Comparison with ASM1 and ASM3

In this study, the experimental data are also simulated with the original ASM1 and ASM3, for comparative evaluation of the results obtained with the established model. The simulation results of the ASM1 are shown in Fig. 6.17 (blue line). A lower consumption rate of S_{NOx} accompanying the lower initial increase in NUR is predicted. The S_{NOx} concentrations are higher in the overall operating time. There are no significant variations of S_{NOx} reduction rates. The simulation is not satisfying for this granule-based denitrification.

The experimental data are also simulated with the original ASM3 model for comparison with the established model. For the ASM3, all biodegradable soluble substrate are initially anoxic stored as internal storage products before it is used for growth and maintenance processes is not taken into consideration. As shown in Fig. 6.17 (red line), ASM3 is capable of simulating the variations of S_{NOx} reduction rate. However, a higher S_{NOx} concentrations and NUR are simulated, which did not agree with the experimental data.

To sort out the problems above, the established model, with considerations of the anoxic storage process, heterotrophic growth on storage polymers, and maintenance, are used to simulate the experimental results (Fig. 6.17, black line).The simulation results show that this model is able to describe the initial NUR response and the subsequent transition to the famine phase. In this model, the initial NUR response is determined dominantly by the rapid storage process and the primary growth process. These two processes compete with each other for substrate. The NUR for the maintenance process after substrate depletion can also be modeled well. Thus, the anoxic simultaneous storage and primary growth processes should be included in ASMs for a better interpretation of the denitrification processes in aerobic granules.

6.3.4 Conclusions

The sensitivity of the NUR towards the system stoichiometric and kinetic coefficients for a wide range of parameter values is analyzed. The NUR is greatly affected by the parameters η_{NOx} and storage and growth parameters. The key affinity constants are found to have a pronounced effect on the system performance. The established model with a significant number of parameters requires adequate methods of parameter estimation. A practical calibration procedure using

NUR, COD, and storage polymer data obtained from batch experiments is proposed and applied successfully to calibrate the model. The parameter values are estimated and calibrated by minimizing the sum of squares of the deviations between the measured data and the model predictions in the context of a selected objective function. The estimation results are demonstrated. The established model gives a good prediction of the anoxic growth yield and the maximum anoxic growth rate of the denitrifiers in aerobic granular sludge, as it accounts for the anoxic storage phenomenon. The established model is also verified with the experimental data and compared with the original ASM1 and ASM3. The verification and comparison results indicate that the model established in this work is applicable to simulating and predicting the performance of an aerobic-granule-based denitrification reactor.

References

Almeida, J.S., Reis, M.A.M., Carrondo, M.J.T.: Competition between nitrate and nitrite reduction in denitrification by Pseudomonas Fluorescens. Biotechnol. Bioeng. **46**, 476–484 (1995)

APHA.: Standard Methods for the Examination of Water and Wastewater, 19th ed. American Public Health Association, Washington, DC (1995)

Arrojo, B., Mosquera-Corral, A., Garrido, J.M., Mendez, R.: Aerobic granulation with industrial wastewater in sequencing batch reactors. Water Res. **38**, 3389–3399 (2004)

Avcioglu, E., Karahan-Gulo, O., Orhon, D.: Estimation of stoichiometric and kinetic coefficients of ASM3 under aerobic and anoxic conditions via respirometry. Water Sci. Technol. **48**, 185–194 (2003)

Beun, J.J., Paletta, F., van Loosdrecht, M.C.M., Heijnen, J.J.: Stoichiometry and kinetics of poly-β-hydroxybutyrate metabolism in aerobic, slow growing activated sludge cultures. Biotechnol. Bioeng. **67**, 379–389 (2000a)

Beun, J.J., Verhoef, E.V., van Loosdrecht, M.C.M., Heijnen, J.J.: Stoichiometry and kineticskinetics of Poly-β Hydroxybutyrate metabolism under denitrifying conditions in activated sludge culture. Biotechnol. Bioeng. **68**, 496–507 (2000b)

Beun, J.J., Heijnen, J.J., van Loosdrecht, M.C.M.: N-removal in a granular sludge sequencing batch airlift reactor. Biotechnol. Bioeng. **75**, 82–92 (2001)

Carucci, A., Dionisi, D., Majone, M., Rolle, E., Smurra, P.: Aerobic storage by activated sludge on real wastewater. Water Res. **35**, 3833–3844 (2001)

Cokgor, E.U., Sözen, S., Orhon, D., Henze, M.: Respirometric analysis of activated sludge behaviour - I. Assessment of the readily biodegradable substrate. Water Res. **32**, 461–475 (1998)

de Kreuk, M.K., van Loosdrecht, M.C.M.: Formation of aerobic granules with domestic sewage. J. Environ. Eng. **132**, 694–697 (2006)

Dionisi, D., Majone, M., Ramadori, R., Beccari, M.: The storage of acetate under anoxic conditions. Water Res. **35**, 2661–2668 (2001)

Dircks, K., Pind, P.F., Mosbæk, H., Henze, M.: Yield determination by respirometry: The possible influence of storage under aerobic conditions in activated sludge. Water SA **25**, 69–74 (1999)

Dircks, K., Henze, M., van Loosdrecht, M.C.M., Mosbaek, H., Aspegren, H.: Storage and degradation of poly-B-hydroxybutyrate in activated sludge under aerobic conditions. Water Res. **35**, 2277–2285 (2001)

Flora, E.M.C.V., Suidan, M.T., Flora, J.R.V., Kim, B.J.: Speciation and chemical interactions in nitrifying biofilms II: Sensitivity analysis. J. Environ. Eng-ASCE **125**, 878–884 (1999)

Gernaey, K., Petersen, B., Dochain, D., Vanrolleghem, P.A.: Modelling aerobic carbon source degradation processes using titrimetric data and combined respirometric-titrimetric data: structural and practical identifiability. Biotechnol. Bioeng. **79**, 754–769 (2002)

Gujer, W., Henze, M., Mino, T., van Loosdrecht, M.C.M.: Activated sludge model NO. 3. Water Sci. Technol. **39**, 183–193 (1999)

Hao, X.D., Heijnen, J.J., van Loosdrecht, M.C.M.: Sensitivity analysis of a biofilm model describing a one-stage completely autotrophic nitrogen removal (CANON) process. Biotechnol. Bioeng. **77**, 266–277 (2002)

Henze, M., Grady, C.P.L. Jr., Gujer, W., Marais, G.V.R., Matsuo, T.: Activated sludge model No. 1. Scientific and Technical Report No. 1, IAWPRC, London (1987)

Insel, G., Orhon, D., Vanrolleghem, P.A.: Identification and modelling of aerobic hydrolysis mechanism—application of optimal experimental design. J. Chem. Tech. Biotechnol. **78**, 437–445 (2003)

Jiang, H.L., Tay, J.H., Maszenan, A.M., Tay, S.T.L.: Enhanced phenol biodegradation and aerobic granulation by two coaggregating bacterial strains. Environ. Sci. Technol. **40**, 6137–6142 (2006)

Karahan, O., Artan, N., Orhon, D., Henze, M., van Loosdrecht, M.C.M.: Experimental assessment of bacterial storage yield. J. Environ. Eng. **128**, 1030–1035 (2002)

Karahan, O., van Loosdrecht, M.C.M., Orhon, D.: Modification of activated sludge NO. 3 considering direct growth on primary substrate. Water Sci. Technol. **47**, 219–225 (2003)

Karahan, O., van Loosdrecht, M.C.M., Orhon, D.: Modeling the utilization of starch by activated sludge for simultaneous substrate storage and microbial growth. Biotechnol. Bioeng. **94**, 43–53 (2006a)

Karahan, O., Martins, M., Orhon, D., van Loosdrecht, M.C.M.: Experimental evaluation of starch utilization mechanism by activated sludge. Biotechnol. Bioeng. **93**, 964–970 (2006b)

Keesman, K.J., Spanjers, H., van Straten, G.: Analysis of endogenous process behaviour in activated sludge. Biotechnol. Bioeng. **57**, 155–163 (1997)

Koch, G., Kuhni, M., Gujer, W., Siegrist, H.: Calibration and validation of activated sludge model no. 3 for Swiss municipal wastewater. Water Res. **34**, 3580–3590 (2000)

Krishna, C., van Loosdrecht, M.C.M.: Substrate flux into storage and growth in relation to activated sludge modelling. Water Res. **33**, 3149–3161 (1999)

Kristensen, H.G., Elberg Jørgensen, P., Henze, M.: Characterisation of functional micro-organism groups and substrate in activated sludge and wastewater by AUR. NUR and OUR Water Sci. Technol. **25**, 43–57 (1992)

Kujawa, K., Klapwijk, B.: A method to estimate denitrification potential for predenitrification systems using NUR batch test. Water Res. **33**, 2291–2300 (1999)

Majone, M., Dircks, K., Beun, J.J.: Aerobic storage under dynamic conditions in activated sludge processes—the state of the art. Water Sci. Technol. **39**, 61–73 (1999)

Malisse, K.: Monitoring van de denitrificatie in actief slib aan de hand van nitraat - en titrimetrische gegevens. Engineers Thesis. Faculty of Agricultural and Applied Biological Sciences. Ghent University, Dutch (2002)

Naidoo, V., Urbain, V., Buckley, C.A.: Characterisation of wastewater and activated sludge from European municipal wastewater treatment plants using the NUR test. Water Sci. Technol. **38**, 303–310 (1998)

Ni, B.J., Yu, H.Q.: Storage and growth of denitrifiers in aerobic granules—Part I: model development. Biotechnol. Bioeng. **99**, 314–323 (2008)

Ni, B.J., Yu, H.Q., Xie, W.M.: Storage and growth of denitrifiers in aerobic granules—Part II: model calibration and verification. Biotechnol. Bioeng. **99**, 324–332 (2008)

Oh, J., Silverstein, J.: Acetate limitation and nitrite accumulation during denitrification. J. Environ. Eng. **125**, 234–242 (1999)

Orhon, D., Söen, S., Artan, N.: The effect of heterotrophic yield on assessment of the correction factor for the anoxic growth. Water Sci. Technol. **34**, 67–74 (1996)

Petersen, B.: Calibration, identifiability and optimal experimental design of activated sludge models. Ph.D. Thesis, Ghent University, Belgium (2000)

Petersen, B., Gernaey, K., Vanrolleghem, P.A.: Anoxic activated sludge monitoring with combined nitrate and titrimetric measurements. Water Sci. Technol. **45**, 181–190 (2002)

Picioreanu, C., Kreft, J.U., van Loosdrecht, M.C.M.: Particle-based multidimensional multispecies biofilm model. Appl. Environ. Microbiol. **70**, 3024–3040 (2004)

Pratt, S., Yuan, Z., Keller, J.: Modelling aerobic carbon oxidation and storage by integrating respirometric, titrimetric, and off-gas CO_2 measurements. Biotechnol. Bioeng. **88**, 135–147 (2004)

Reichert, P.: AQUASIM 2.0—User Manual, Computer Program for the Identification and Simulation of Aquatic Systems. Swiss Federal Institute for Environmental Science and Technology (EAWAG) (1998)

Schwarzenbeck, N., Erley, R., McSwain, B.S., Wilderer, P.A., Irvine, R.L.: Treatment of malting wastewater in a granular sludge sequencing batch reactor (SBRSBR). Acta Hydrochim. Hydrobiol. **32**, 16–24 (2004)

Shahalam, A.B., Elsamra, R., Ayoub, G.M., Acra, A.: Parametric sensitivity of comprehensive model of aerobic fluidized-bed biofilm process. J Environ. Eng-ASCE **122**, 1085–1093 (1996)

Siegrist, H., Vogt, D., Garcia-Heras, J.L., Gujer, W.: Mathematical model for meso- and thermophilic anaerobic sewage sludge digestion. Environ. Sci. Technol. **36**, 1113–1123 (2002)

Sin, G.: Systematic calibration of activated sludge model. Thesis for the Degree of Doctor (Ph.D) in Applied Biological Sciences: Environmental Technology. Ghent University (2004)

Sin, G., Guisasola, A., De Pauw, D.J.W., Baeza, J.A., Carrera, J., Vanrolleghem, P.A.: A new approach for modelling simultaneous storage and growth processes for activated sludge systems under aerobic conditions. Biotechnol. Bioeng. **92**, 600–613 (2005)

Soen, A., Ubay, Cokgor E., Henze, Orhon D., M., : Respirometric analysis of activated sludge behaviour—II. Heterotrophic growth under aerobic and anoxic conditions. Water Sci. Technol. **32**, 476–488 (1998)

Sozen, S., Orhon, D.: The effect of nitrite correction on the evaluation of the rate of nitrate utilization under anoxic conditions. J. Chem. Technol. Biotechnol. **74**, 790–800 (1999)

Su, K.Z., Yu, H.Q.: Formation and characterization of aerobic granules in a sequencing batch reactor treating soybean processing wastewater. Environ. Sci. Technol. **39**, 2818–2827 (2005)

Su, K.Z., Yu, H.Q.: A generalized model for aerobic granule-based sequencing batch reactor 1. Model development. Environ. Sci. Technol. **40**, 4703–4708 (2006)

Tay, S.T.L., Zhang, W.Q., Tay, J.H.: Start-up, microbial community analysis and formation of aerobic granules in a tert-butyl alcohol degrading sequencing batch reactor. Environ. Sci. Technol. **39**, 5774–5780 (2005)

Thomsen, J.K., Geest, T., Cox, R.P.: Mass spectrometric studies of the effect of pH on the accumulation of intermediates in denitrification by *Paracoccus denitrificans*. Appl. Environ. Microbiol. **60**, 536–541 (1994)

van Loosdrecht, M.C.M., Heijnen, J.J.: Modelling of activated sludge processes with structured biomass. Water Sci. Technol. **45**, 12–23 (2002)

van Loosdrecht, M.C.M., Henze, M.: Maintenance, endogeneous respiration, lysis, decay and predation. Water Sci. Technol. **39**, 107–117 (1999)

van Loosdrecht, M.C.M., Pot, M.A., Heijnen, J.J.: Importance of bacterial storage polymers in bioprocesses. Water Sci. Technol. **35**, 41–47 (1997)

Vanrolleghem, P.A., Van Daele, M., Dochain, D.: Practical identifiability of a biokinetic model of activated sludge respiration. Water Res. **29**, 2561–2570 (1995)

Vanrolleghem, P.A., Spanjers, H., Petersen, B., Ginestet, P., Takács, I.: Estimating (combinations of) activated sludge model No.1 parameters and components by respirometry. Water Sci. Technol. **39**, 195–215 (1999)

Vanrolleghem, P., Sin, G., Gernaey, K.: Transient response of aerobic and anoxic activated sludge activities to sudden concentration changes. Biotechnol. Bioeng. **86**, 277–290 (2004)

Wanner, O., Reichert, P.: Mathematical modelling of mixed-culture bioflms. Biotechnol. Bioeng. **49**, 172–184 (1996)

Wild, D., Kisliakova, A., Siegrist, H.: P-fixation by Mg, Ca and zeolite A during stabilization of excess sludge from enhanced biological P-removal. Water Sci. Technol. **34**, 391–398 (1996)

Wilderer, P.A., Warren, L.J., Dau, U.: Competition in denitrification systems affecting reduction rate and accumulation of nitrite. Water Res. **21**, 239–245 (1987)

Xu, S.L., Hultman, B.: Experiences in wastewater characterization and model calibration for the activated sludge process. Water Sci. Technol. **33**, 89–98 (1996)

Yi, S., Zhang, W.Q., Wu, B., Tay, S.T.L., Tay, J.H.: Biodegradation of p-nitrophenol by aerobic granules in a sequencing batch reactor. Environ. Sci. Technol. **40**, 2396–2401 (2006)

Chapter 7
Formation Processes of Extracellular Polymeric Substances

Extracellular polymeric substances (EPS) produced by mixed microbial community were characterized using gel-permeating chromatography (GPC) and three-dimensional excitation emission matrix (EEM) fluorescence spectroscopy measurement. The production of EPS, as well as its molecular weight (MW), depended on the external substrate utilization. A novel and convenient approach to evaluate the EPS production kinetics was then developed. Electrons from the external substrate were found to be distributed in the following order: new biomass synthesis of 61 %, oxygen for respiration of 21 %, and EPS formation of 18 %.

7.1 Characterization of Extracellular Polymeric Substances

7.1.1 Introduction

In biological wastewater treatment, EPS are produced by the microorganisms in bioreactors when organic materials present in wastewater are consumed. The accumulation of EPS occurs via a number of different mechanisms including excretion, secretion, cell lysis, and sorption of constituents from wastewater, etc. However, the production of EPS involves a significant investment of carbon and electrons by microorganisms and the flow of energy into EPS during the microbial growth (Laspidou and Rittmann 2002a, b). In bioreactors microorganisms are usually in the form of aggregates, such as sludge flocs and granules (Zartarian et al. 1997; Liao et al. 2002). The presence of EPS in pure cultures, aerobic, and anaerobic sludge all have been confirmed and observed using various electron microscopy techniques (Kawaguchi and Decho 2002). EPS on the cell surface help to keep microbial aggregates together in a three-dimensional gel-like hydrated matrix through weak physicochemical interactions (electrostatic, hydrophobic, van der Waals, and hydrogen-bond interactions) (Zartarian et al. 1997; Leone et al.

B.-J. Ni, *Formation, Characterization and Mathematical Modeling of the Aerobic Granular Sludge*, Springer Theses, DOI: 10.1007/978-3-642-31281-6_7,

2006; González-Brambila et al. 2006). EPS have a close relationship with the structure, composition, function, and microbial community of such aggregates (Urbain et al. 1993; Frolund et al. 1996). The content and composition of EPS of the microbial aggregates play a crucial role in biological wastewater treatment.

EPS essentially consist of polysaccharides, proteins, and smaller amounts of DNA and lipids, etc. (Jorand et al. 1995; Liao et al. 2001; Sheng et al. 2008), which fill and form the space between microbial cells. The extent of each bond contribution to the cumulative binding force depends strongly on the nature of EPS. The flocculating ability and settling of activated sludge are greatly related to its EPS properties. Thus, it is essential to understand the roles of EPS in activated sludge thorough characterizing EPS. The MW, chemical nature, and kinetics are of particular importance (Jorand et al. 1998; Gorner et al. 2003).

The GPC has been used to characterize the metabolic products of activated sludge (Frolund and Keiding 1994; Gorner et al. 2003; Garnier et al. 2006). EPS macromolecules can be separated on the basis of MW analyzed by GPC. The method was successful in showing differences and similarities between microbial products from two different activated sludge treatment plants, and showing the degradation of compounds (Gorner et al. 2003). Fluorescence spectroscopy has been used in biotechnology for different applications in recent years (Chen et al. 2003; Farabegoli et al. 2003; Arunachalam et al. 2005). Additional information can be obtained by using more different filters or reflection gratings for selecting excitation and emission wavelengths, which leads to fluorescence spectroscopy (Li et al. 1991). With this technique it is possible to measure several intra- and extracellular fluorophors simultaneously. Three-dimensional EEM fluorescence spectroscopy is a rapid, selective, and sensitive technique. Its advantage is that information regarding the fluorescence characteristics can be acquired entirely (Sheng and Yu 2006). Thus, considering the complex compositions in bioreactors, EEM fluorescence spectroscopy might be appropriate for exploring the physico-chemical properties of sludge EPS and distinguishing the fluorescence compounds present in EPS. Fluorescent intensities of characteristic peaks on EEM plot can quantitatively evaluate organic matter. Chen et al. (2003) delineated the EEM plot into distinct excitation-emission regions: tryptophan-like, tyrosin-, microbial byproduct-, humic- and fulvic-like compounds. Their results demonstrate that the EEM is a useful toll to characterize complex environmental samples.

Kinetic modeling has been proven to be an important tool to understand the microbial growth (Sheintuch and Tartakovsky 1997; Ding et al. 2006; Curvers et al. 2007; Ni and Yu 2008). However, so far most models have focused on the heterogeneity of substrate concentration and cellular biomass growth (Henze et al. 2000). Little attention has been paid to the EPS profiles in activated sludge process. Mathematical modeling of EPS production is an important issue, as it is closely relates to microbial growth and has a great effect on the substrate dynamics. Modeling can be used to quantify the relationships between the particulate components in activated sludge, such as active microorganisms, inert residues and EPS, and the total energy conversion into biomass and EPS production (Laspidou and Rittmann 2002b). Model simulation and prediction for

EPS formation can play an important role in design and operation of biological treatment systems (Laspidou and Rittmann 2002a). Thus, a kinetic analysis for the formation of EPS in activated sludge processes is essential.

In this work, GPC and fluorescence spectroscopy are combined to provide information about both MW and the chemical nature of EPS in activated sludge. In addition, a kinetic modeling on EPS production is also performed to better understand the dynamics of this important microbial product.

7.1.2 Materials and Methods

7.1.2.1 Sludge

Sludge was taken from a bench-scale SBR which was operated similar to that described in Chap. 3. The reactor was operated at 20 °C and fed with a synthetic wastewater at COD of 1,000 mg L^{-1}. The criteria for reaching steady state were constancy of the biomass concentration, and of the dynamic patterns of substrate utilization and EPS production in repetitive cycles. The composition of the synthetic wastewater used in the reactor was as follows: Acetate, 1,000 mg L^{-1}; NH_4Cl, 190 mg L^{-1}; KH_2PO_4, 224 mg L^{-1}; $MgSO_4$, 90 mg L^{-1}; KCl, 37 mg L^{-1} and trace element solution (in mg L^{-1}): EDTA, 50; $ZnSO_4 \cdot 7H_2O$, 22; $CaCl_2 \cdot 2H_2O$, 8.2; $MnCl_2 \cdot 4H_2O$, 5.1; $FeSO_4 \cdot 7H_2O$, 5.0; $(NH_4)_6Mo_7O_{24} \cdot 4H_2O$, 1.1; $CuSO_4 \cdot 5H_2O$, 1.8; $CoCl_2 \cdot 6H_2O$, 1.6. The influence pH value was adjusted to 7.0 through the dose of 1 M HCl and NaOH.

7.1.2.2 Experimental Design

A set of batch experiments was conducted at known initial conditions to evaluate the formation and dynamics of EPS, in terms of the MW and chemical nature during the biological process. The seeding sludge was sampled from the SBR after its reaching steady state when no substrate was present in the medium, and was washed twice with distilled water to remove external soluble organic material. A batch reactor was inoculated with 1 L of diluted activated sludge and aerated continuously so that the DO level was above 2 mg O_2 L^{-1}. Then, external substrate (the synthetic wastewater) with a concentration of approximately 700 mg COD L^{-1} was added. Samples were taken at given intervals for the analysis of COD, EPS, and MLVSS. The experiments were conducted at 20 ± 1 °C, pH of 7.0 ± 0.05, and MLVSS of 3,400 mg L^{-1}. All extracted EPS samples were characterized for measurement of polysaccharides, proteins, nucleic acids, and also for GPC and EEM fluorescence spectral analysis.

7.1.2.3 Analysis

The EPS of the sludge samples were extracted using the cation exchange resin (CER) technique (Dowex Marathon C, 20–50 mesh, sodium form, Fluka 91973) according to Frolund et al. (1996). Sludge samples were harvested by centrifugation at 600 × *g* for 15 min, and then the pellets were washed twice with 100 mM NaCl solution. After that, the sludge pellets were resuspended to a predetermined volume and the solution was transferred to an extraction beaker, followed by the CER addition with a dosage of 60 g g^{-1} SS. These suspensions were then stirred for 12 h at 200 rpm and 4 °C. Afterwards, the CER/sludge suspensions were settled for 3 min in order to remove CER, and the EPS were harvested by centrifugation at 9560 × *g* and 4 °C for 30 min to remove remaining sludge components. The supernatants were then filtrated through 0.45-μm acetate cellulose membranes and were used as the EPS fraction for chemical, GPC, and EEM fluorescence spectral analyses.

The peak molar masses (M_p) were estimated by a GPC (Waters Co., USA). Series of ultrahydrogel 250, 500, and 2000 columns, which were heated to 40 °C and maintained by thermostat control, were used in this work with deionized water as eluent at a flow rate of 1.0 ml min^{-1}. The detection was carried out at 35 °C with a diode array UV detector at 254 nm and simultaneously with a refractive index detector. The samples were filtered with a 0.45 μm hydrophilic filtration membrane prior to injection (100 μl). The column was calibrated by standard polysaccharides of molecular mass 180, 738, 5900, 1180, 2280, 47,300, 112,000, 212,000, 404,000, and 788,000 Da (Polymers Labs, EasiCal; provided by Phenomenex Inc.) and standard proteins of molecular mass 13,700, 45,000, 67,000, 200,000, and 670,000 Da (ribonuclease A: R-4875, bamylase: A-7130, chicken egg albumin: A-5378 and bovine serum albumin: A-7906), provided by Sigma Co. (France).

All EEM spectra were measured using a luminescence spectrometry (LS-55, Perkin-Elmer Co., USA). The spectrometer displayed a maximum emission intensity of 1000 arbitrary units (AU). EEM spectra are a collection of a series of emission spectra over a range of excitation wavelengths, and they can be used to identify fluorescent compounds present in complex mixtures. In this study, extracellular EEM spectra were collected with subsequent scanning emission spectra from 300 to 550 nm at 0.5 nm increments by varying the excitation wavelength from 250 to 450 nm at 10 nm increments. Excitation and emission slits were both maintained at 10 nm and the scanning speed was set at 1,500 nm min^{-1} for all the measurements. A 290 nm emission cutoff filter was used in scanning to eliminate second order Raleigh light scattering. The spectrum of double distilled water was recorded as the blank. The software MatLab 7.0 (MathWorks Inc., USA) was employed for handling EEM data. EEM spectra are illustrated as the elliptical shape of contours. The X-axis is the emission spectra from 300 to 550 nm, whereas the Y-axis is the excitation wavelength from 250 to 450 nm.

Parallel factor analysis (PARAFAC) is a decomposition method, which decomposes higher order arrays into tri-linear components. PARAFAC was used to model the EEM fluorescence data in this work. The algorithm used in this paper was proposed by Bro (1997). As validation of the models, core consistency diagnostic is used. PARAFAC decompose the data X into a structured part, where the information is contained, and into a noise part. For each component in the structural part one score vector and two loading vectors, emission and excitation, are generated.

The polysaccharides in EPS were determined as glucose equivalent using anthrone-sulfuric acid method (Dubois et al. 1956), whereas the protein was measured as bovine albumin equivalent using the Lowry method (Lowry et al. 1951). The content of nucleic acids in EPS was determined using a UV spectrophotometer at 260 nm (UV751GD, Analytical Instrument, Shanghai). Measurement of MLVSS and COD was performed according to the Standard Methods (APHA 1995).

7.1.3 Kinetic Modeling

A kinetic model for EPS formation is developed in this work. For the flow of electrons from the external substrate, part of the external substrate is used for biomass synthesis. Other substrate electrons are diverted to the formation of EPS. EPS are released as a solid to form the aggregate matrix. Substrate oxidation and respiration of the electrons to reduce oxygen and generate the energy needed to fuel the formation of active biomass and EPS. The model describes the relationships among the three solid species: active bacteria (X_a), EPS (X_{EPS}), and residual inert biomass (X_{I}); one soluble species: external substrate (S_{S}); and an electron acceptor, which is dissolved oxygen (DO, S_{O}). The units for all species are oxygen demand or oxygen (for DO), which is directly proportional to electron equivalents (8 g O_2 per e^- equivalent).

The rate of EPS formation is proportional to the rate of substrate utilization (Laspidou and Rittmann 2002a). Decay of EPS leads to EPS loss. The decay rate is proportional to the EPS concentration, with k_{decay} being the first-order rate coefficient. The kinetic equation for EPS is described as follows:

$$\frac{\mathrm{d}X_{\mathrm{EPS}}}{\mathrm{d}t} = k_{\mathrm{EPS}}\frac{\mu}{Y}\frac{S_{\mathrm{S}}}{K_{\mathrm{S}}+S_{\mathrm{S}}}\frac{S_{\mathrm{O}}}{K_{\mathrm{O}}+S_{\mathrm{O}}}X_a - k_{\mathrm{decay}}X_{\mathrm{EPS}} \tag{7.1}$$

The first term on the right-hand side is for EPS formation, while the second term is EPS loss. k_{EPS} is the EPS formation coefficient. μ is the maximum specific growth rate of active biomass, Y is its yield, and K_{S} and K_{O} are the half maximum-rate concentrations for S_{S} and S_{O}, respectively.

Equation 7.2 is the kinetic equation for active biomass, while the kinetics for external substrate is described in Eq. 7.3.

$$\frac{dX_a}{dt} = (1 - k_{EPS})\mu \frac{S_S}{K_S + S_S} \frac{S_O}{K_O + S_O} X_a - b \frac{S_O}{K_O + S_O} X_a \tag{7.2}$$

$$\frac{dS_S}{dt} = -\frac{\mu}{Y} \frac{S_S}{K_S + S_S} \frac{S_O}{K_O + S_O} X_a \tag{7.3}$$

The last term of Eq. 7.2 represents the loss of active biomass attributed to endogenous respiration which produces X_I.

$$\frac{dX_I}{dt} = f_I b \frac{S_O}{K_O + S_O} X_a \tag{7.4}$$

where f_I is the fraction of active biomass converted into inert biomass in endogenous decay.

7.1.4 Results

7.1.4.1 EPS Formation

The time courses of substrate consumption and EPS formation are similar (Fig. 7.1). External substrate consumption results in an increase in the amount of total EPS. After the external substrate becomes depleted, the total EPS concentration slightly decreases. This indicates that EPS formation mainly occur during the substrate consumption period. The substantial decrease in EPS concentration after the depletion of the external substrate might be attributed to the EPS decay.

As shown in Fig. 7.1a, the total EPS concentration increases with the biomass growth. The contents of carbohydrates, proteins, and nucleic acids in EPS all increase slowly in the substrate consumption process. After the depletion of the external substrate, however, there are only marginal increases in the amount of carbohydrates, while the contents of proteins and nucleic acids in EPS slightly decrease. The three components have different formation rates. The content of carbohydrates is higher than that of proteins and nucleic acids, and carbohydrates are the predominant components in the produced EPS.

To have a better insight into the EPS production, the kinetic EPS formation was modeled with regard to their MW and chemical nature. Figure 7.1b compares the simulated and measured results for COD and EPS. Model parameters used in model simulation as well as the values are shown in Table 7.1. The simulated and measured data match very well for the two parameters. COD decreases rapidly and continuously for the initial ~2 h, as the external substrate is consumed. The modeling results show that COD is almost constant in the period from 2 to 4 h. In the period of rapid microbial growth, EPS increase rapidly. However, once the external substrate is consumed and the rapid microbial growth ceases, EPS decrease slightly. These model predictions are in qualitative agreement with the experimental observations.

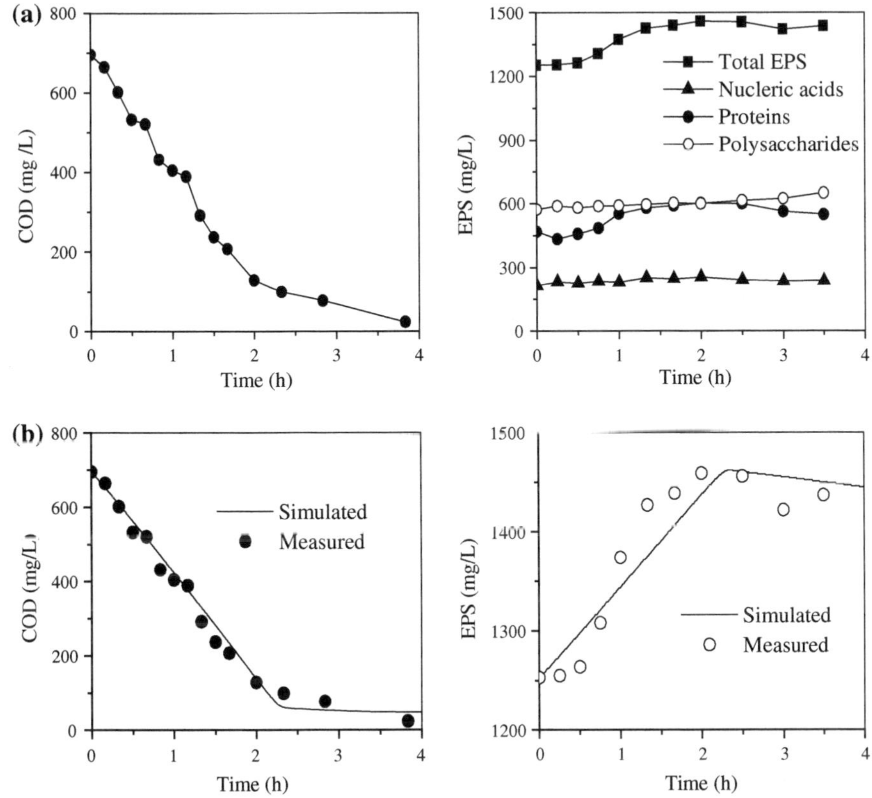

Fig. 7.1 **a** External substrate consumption and EPS formation in activated sludge process: COD concentrations and EPS profiles (Reprinted from Ni et al. (2009), with permission from Elsevier) **b** Kinetic modeling of the measured COD and EPS profiles of the activated sludge with the developed model (Reprinted from Ni et al. (2009), with permission from Elsevier)

Table 7.1 Kinetic and stoichiometric parameters used in models

Parameter	Definition	Value	Units
Y	Yield coefficient for growth on S_S	0.61	g COD_X g^{-1} COD_S
f_I	Fraction of X_I in decay	0.20	g COD_X g^{-1} COD_X
μ	Maximum growth rate on S_S	0.40	h^{-1}
K_S	Substrate affinity constant	11.38	g COD_S m^{-3}
K_O	Dissolve oxygen affinity constant	0.20	g COD_O m^{-3}
b	Decay rate coefficient of X_a	0.021	h^{-1}
k_{EPS}	EPS formation coefficient	0.18	g COD_{EPS} g^{-1} COD_S
k_{decay}	EPS hydrolysis rate coefficient	0.0071	h^{-1}

7.1.4.2 GPC Profiles

The GPC chromatogram of the EPS extracted from the sampled sludge at various times during the reactor cycle is shown on Fig. 7.2. Based on the GPC chromatograms, different molecular size fractions are identified for EPS samples. There are seven peaks at retention times of $t_{R1} = 19$ min, $t_{R2} = 21.5$ min, $t_{R3} = 22$ min, $t_{R4} = 23$ min, $t_{R5} = 26$ min, $t_{R6} = 26.5$ min, and $t_{R7} = 33.5$ min. The MWs corresponding to the compounds with peaks at retention times from t_{R1} to t_{R7} calibrated with polysaccharides and proteins are given in Table 7.2.

Among them, three important peaks at 22, 23, and 33.5 min are observed. The solid line represents UV/Vis diode array detector signal at $\lambda = 280$ nm. This wavelength detection represents the proteins in EPS. The hatched line corresponds to the refractive index detection, which represents the polysaccharides in EPS. The MW of the fraction at retention time of $t_{R1} = 19$ min is around 800,000 Da for all samples, which is recognized as polysaccharides. The chromatographic peaks at 21.5 and 22 min contain both proteins and polysaccharides, while the peaks at 33.5 min are mainly composed of polysaccharides. However, these fraction components show different EPS content in terms of proteins and polysaccharides among the different samples. With the consumption of the external substrate at initial 1 h, the peak area of the GPC chromatogram increase to a great content. Thus, the proteins and polysaccharides content increase with the utilization of substrate, which is in accordance with the observations shown in Fig. 7.1.

The GPC tests suggest a shift in EPS composition and EPS MW distribution. The chromatogram of EPS extracted after 5 min of cultivation, has three main peaks with a retention time of about 21.5, 22, and 33.5 min at MW of 250,000 Da, 200,000 Da, and 300 Da, respectively. Over 75 % EPS compounds is in this MW range for all samples. With the consumption of the external substrate, a new class of macromolecules, most of which are proteins, with larger sizes and higher MWs at a retention time of 16 min appears. The peak at 21.5 min corresponds to a mixture of proteins and polysaccharides, while the peak at 22.5 and 33.5 min mainly corresponds to polysaccharides. The peak with 33.5 min remains unchanged in the entire activated sludge cultivation process. In the initial 60 min (Fig. 7.2a–d), the external substrate is transformed to proteins and polysaccharides in EPS with a broad MW range. The GPC peaks of MWs of 200 and 250 kDa appear, reflecting the formation of EPS related to the substrate utilization or microbial growth. The EPS is likely to be partially hydrolyzed, thus the GPC peaks decrease and are transformed into the compounds with smaller MWs at 170 min.

7.1.4.3 EEM Spectra

The three-dimensional EEM spectroscopy is also applied to characterize the extracted EPS. Each EEM gives spectral information about the chemical compositions of EPS samples. Nine EPS-EEM fluorescence spectra are acquired and shown in Fig. 7.3. They are similar in the peak locations, but have different

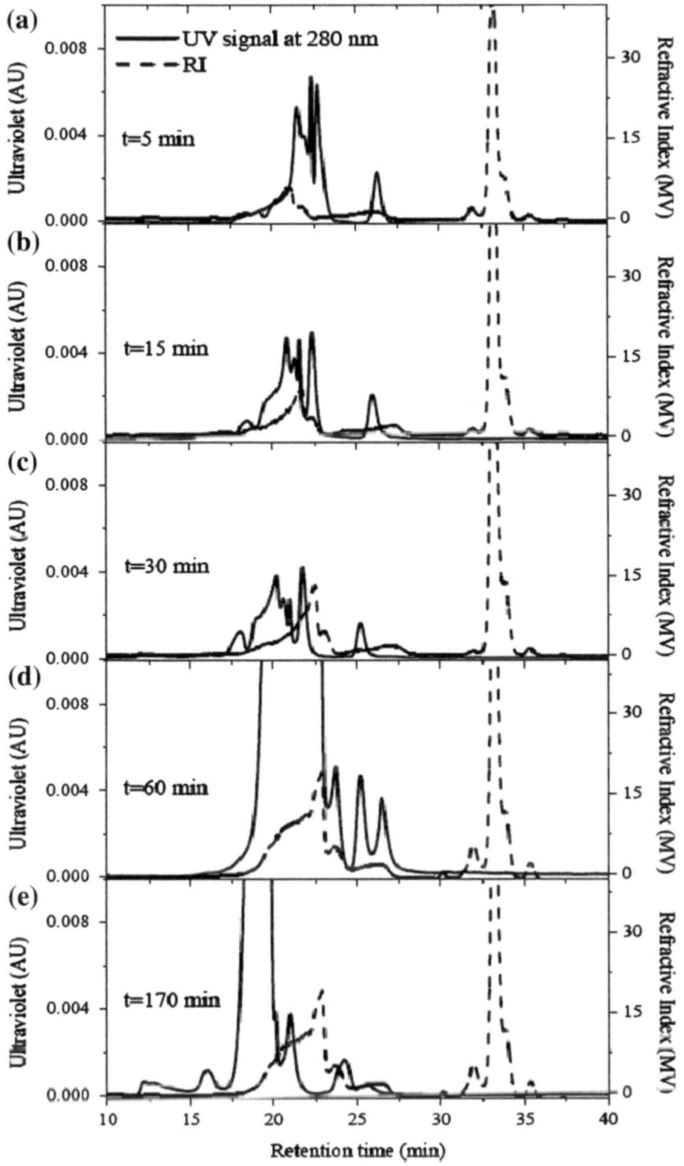

Fig. 7.2 GPC chromatograms of extracted EPS at different times of activated sludge cultivation process (*solid line*: UV detection signal at 280 nm; hatched line: refractive index detection): **a** $t = 5$ min; **b** $t = 15$ min; **c** $t = 30$ min; **d** $t = 60$ min; and **e** $t = 170$ min (Reprinted from Ni et al. (2009), with permission from Elsevier)

fluorescence intensities. To find out the appropriate number of PARAFAC components, core consistency diagnostic is also used to determine the correct number of components in the PARAFAC solution. Three components are found to be

Table 7.2 Calibrated MWs and corresponding retention times t_R

t_R (min)	19	21.5	22	23	26	26.5	33.5
MW (Da)	800,000	250,000	200,000	125,000	50,000	20,000	300

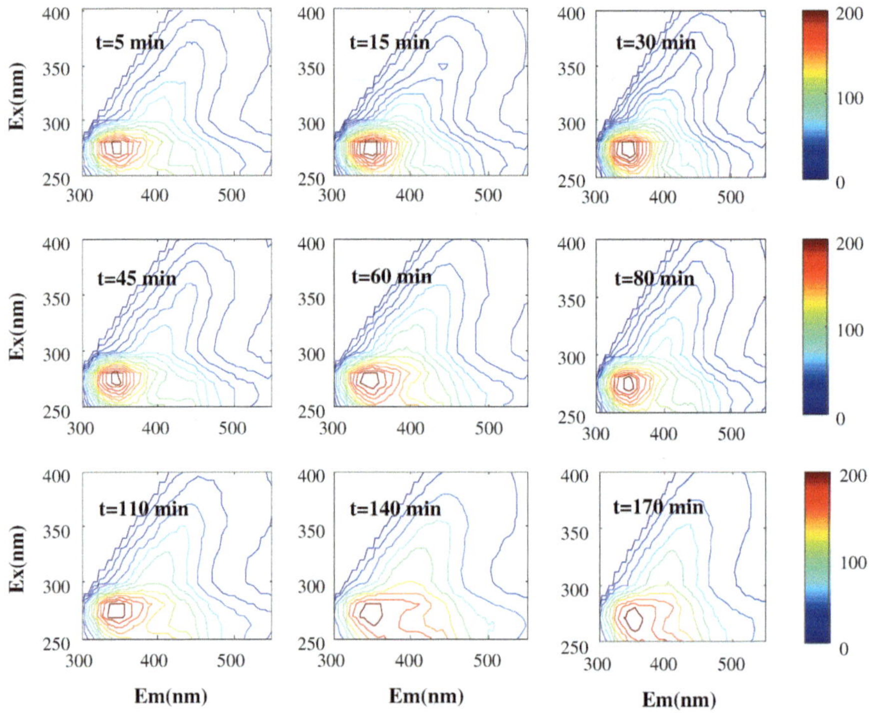

Fig. 7.3 EEM spectra of the EPS samples collected at different times in activated sludge cultivation process (Reprinted from Ni et al. (2009), with permission from Elsevier)

appropriate. Using the PARAFAC, the matrix X is decomposed into three vectors: two loading vectors and one score vector. The two loading vectors correspond to the excitation and emission characters of the fluorescence components. They are multiplied and shown in Fig. 7.4. Two components of the extracellular products are identified as proteins and fulvic-acid-like substances. The fluorescence peak of proteins is at Ex/Em 280/340 nm, and that of the fulvic-acid-like components is at 320/400 nm.

7.1.5 Discussion

In activated sludge processes, the EPS composition is a contribution of various responses, which arise from microbial activity, cell death, lysis, or accumulation

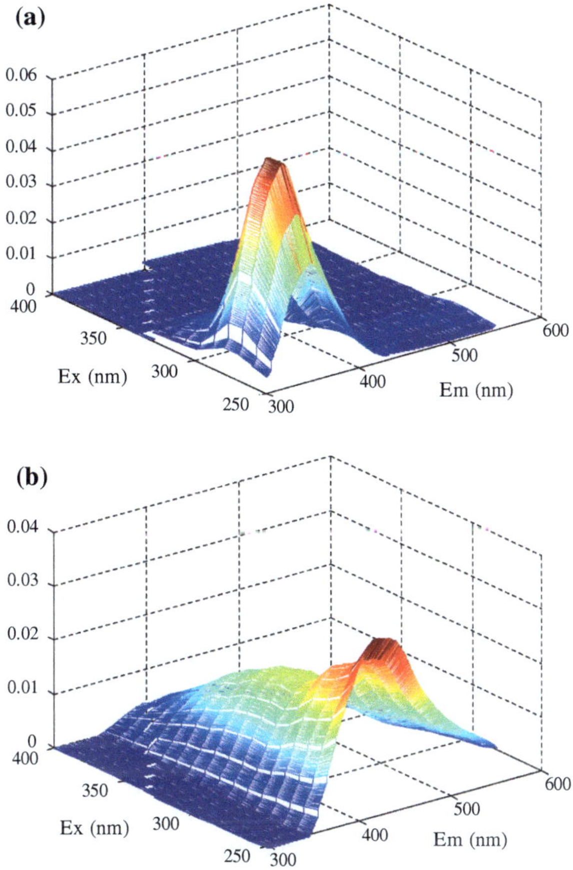

Fig. 7.4 Two components of EPS decomposed using the PARAFAC approach: **a** proteins; and **b** fulvic-acid-like substances (Reprinted from Ni et al. (2009), with permission from Elsevier)

from components in wastewater. The extracellular polymeric matrix associated with flocs and films represents an essential structure and contributes significantly to the structure–function relationship and bioactivities of activated sludge. Their composition and physicochemical properties play an important role in biological wastewater treatment. Fundamentally, EPS represent a significant sink for carbon and energy. EPS might represent a stored energy supply and can provide a reservoir for nutrients. Their complicated nature is reflected by the wide range of proteins, polysaccharides, lipids, and DNA found in EPS (Jorand et al. 1995; Liao et al. 2001; Sheng and Yu 2006; Sheng et al. 2008). The composition and role of proteins and polysaccharides (the two main constituents) and their formation dynamics in a particular system is of great importance. In this paper the characterization of EPS in activated sludge is realized using several techniques including GPC, EEM fluorescence spectroscopy, and kinetic modeling.

The ratio of proteins to carbohydrates of the sludge EPS reported in this work is relatively low (Fig. 7.1a). This ratio has a significant influence on the sludge stability (Mikkelsen and Nielsen 2001). Carbohydrates were found to be the predominant component in the EPS fraction for the activated sludge, but for the anaerobic flocs proteins were the major component (Sheng et al. 2006). In addition, the COD loading has a significant effect on the total EPS composition (Laspidou and Rittmann 2002b). As a result, the relatively high COD loading to the SBR in this work might result in more EPS production than that at a lower COD loading (200–300 mg L^{-1}).

In a previous study (Sheng and Yu 2006), the EEM fluorescence spectroscopy was applied to characterize effects of pH, EPS concentration, and ionic strength on EEM fluorescence spectra of the EPS extracted from both aerobic and anaerobic sludge in wastewater treatment. The EEM spectroscopy was proven to be an appropriate and effective method to characterize the EPS. However, the composition of the EPS from biological wastewater treatment reactors is very complex. There are many fluorescent constituents in the polymer matrix, some of which may originate from humic acids, proteins and coenzymes, etc. Their fluorescence peaks are always wide and overlapping. In this case, it is difficult to acquire the actual fluorescence peak intensity. As a tri-linear method, the PARAFAC approach is an ideal tool for EEM decomposition. It can extract the principle components and discard the errors. It can also solve the problem of spectral overlapping. The EEM peaks of EPS are separated clearly and the fluorescence intensity scores are used to precisely reflect their fluorophore concentration with the PARAFAC analysis. Furthermore, in this study GPC chromatographic separation is used to obtain the MW-based fractions with a limited number of compounds. Another feature of this work is the exploration of the EPS formation dynamics by using the kinetic model evaluation.

The EEM of the collected fractions provides additional information about the chemical nature of the biological macromolecules. The fluorescence peak locations do not change significantly in the cultivation process of activated sludge, while the fluorescence peak intensities do change. EEM results reveal that the relative concentrations of proteins and fulvic-acid-like substances increase in the substrate consumption phase, but decrease in the endogenous phase. Extracellular proteins are mainly excreted by microorganisms. Tryptophane fluorescence is the dominant part of the protein fluorescence, which has a fluorescence maximum at excitation/emission of 280/350 nm. The decrease in the extracellular protein concentration may be attributed to their hydrolysis and partial utilization by microorganisms, especially in the endogenous decay phase. The fulvic-acid-like component contained in the polymer matrix might be attributed to either fulvic acids directly or NADH and pyridoxine (Pons et al. 2004; Sheng and Yu 2006). In addition, the extracellular microbial compounds showing a fluorescent signal maybe excreted by the microorganisms whereas others are derived from their decay. The extracellular proteins are mainly substrate-utilization-associated, while the fulvic-acid-like substances are non-growth-associated (Li et al. 2008). In a way the extracellular fluorescent signals of these constituents can be related to the reactor performance and have the potential of being applied for bioreactor monitoring.

With the three-component model in the PARAFAC analysis (Bro and Kiers 2003), the matrix of EEM fluorescence spectra is decomposed into three vectors. As shown in Fig. 7.4, two components of the extracellular products are identified as the proteins and fulvic-acid-like substances.

In addition to the fluorescence locations, the score vector of the individual component could also be obtained from the EEM spectra using the PARAFAC analysis. The excitation wavelengths of the EEM fluorescence spectra exceed 250 nm, where the absorbance of the samples is lower than 0.05. In this case, the inner filter effect can be neglected. The score vector is correlated to the relative concentration of proteins (Baunsgaard 1999). The relative intensity of the extracellular proteins increases in the substrate consumption phase, which has a similar variation trend to the protein concentration in the supernatant determined using the Folin method (Fig. 7.1). The concentration of the extracellular fulvic-acid-like substances does not vary significantly in the cultivation period.

Several separated EPS families with different MWs have been obtained based on GPC chromatographic separation. A qualitative description of two main components, proteins and polysaccharides, is possible. The EPS fractions vary in MWs from 300 to 800,000 Da. Evolution of the GPC chromatogram area indicates that the quantity of produced EPS are formed substantially in the substrate utilization process, suggesting that the EPS production is related to the substrate consumption.

Kinetic modeling results show that the activated sludge consumes electron-donor substrate to build active biomass. At the same time, they produce EPS. Electrons from the external substrate are distributed in this order when EPS are produced: new biomass synthesis 61 %, oxygen for respiration 21 %, and EPS 18 %. Thus, more substrate electrons are diverted to the active cells, but less is distributed to produce EPS when ample substrate is available. EPS decay when active biomass undergoes endogenous respiration to form residual dead cells. The decay rate for active biomass (b) is 0.021 h^{-1}, which is about three times greater than the decay rate for EPS (k_{decay}) of 0.0071 h^{-1}. Such a difference suggests that EPS are not used by the microorganisms as an energy- and electron-storage material.

In this study, the developed kinetic model for describing the production and consumption of EPS has different parameter values for the substrate electrons diverting to oxygen for respiration. The obtained oxygen yield on COD consumed in our work, i.e., 0.21, is substantially lower than the corresponding value of 0.4, which is often used in the conventional activated sludge modeling approaches (Henze et al. 2000). This reflects the extended structure of our model for incorporation of EPS and subsequent parameter evaluation, as well as the different microbial communities. The lower oxygen yield in this work is partially attributed to the fact that in our model EPS could be synthesized from the utilization of substrate, which is not taken into account in the conventional activated sludge models. In conventional approaches, the total substrate utilization kinetics is lumped into the yields for cell synthesis and oxygen respiration. In this case, all the substrate electrons for the synthesis of EPS are accounted for the oxygen yield, resulting in a higher oxygen yield value. Additionally, different wastewater characteristics, reactor operating conditions, and microbial communities may also be responsible for such a difference.

Since the formation of EPS by the mixed microorganisms in activated sludge is a very complex bioprocess, it is difficult to elucidate such a process and accordingly the characteristics of EPS formed. In the present work, combination of EEM fluorescence spectroscopy and gel-permeating chromatography measurement with kinetic modeling is successfully used to reveal the complex nature of EPS. Such a combination is able to characterize the changes in EPS content and composition during the growth of activated sludge and provides unique and useful information on better understanding the characteristics of EPS formed in activated sludge process.

7.1.6 Conclusions

In this work three-dimensional EEM fluorescence spectroscopy, gel-permeating chromatography measurement, and kinetic modeling are used to elucidate the characteristics of EPS produced by activated sludge. The amount of produced EPS and their molecular weight distribution depend on the utilization of external substrate. Evolution of the chromatogram area indicates that the quantity of produced EPS increases in the substrate utilization process. With the PARAFAC approach, two components of EPS are identified by the EEM analysis: proteins at Ex/Em 280/340 nm, and the fulvic-acid-like component at 320/400 nm. The kinetic modeling of EPS formation was performed in relation to their molecular weight and chemical nature identified by GPC and EEM. The model predictions are in qualitative agreement with the experimental observations. Results show that EPS increase rapidly with the substrate consumption, but decrease slightly after the external substrate is completely consumed. Electrons from the external substrate are distributed in the following order: new biomass synthesis of 61 %, oxygen for respiration of 21 %, and EPS of 18 %.

7.2 A Novel Approach to Evaluate the Production Kinetics of EPS

7.2.1 Introduction

EPS come from the natural secretions of microorganisms which refer to a rich matrix of biopolymers, such as polysaccharides, proteins, nucleic acids, and others (McSwain et al. 2005; Liu et al. 2007; Haberkamp et al. 2008). They play an important role in the aggregation of microbial cells, the formation of an active gel-like structure that maintains cell cohesion, governing the microbial physiology, stabilization of sludge structure, and protection of microbes against noxious environmental conditions (McSwain et al. 2005; Fonseca et al. 2007; Laspidou and Rittmann 2002a). It is also found that EPS are important sinks for electrons and carbon derived from the original substrate (Laspidou and Rittmann 2002b; Aquino and Stuckey 2008). The diversion of electrons and carbon for EPS production could otherwise be invested in the cell yield and the growth rate (Laspidou and Rittmann 2002a, b). Ignoring EPS formation could lead to an overestimation of true cellular growth rates, which should have significant implications on the performance of activated sludge systems. Therefore, the formation kinetics of EPS in biological wastewater treatment processes should be explored.

Several approaches have been used to model EPS production kinetics. Kreft and Wimpenny (2001) developed a model of biofilm in which the EPS production was described by a growth rate-dependent and an independent term. Laspidou and Rittmann (2002a, b) formulated a model that EPS production was proportional to the substrate utilization rate. The biochemical reactions in these models are expressed using nonlinear equations, resulting in complexity of their models and difficulty in estimating kinetic parameters for EPS production.

The evaluation of microbial and enzymatic reaction rates requires representative rate data and an effective method for fitting rate equations to the experimental data. The integrated Monod-based equation is found to be useful for evaluation of kinetic parameters through progress curves from a few batch experiments or even one batch experiment (Smith et al. 1998). Weighted nonlinear least-squares analysis is an approach that can be used to minimize differences between experimental data and model predictions when it is necessary to use an implicit expression in the model (Smith et al. 1998).

On the other hand, the Monte Carlo simulation is a practical way of imitating the inherent randomness in biological systems using deterministic models (Goovaerts et al. 2001; Miller et al. 2006; Huijbregts et al. 2003; Luo and Farrell 2003; Leterm et al. 2007; Saloranta et al. 2008). This simulation technique has been used for the design and upgrade of wastewater treatment plants, as well as generating different wastewater compositions for posterior process performance evaluation. It also has a wide application in the hydrological modeling field (Sin et al. 2008).

This paper aims to develop a novel and convenient approach for determining the best-fit values for kinetic coefficients of EPS production by activated sludge and their practical identifiability using a combination of weighted least-squares analysis and Monte Carlo method. Then experimental and literature data are used to validate this approach. Finally, the comparison between this novel approach and others is performed to obtain a better understanding of EPS production kinetics overall. To the best of our knowledge, this might be the first attempt to evaluate the EPS production kinetics by activated sludge using the weighted nonlinear least-squares and the Monte Carlo method.

7.2.2 Materials and Methods

7.2.2.1 Monod-Based Kinetics for EPS Production

A Monod-based kinetics for EPS formation by activated sludge is evaluated in this work. The rate of EPS formation is proportional to the rate of substrate utilization (Laspidou and Rittmann 2002a), and the kinetic equation is expressed as follows:

$$\frac{dX_{EPS}}{dt} = k_{EPS}\frac{\mu_H}{Y_H}\frac{S_S}{K_S + S_S}X_H \tag{7.5}$$

where X_H is the active bacteria (mg COD L^{-1}), X_{EPS} is the produced EPS (mg COD L^{-1}), S_S is the rate-limiting external substrate (mg COD L^{-1}), t is time (h), k_{EPS} is the EPS formation coefficient (g COD_{EPS} g^{-1} COD_S), μ_H is the maximum specific growth rate of active biomass (h^{-1}), Y_H is the yield for active cell synthesis (g COD_X g^{-1} COD_S), and K_S is the half-maximum-rate concentration for S_S (g COD_S m^{-3}).

Equation 7.6 is the kinetic equation for active biomass, while the kinetics for external substrate is described with Eq. 7.7.

$$\frac{dX_H}{dt} = (1 - k_{EPS})\mu_H\frac{S_S}{K_S + S_S}X_H \tag{7.6}$$

$$\frac{dS_S}{dt} = -\frac{\mu_H}{Y_H}\frac{S_S}{K_S + S_S}X_H \tag{7.7}$$

7.2.2.2 Integrated EPS Production Equation

According to Eqs. 7.6 and 7.7, the active cell and EPS production associated with growth substrate utilization can be described below:

$$X_H = X_{Ho} + Y_H(1 - k_{EPS})(S_{S0} - S_S) \tag{7.8}$$

$$X_{EPS} = X_{EPSo} + k_{EPS}(S_{S0} - S_S) \tag{7.9}$$

where the subscript 0 denotes time zero.

Equations 7.5, 7.8 and 7.9 are combined into the following equation:

$$\frac{K_S + \left(S_{S0} - \frac{X_{EPS} - X_{EPSo}}{k_{EPS}}\right)}{\left(S_{S0} - \frac{X_{EPS} - X_{EPSo}}{k_{EPS}}\right)\left(X_{Ho} + Y_H(1 - k_{EPS})\frac{X_{EPS} - X_{EPSo}}{k_{EPS}}\right)}dX_{EPS} = k_{EPS}\frac{\mu_H}{Y_H}dt \tag{7.10}$$

Then, Eq. 7.10 can be integrated to calculate the integrated EPS production by activated sludge:

$$t = \frac{Y_{\mathrm{H}}}{\mu_{\mathrm{H}}} \left\{ \begin{array}{l} \dfrac{K_{\mathrm{S}}}{X_{\mathrm{Ho}} + Y_{\mathrm{H}}(1 - k_{\mathrm{EPS}})S_{\mathrm{S0}}} \ln\left(\dfrac{k_{\mathrm{EPS}} X_{\mathrm{Ho}} + Y_{\mathrm{H}}(1 - k_{\mathrm{EPS}})(X_{\mathrm{EPS}} - X_{\mathrm{EPSo}})}{k_{\mathrm{EPS}} K_{\mathrm{S}} S_{\mathrm{S0}} - K_{\mathrm{S}}(X_{\mathrm{EPS}} - X_{\mathrm{EPSo}})} \right) \\ \quad + \dfrac{1}{Y_{\mathrm{H}}(1 - k_{\mathrm{EPS}})} \ln\left(X_{\mathrm{Ho}} + \dfrac{Y_{\mathrm{H}}(1 - k_{\mathrm{EPS}})(X_{\mathrm{EPS}} - X_{\mathrm{EPSo}})}{k_{\mathrm{EPS}}} \right) \\ \quad - \dfrac{K_{\mathrm{S}}}{X_{\mathrm{Ho}} + Y_{\mathrm{H}}(1 - k_{\mathrm{EPS}})S_{\mathrm{S0}}} \ln\left(\dfrac{X_{\mathrm{Ho}}}{K_{\mathrm{S}} S_{\mathrm{S0}}} \right) - \dfrac{1}{Y_{\mathrm{H}}(1 - k_{\mathrm{EPS}})} \ln(X_{\mathrm{Ho}}) \end{array} \right\} \tag{7.11}$$

Thereafter, the weighted nonlinear least-square analysis is used to construct the objective function and to determine the kinetic parameter for EPS production by comparing the model predictions both with the observed EPS concentrations and t by using the known values. Thus, the estimation of the kinetic parameters is successively performed by minimizing the sum of the squared weighted errors (SSWE) based on n observations using the Monte Carlo method (Eq. 7.12):

$$\mathrm{SSWE} = \sum_{i=1}^{n} \left[w_i \left(t_i^{\mathrm{obs}} - t_i^{\mathrm{pred}} \right) \right]^2 \tag{7.12}$$

where w_{i} is an appropriate weighting factor; $t_i^{\mathrm{obs}}$is the time of the ith observation; t_i^{pred} is the t value predicted by the model for the measured X_{EPS} value.

The differences between the predicted and the observed EPS values can be estimated by multiplying $t_i^{\mathrm{obs}} - t_i^{\mathrm{pred}}$ by the local slope of the EPS production curve, $\Delta X_{\mathrm{EPS}}/\Delta t$. Therefore, the logical weighting factor is the local slope of the EPS production curve (Eq. 7.13):

$$\mathrm{SSWE} = \sum_{i=1}^{n} \left[\frac{\Delta X_{\mathrm{EPS}}}{\Delta t} \left(t_i^{\mathrm{obs}} - t_i^{\mathrm{pred}} \right) \right]^2 \approx \sum_{i=1}^{n} \left(X_{\mathrm{EPS}i}^{\mathrm{obs}} - X_{\mathrm{EPS}i}^{\mathrm{pred}} \right)^2 \tag{7.13}$$

where $X_{\mathrm{EPS}i}^{\mathrm{obs}}$ and $X_{\mathrm{EPS}i}^{\mathrm{pre}}$ are the ith measured and predicted EPS concentrations, respectively.

7.2.2.3 Monte Carlo Method

Figure 7.5 shows a flowchart of the calculation steps of the Monte Carlo method. In our work random uniform distributions were applied to simplify its complexity. In our approach, we used uniform random distributions, with a random number generation, to generate the parameter vector at each of the 1,000,000 calculations. To do this, the objective function for the entire parameter space was mapped. In this case, the random number generation was used to select parameter values at each step instead of over stepping through the parameter combinations in a systematic manner, because the later approach might lead to a loss of the global optima and the solution accuracy was limited. Since the EPS production kinetic estimation has four optimizing variables, four random numbers need to be

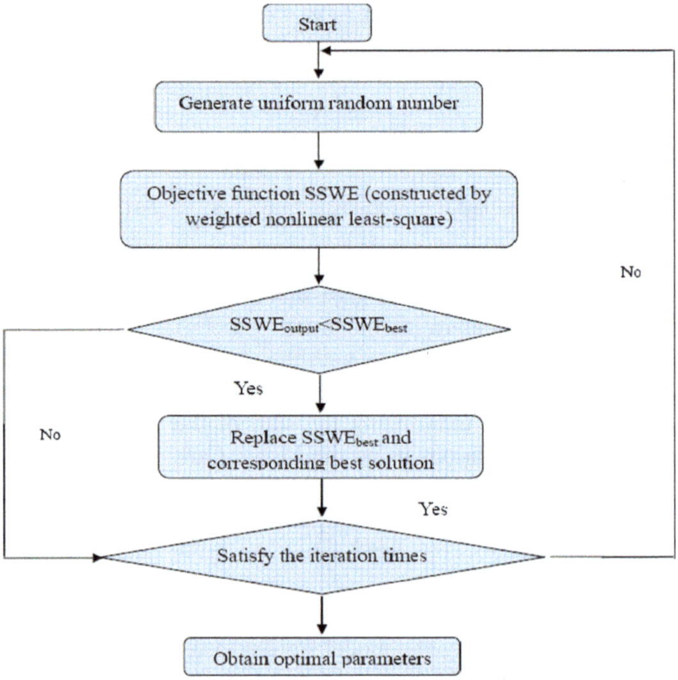

Fig. 7.5 A flowchart of the calculation steps of the Monte Carlo method (Reprinted from Ni et al. (2009), with permission from American Chemical Society)

generated at each step. The uniform distribution is a widely used random number generation method when the distribution of the kinetic parameters is unclear.

In this work, each parameter is assumed to distribute uniformly over the entire parameter space. Thus, the uniform distribution approach to generate four random numbers at each step for the selection of parameter is applied in our work. The comparison between this method of testing parameter value generation and other options, which can generate similar combinations of parameters without needing to generate four random numbers at each step, indicates that it is appropriate for the EPS kinetics estimation.

The computation for the Monte Carlo method is performed using software MatLab 7.0 (Mathworks, Natick, USA).

7.2.2.4 Self-Organizing Map

SOM is a neural network model and algorithm that implements a characteristic nonlinear projection from the high-dimensional space of sensory or other input signals onto a low-dimensional array of neurons. SOM was applied to visualize the relationships between the EPS production and the kinetic parameters.

7.2.2.5 Confidence Intervals for Parameter Estimation

The mean square error (MSE) of a fitting parameter, which is used to calculate the 95 % confidence interval of a given parameter estimate, is calculated from the mean square fitting error and the sensitivity of the model to the parameter (Smith et al. 1998). The model sensitivity to the parameters is evaluated through calculating the approximate first derivatives of the model predictions with respect to the parameters. Two sets of model predictions are compared, in which one parameter is varied by a small step. For the cases involving more than one fitting parameter as applied in this work, the sensitivity of the model predictions to the parameters is represented by a $p \times p$ matrix, A, where p is the number of fitting parameters.

7.2.2.6 Collection of Kinetic Data

Three different sets of batch experiments (I, II, and III) were conducted under different substrate concentrations (850, 1,150, 1,600 mg COD L^{-1}, respectively) and sludge concentrations (2,350, 3,050, 3,700 mg VSS L^{-1}, respectively) to evaluate the EPS production by activated sludge. In each batch test, the ratio of substrate to VSS was selected to be of a similar level and to be close to that in the SBR (approximately 0.4), in order to simulate the SBR operation better. Each batch experiment was performed in triplicates. The seeding sludge was sampled from the SBR when no substrate was present in the system, and was washed twice with distilled water to remove external soluble organic material. A batch reactor with working volume of 2 L was inoculated with 1 L of predetermined content of activated sludge and aerated it continuously so that the DO concentration exceeded 2 mg O_2 L^{-1}. Then, the 1-L synthetic wastewater with predetermined concentrations was added. Samples were taken at given intervals for the analysis of COD, EPS, and VSS. The experiments were conducted at pH of 7.0 ± 0.05. The measurement of VSS and COD was performed according to the Standard Methods (APHA 1995).

7.2.3 Results and Discussion

7.2.3.1 EPS Production by Activated Sludge

The profiles of the measured substrate consumption and EPS formation in the three sets of batch tests were similar (Figs 7.6, 7.10a, b). External substrate consumption resulted in an increase in the amount of total EPS. After the external substrate became depleted, the total EPS concentration did not change significantly. This indicates that EPS formation mainly occurred during the substrate consumption period. This observation was in accordance with the unified theory for EPS production proposed by Laspidou and Rittmann (2002a, b). They proposed

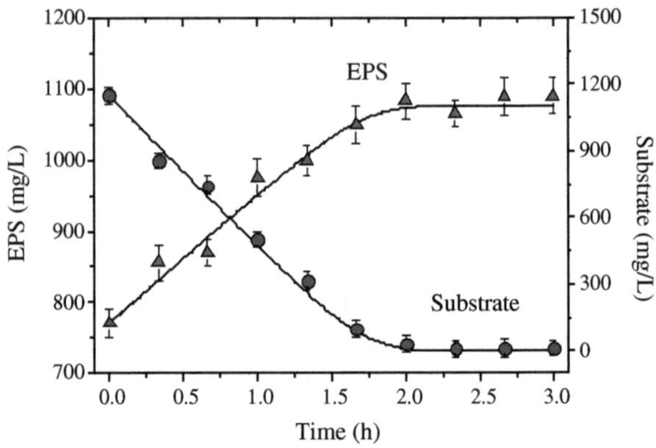

Fig. 7.6 Fitting results of the integrated EPS production equations to the EPS formation rate data. The model (*solid line*) is fitted to one data set I of EPS concentrations (*scatter*), which result in the parameter values shown in Table 7.3 (Reprinted from Ni et al. (2009), with permission of American Chemical Society)

that the formation of EPS was growth-associated and was produced in direct proportion to substrate utilization. Furthermore, Turakhia and Characklis (1988) also reported that there was a linear relationship between the EPS production rate and the biomass growth rate. The total EPS concentration in activated sludge increased with the microbial growth. The contents of carbohydrates, proteins, and nucleic acids in EPS all increased slowly in the substrate consumption process (data not shown), although those three components had different formation rates. The content of carbohydrates was higher than those of proteins and nucleic acids, and carbohydrates were the predominant components in the EPS formed.

7.2.3.2 Application of the Kinetic Approach

The experimental data for EPS production were analyzed to determine the rate coefficients μ_H and K_S and the two yield coefficients for active bacteria (Y_H) and EPS (k_{EPS}) by fitting Eq. 7.11 to the EPS production data by using the Monte Carlo method and weighted nonlinear least-squares analysis (Eqs. 7.12 and 7.13). The μ_H was used as a fitting parameter, because it was not known after incorporation EPS production. Ignoring EPS formation may lead to an overestimation of true cellular growth rates and it would not be appropriate to force the best-fit curve through the defaulted value of μ_H. The input value for X_{H0} was estimated based on the results of several experiments.

The kinetic parameter values and their standard errors are summarized in Table 7.3 along with their 90 % likelihood intervals for nonlinear systems, suggesting that the estimated kinetic parameters were reasonable. The best estimates of the rate coefficients μ_H (0.32 h^{-1}) and K_S (11.68 g COD_S m^{-3}), obtained from fitting Eq. 7.11 to the data along with the two yield coefficients of Y_H (0.49 g COD_X g^{-1} COD_S), and k_{EPS} (0.23 g COD_{EPS} g^{-1} COD_S), yielded a simulation curve that matched data set I shown in Fig. 7.6. It should be noted that the substrate-type, reactor operating conditions and microbial communities are important factors affecting the EPS production. Since, the data were obtained from

Table 7.3 Parameters for the optimum fitting to the experimental data

Parameter	Definition	Values	Unit
k_{EPS}	Yield coefficient for EPS	0.23 ± 0.01	g COD_{EPS} g^{-1} COD_S
Y_H	Yield coefficient for bacteria	0.49 ± 0.02	g COD_X g^{-1} COD_S
μ_H	Maximum growth rate of X_H	0.32 ± 0.02	h^{-1}
K_S	Half-saturation constant	11.68 ± 0.15	g COD_S m^{-3}

Table 7.4 Comparison among the values of the kinetic parameters for EPS formation reported in literatures

Sources	k_{EPS} (g COD_{EPS} g^{-1} COD_S)	Y_H (g COD_X g^{-1} COD_S)	μ_H (h^{-1})	K_S (g COD_S m^{-3})
This study	0.23	0.49	0.32	11.68
Laspidou and Rittmann (2002b)	0.18	0.34	0.403	0.7
Kommedal et al. (2001)	0.18	0.44	0.39	1.7
Horn et al. (2001)	0.182	0.13	0.05	2.0
Ni et al. (2009)	0.18	0.61	0.40	11.38
Laspidou and Rittmann (2004)	0.18	0.34	0.403	20.0

mixed-culture systems (activated sludge), the parameters reported here were highly dependent on these factors. The obtained parameter coefficients in Table 7.3 represent the overall values for the mixed-culture systems.

Table 7.4 lists the values of the kinetic parameters for EPS production reported in literatures for comparison. The value of k_{EPS} (0.23 mg COD_{EPS} mg^{-1} COD_S) estimated from our approach was generally in accord with those reported in other studies. The Y_H estimated from our approach, 0.49 g COD_X g^{-1} COD_S, fell in the range of the values reported in literatures (Table 7.4). The heterotrophs in activated sludge have a wide range of Y_H value, depending on the cultivation conditions such as substrate type, nutrient concentration, and many other factors. The value of the estimated K_S, 11.68 g COD_S m^{-3}, was between 0.7 and 20.0 g COD_S m^{-3} shown in Table 7.4. The μ_H reported in literatures varied largely between 0.05 and 0.403 h^{-1} (Table 7.4), whereas the μ_H value estimated in our work, 0.32 h^{-1}, was also in the range of these values.

The cells loss attributed to the endogenous decay over the experimental course was much smaller compared with the biomass growth. Therefore, the biomass lose in the decay was neglected in our work. EPS might also undergo the decay process and results in an EPS lose. However, the decay rate for EPS reported in previous work was approximately 0.0071 h^{-1}, which was significantly lower than its production rate (0.21 h^{-1}). Such a difference suggests that the EPS lose could be neglected in the EPS production process, when the substrate was being consumed. In addition, since the main objective of this work was to develop a convenient approach to evaluate the EPS production kinetics, the EPS decay was not taken into account in our approach.

7.2.3.3 Practical Parameter Identifiability

Practical parameter identifiability of a model structure is important as it tells which parameter combinations can be estimated under given measurement accuracy and quantity. Sensitivity analysis is a useful tool to evaluate whether the values of theoretical identifiable parameters can be reliably obtained from experimental data. The four kinetic parameters for EPS production (Table 7.3) were changed in the simulation one by one. The output sensitivities of model parameters calculated using best-fit parameters for the EPS production are shown in Fig. 7.7.

The base value for μ_H was 0.32 h^{-1} with a range of $0.12 \leq \mu_H \leq 0.62$ (Fig. 7.7a). The effect of μ_H on the EPS output was minimal at high values of μ_H. However, as μ_H was reduced, the EPS production process changed significantly. During the reaction period, at a μ_H of 0.12, the EPS production rate was much lower than the corresponding values at high μ_H values. Figure 7.7b shows that all curves had a similar mode in a range of $1.68 \leq K_S \leq 51.68$. However, as K_S was increased, the sudden inflection curve became much smooth. This was attributed to the fact that the substrate affinity constant is a biological characteristic parameter of the particular microbial species. It represents the appetency between the substrate and microbial species.

An increase in the affinity constant will reduce such an appetency. An increase in k_{EPS} resulted in an increase in the EPS production rate and EPS content (Fig. 7.7d). Such an effect was attributed to the fact that k_{EPS} is the yield coefficient for EPS, which is a multiplicator of the total EPS production rate. However, the yield coefficient of microbial Y_H had an influence on the model output in an opposite way (Fig. 7.7c). A lower Y_H value resulted in a higher EPS production rate. This was associated with the indirect effect of this yield on the substrate utilization rate by affecting the substrate electron distribution.

The contour plots of the objective function shown in Fig. 7.8 were calculated around the optimum for different combinations of parameters. The contours of the objective function were large in all planes of the two-parameter subsets. This implies that several combinations of parameters would give almost equally good fits to the data, leading to large confidence intervals of the parameter estimates. The sensitivity analysis results show that the parameters of μ_H and k_{EPS} were more sensitive than those of K_s and Y_H. In the plane K_s and Y_H (Fig. 7.8), the objective function was observed to be a valley-like shape with a flat bottom. This shape shows a valley where the optimum values reside approximately within the center, indicating a good identifiability of these parameters.

The component plane calculated using SOM to visualize these overall relationships is shown in Fig. 7.9. A higher EPS production was linked to a higher k_{EPS}, while the total EPS production had no such a strong relationship with μ_H, K_S, and Y_H. The results from a correlation analysis comprising the input parameters μ_H, K_S, Y_H and k_{EPS}, and the target variable (EPS production) were in agreement with the results shown in Fig. 7.9, highlighting the key relationships revealed by the SOM. The k_{EPS}, as key parameter, correlated positively with the EPS

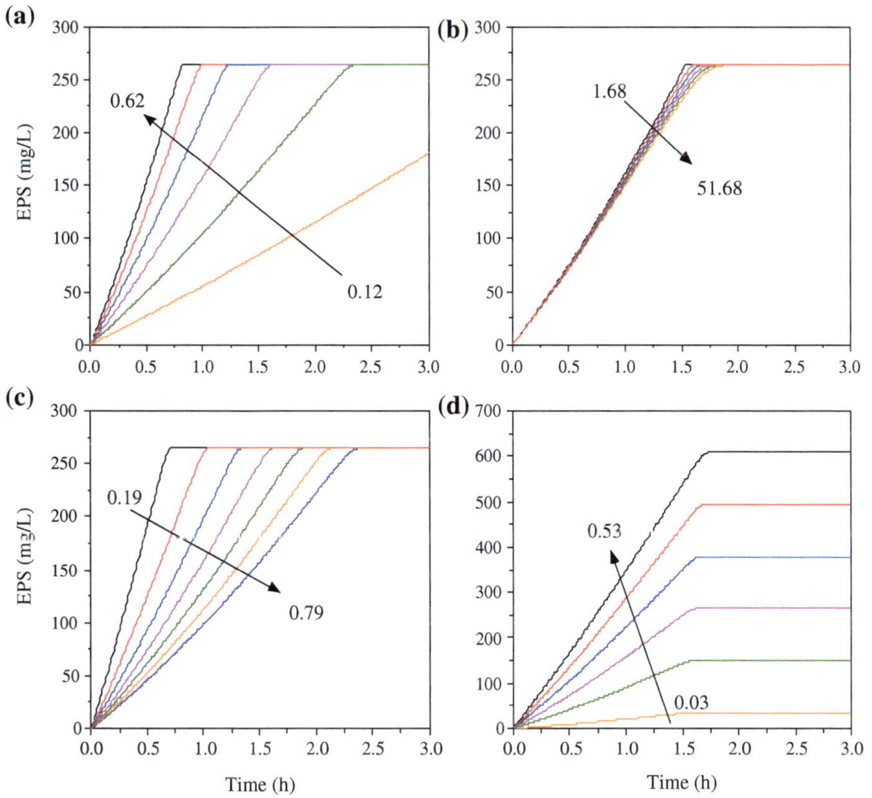

Fig. 7.7 Sensitivity of parameters to EPS production: **a** μ_H; **b** K_S; **c** Y_H; and **d** k_{EPS} (Reprinted from Ni et al. (2009), with permission of American Chemical Society)

production ($R = 0.970$, see Fig. 7.9), indicating a higher k_{EPS} would result in more EPS production.

7.2.3.4 Verification of the Kinetic Approach

Verification is performed through comparing the measured and simulated results. In our work, one data set (set I) was used to estimate the kinetic parameters, and the resulting values for the coefficients were used to generate the two additional curves for comparison with the other two data sets (Fig. 7.10a, b). The maximum difference between the measured and calculated values was 10 %, and 70 % of the results had a difference of less than 5 %. Furthermore, the simulation shows no systematic deviations, suggesting the validity of the estimated kinetic parameters.

It should be noted that the kinetic information obtained from these batch tests was heavily dependent upon the ratio of the initial substrate concentration to the initial biomass concentration S_o/X_o. The use of a low S_o/X_o ratio (e.g., < 0.2)

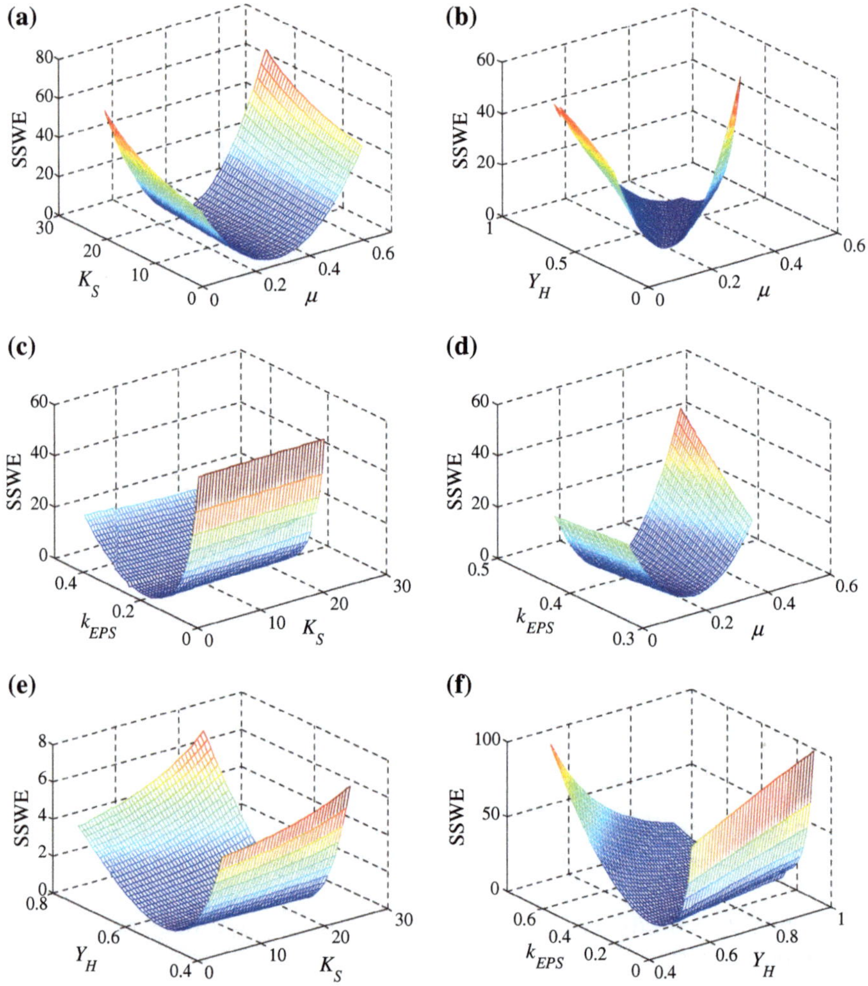

Fig. 7.8 Contour plots of the objective functional used for parameter estimation as a function of different parameter combinations: **a** μ and K_S; **b** μ and Y_H; **c** k_{EPS} and K_S **d** μ and k_{EPS} **e** K_S and Y_H; and **f** Y_H and k_{EPS} (Reprinted from Ni et al. (2009), with permission of American Chemical Society)

would lead to a short-term experiment, as the substrate was rapidly assimilated and no significant biomass growth and EPS production occurred. On the other hand, at a high S_o/X_o (e.g., > 1.0), the degradation of these components was readily achieved, but a significant microbial growth occurred. This would result in a complex interpretation of the multi-component kinetics, such as incorporation of EPS production. Therefore, in the procedures proposed in this work, accurate and reliable EPS production kinetics estimation requires an optimal S_o/X_o ratio. This optimal value was greatly dependent on the origins and characteristics of the

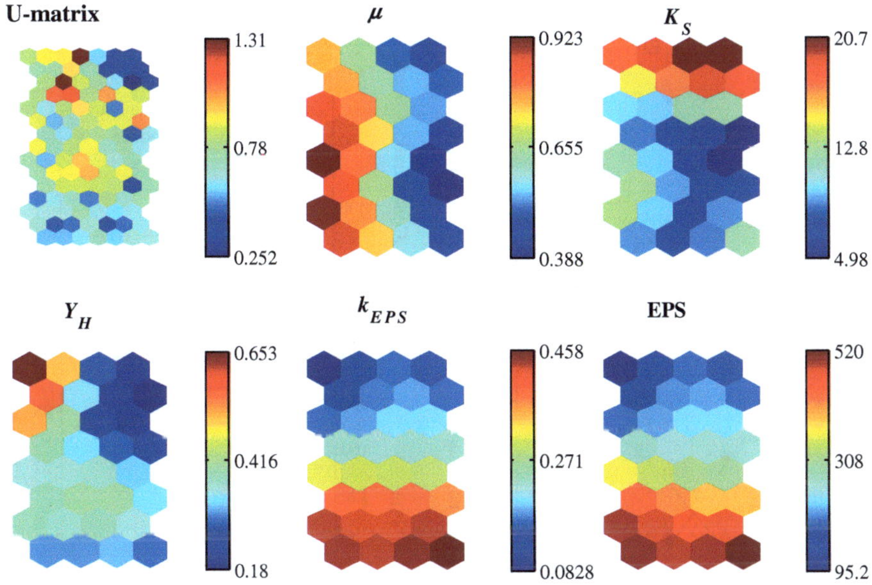

Fig. 7.9 Abstract visualization of the relationships between EPS (mg COD L^{-1}) and μ_H (h^{-1}), K_S (mg COD L^{-1}), Y_H (g COD_X g^{-1} COD_S), and k_{EPS} (g COD_{EPS} g^{-1} COD_S) using an SOM model (Reprinted from Ni et al. (2009), with permission of American Chemical Society)

wastewater and the sludge. Therefore, the steady-state operation conditions for the SBR can be considered to determine the optimal batch experimental conditions, allowing the accuracy of the estimation to be maximized. In this work, the ratios of substrate to VSS were selected to be at a level close to that in the original SBR, in order to better verify the estimated parameters.

To further validate our approach, the predicted EPS and external substrate profiles in Fig. 7.10c were compared with the corresponding experimental data from Hsieh et al. (1994) for a transient experiment under batch conditions. Overall, the model outputs captured all the experimental trends for this experiment in terms of growth and EPS production. The most appropriate estimate of the k_{EPS} was 0.16 g COD_{EPS} g^{-1} COD_S. The external substrate declined at an increasing rate over the initial 10 h. During the period of rapid microbial growth, the EPS increased gradually. Laspidou and Rittmann (2002b) also obtained EPS production kinetic parameters using a complex mathematical model to describe the same set of experimental data from Hsieh et al. (1994). The k_{EPS} was estimated to be 0.18 g COD_{EPS} g^{-1} COD_S, which was comparable to that of estimation using our approach. However, the model structure of Laspidou and Rittmann (2002b) included eight complex mass balance equations with 14 model parameters, which needed to be calibrated using extensive experimental data and complex model solution method. In our approach, only one data set was needed to obtain similar

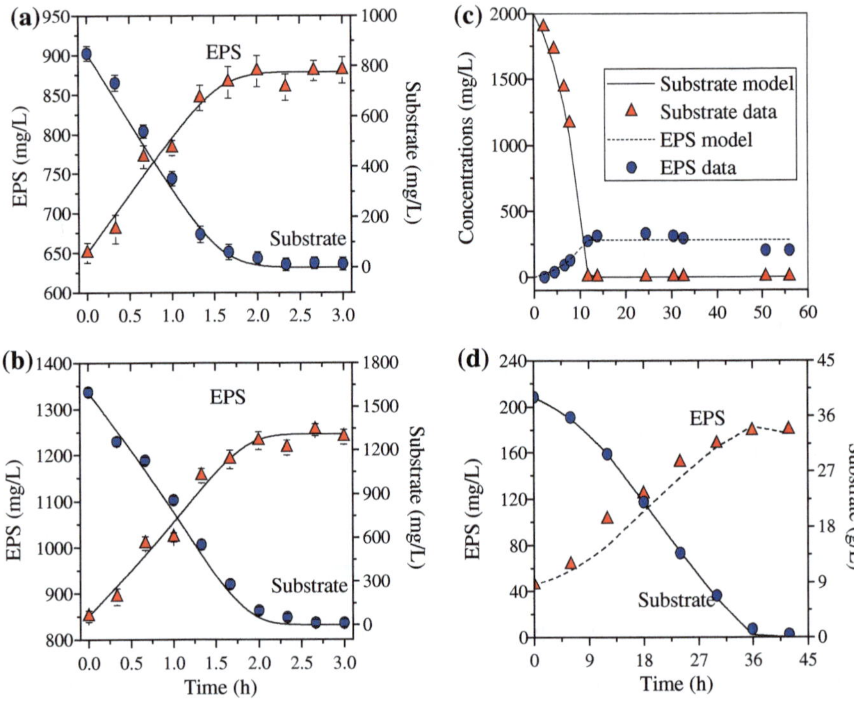

Fig. 7.10 Comparisons between the model simulations and the experimental data for the verification of the approach: **a** data set II of this work; **b** data set III of this work; **c** data set of Hsieh et al. (1994); and **d** data set of Wang and Yu (2007) (Reprinted from Ni et al. (2009), with permission of American Chemical Society)

accurate kinetic parameters by using a convenient kinetic approach, as demonstrated in this work.

In addition, the model was further evaluated by using the experimental results of Wang and Yu (2007). They investigated the production of EPS in a glucose medium by *Ralstonia eutropha* ATCC 17699 in batch cultures. The good agreement between the model predictions and the experimental data shown in Fig. 7.10d suggests that our approach was appropriate to evaluate the production kinetics of EPS by a pure culture. Again, only one data set was used to obtain the kinetic parameters of EPS production for this pure culture, which still accurately described the substrate utilization and EPS formation (Fig. 7.10d).

7.2.3.5 Comparison with Other Approaches

Parameter estimation consists of the determination of their "optimal" values from the measured data. The parameter estimation routine basically consists of minimizing an objective function, which for example, can be defined as the weighted sum of squared errors between the model output and the data. When the objective function reaches a minimum value with a given accuracy, the optimal parameter values are obtained. Thus, the parameter estimation is carried out using the mathematical search algorithms (Smith et al. 1998; Deiveegan et al. 2006).

A comparison among the several evaluation methods for biokinetics parameters was performed to evaluate our approach. Smith et al. (1998) used a computer spreadsheet program to estimate the kinetic parameters. However, this method was based on the simplex for linear programming and generalized reduced gradient for nonlinear programming. This might result in a miss finding of the global minimum (Behzadi et al. 2005). Gradient descent algorithm is a widely used search algorithm to find the minimum for parameter estimation through measuring the errors, calculating the gradient of the errors, and adjusting the parameters in the descending gradient direction. However, the algorithm inherently has the problem of being trapped into local optima (Salman and Al-Shammiri 2007). Levenberg–Marquardt algorithm is the most popular tool for solving the nonlinear minimum problems. Attributed to the properties of rapid convergence and stability, this method has been employed in many modeling works (Sakamoto et al. 2005). But its accuracy depends highly on the selection of the initial values of parameters and also has the problem of being trapped into local optima (Deiveegan et al. 2006). In this case, the problem of being trapped into local optima would not be avoided by increasing the overall iteration times. To sort out these problems, the Monte Carlo method, which is a technique to quantify the variability by using probability distribution, is applied in our approach. It uses an iterative search for the parameter values that yield the minimum SSWE, and the initial estimates of the parameters are automatically replaced by the best estimates. It can quickly search for a minimum value for a selected objective by varying the values for one or more selected parameters. It provides a flexible simulation tool capable of accounting in a straightforward manner. Although long CPU times are needed in the Monte Carlo simulation, it provides better simulation accuracy and avoids trapping into local optimum through increasing the overall iteration times, compared with the evaluation algorithms mentioned above (Mantoglou et al. 2007). This has been demonstrated in this work.

The comparison results among the three other approaches, i.e. (i) spreadsheet program method (ii) gradient descent algorithm, and (iii) Levenberg–Marquardt algorithm, are given in Table 7.5. The values of the parameters estimated with the different approaches listed in Table 7.5 show that the gradient descent algorithm and the Levenberg–Marquardt algorithm might trap into local optima, and that the parameter values obtained from these methods are unreasonable. In addition, these approaches exhibit different optimal minimum values. The parameters values estimated with the spreadsheet program method and this work are nearly equal, but the optimal solution could be obtained more rapidly and accurately by the approach proposed in our work. Such a comparison demonstrates that the approach of our work provides greater simulation accuracy and avoids trapping into local optimum, compared with the evaluation algorithms mentioned above.

Table 7.5 Comparison between our approach with other approaches

Method	Estimated parameters				Testing function	
	k_{EPS}	Y_H	μ_H	K_S	Global/Local optima	Optimal minimum
(i)	0.24	0.46	0.33	14.55	Global	1,593
(ii)	0.11	0.26	1.26	4.37	Local	962
(iii)	0.39	0.79	2.41	25.68	Local	894
This work	0.23	0.49	0.32	11.68	Global	878

7.2.3.6 Correction for the Overestimation of Cellular Production

The estimated kinetic parameters showed that the activated sludge consumed the substrate (electron donor) to produce active biomass and EPS. The electron distribution was: new biomass synthesis 49 %, oxygen for respiration 28 %, and EPS 23 %. Thus, a significant amount of substrate electrons was distributed to produce EPS besides the active cells. The obtained biomass yield on substrate consumed in our work, i.e., 0.49, was lower than the corresponding value of 0.67, which is often used in the conventional activated sludge modeling approaches (Sin et al. 2005). In the conventional approaches, the EPS production kinetics is lumped into the yields for cell synthesis. In this case, the substrate electrons for the synthesis of EPS are accounted for the biomass yield, resulting in a higher biomass yield value. This suggests that the improved calculation of EPS kinetics by our approach could correct the overestimation of the cellular production.

7.2.4 Conclusions

The integrated EPS production method is a convenient procedure to obtain reliable estimates of kinetic coefficients of EPS formation by activated sludge. The fraction of EPS lost over the experimental course because of hydrolysis of EPS must be small. Changes in EPS production attributed to toxicity or other conditions do not limit the applicability of this method, as they can be included in the integrated form of the equation. The weighting method and Monte Carlo method applied in this work has a rational statistical basis. The difference between the observed and predicted EPS concentrations is estimated by multiplying the slope of the EPS production curve by the difference between the measured and predicted sample times. This approach is appropriate because it approximates the error in the measurement of EPS concentration and such errors are usually larger than errors in time measurement in biological wastewater treatment experiments. In addition, the approach described above for calculating the practical parameter identifiability makes great use of the information available from the data without requiring the use of sophisticated mathematics. Therefore, it successfully corrects the overestimation of cellular production and identifies that k_{EPS} is the key parameter in EPS production kinetics. Furthermore, this study demonstrates that this approach could estimate the kinetic parameters accurately using few data sets or even one set, which becomes very attractive for the processes where data are obtained costly.

References

Arunachalam, R., Shah, H.K., Ju, L.K.: Monitoring aerobic sludge digestion by online scanning fluorometry. Water Res. **39**, 1205–1214 (2005)

APHA: Standard methods for the examination of water and wastewater, 19th edn. American Public Health Association, New York (1995)

Aquino, S.F., Stuckey, D.C.: Integrated model of the production of soluble microbial products (SMP) and extracellular polymeric substances (EPS) in anaerobic chemostats during transient conditions. Biochem. Eng. J. **38**, 138–146 (2008)

Baunsgaard, D.: Factors affecting 3-way modeling (PARAFAC) of fluorescence landscapes. http://www.models.kvl.dk(1999)

Behzadi, B., Ghotbi, C., Galindo, A.: Application of the simplex simulated annealing technique to nonlinear parameter optimization for the SAFT-VR equation of state. Chem. Eng. Sci. **60**, 6607–6621 (2005)

Bro, R.: PARAFAC. Tutorial and applications. Chemom. Intell. Lab. Syst. **38**, 149–171 (1997)

Bro, R., Kiers, H.A.L.: A new efficient method for determining the number of components in PARAFAC models. J. Chemom. **17**, 274–286 (2003)

Chen, W., Westerhoff, P., Leenheer, J.A., Booksh, K.: Fluorescence excitation—Emission matrix regional integration to quantify spectra for dissolved organic matter. Environ. Sci. Technol. **37**, 5701–5710 (2003)

Curvers, D., Maes, K.C., Saveyn, H., De Baets, B., Miller, S., Van der Meeren, P.: Modelling the electro-osmotically enhanced pressure dewatering of activated sludge. Chem. Eng. Sci. **62**, 2267–2276 (2007)

Deiveegan, M., Balaji, C., Venkateshan, S.P.: Comparison of various methods for simultaneous retrieval of surface emissivities and gas properties in gray participating media. J. Heat Transf.-Trans. ASME. **128**, 829–837 (2006)

Ding, A., Hounslow, M.J., Biggs, C.A.: Population balance modelling of activated sludge flocculation: investigating the size dependence of aggregation, breakage and collision efficiency. Chem. Eng. Sci. **61**, 63–74 (2006)

Dubois, M., Gilles, K.A., Hamilton, J.K., Rebers, P.A., Smith, F.: Colorimetric method for determination sugars and related substance. Anal. Chem. **28**, 350–356 (1956)

Farabegoli, G., Hellinga, C., Heijnen, J.J., van Loosdrecht, M.C.M.: Study on the use of NADH fluorescence measurements for monitoring wastewater treatment systems. Water Res. **37**, 2732–2738 (2003)

Fonseca, A.C., Summers, R.S., Greenberg, A.R., Hernandez, M.T.: Extra-cellular polysaccharides, soluble microbial products, and natural organic matter impact on nanofiltration membranes flux decline. Environ. Sci. Technol. **41**, 2491–2497 (2007)

Frolund, B., Keiding, K.: Implementation of an HPLC polystyrene divinylbenzene column for separation of activated sludge exopolymers. Appl. Microbiol. Biotechnol. **4**, 708–716 (1994)

Frolund, B., Palmgren, R., Keiding, K., Nielsen, P.H.: Extraction of extracellular polymers from activated sludge using a cation exchange resin. Water Res. **30**, 1749–1758 (1996)

Garnier, Ch., Gorner, T., Guinot-Thomas, P., Chappe, P., de Donato, P.: Exopolymeric production by bacterial strains isolated from activated sludge of paper industry. Water Res. **40**, 3115–3122 (2006)

González-Brambila, M., Monroy, O., López-Isunza, F.: Experimental and theoretical study of membrane-aerated biofilm reactor behavior under different modes of oxygen supply for the treatment of synthetic wastewater. Chem. Eng. Sci. **61**, 5268–5281 (2006)

Gorner, T., de Donato, P., Ameil, M.-H., Montarges-Pelletier, E., Lartiges, B.S.: Activated sludge exopolymers: separation and identification using size exclusion chromatography and infrared micro-spectroscopy. Water Res. **37**, 2388–2393 (2003)

Goovaerts, P., Semrau, J., Lontoh, S.: Monte Carlo analysis of uncertainty attached to microbial pollutant degradation rates. Environ. Sci. Technol. **35**, 3924–3930 (2001)

Haberkamp, J., Ernst, M., Bockelmann, U., Szewzyk, U., Jekel, M.: Complexity of ultrafiltration membrane fouling caused by macromolecular dissolved organic compounds in secondary effluents. Water Res. **42**, 3153–3161 (2008)

Henze, M., Gujer, W., Mino, T., van Loosdrecht, M.C.M.: Activated sludge models ASM1, ASM2, ASM2d, and ASM3. IWA Scientific and technical report No. 9. IWA Publishing, London (2000)

Horn, H., Neu, T.R., Wulkow, M.: Modelling the structure and function of extracellular polymeric substances in biofilms with new numerical techniques. Water Sci. Technol. **43**, 121–127 (2001)

Hsieh, K. M., Murgel, G. A., Lion, L. W., Schuller, M. L.: Interactions of microbial biofilms with toxic trace metals 1. Observation and modeling of cell growth, attachment, and production of extracellular polymer. Biotechnol. Bioeng. **44**, 219–231 (1994)

Huijbregts, M.A.J., Gilijamse, W., Ragas, A.M.J., Reijnders, L.: Evaluating uncertainty in environmental life-cycle assessment. A case study comparing two insulation options for a Dutch one-family dwelling. Environ. Sci. Technol. **37**, 2600–2608 (2003)

Jorand, F., Boue-Bigne, F., Block, J.C., Urbain, V.: Hydrophobic/hydrophilic properties of activated sludge exopolymeric substances. Water Sci. Technol. **37**, 307–315 (1998)

Jorand, F., Zartarian, F., Thomas, F., Block, J.C., Bottero, J.Y., Villemin, G., Urbain, V., Manem, J.: Chemical and structural (2D) linkage between bacteria within activated sludge flocs. Water Res. **29**, 1639–1647 (1995)

Kawaguchi, T., Decho, A.W.: In situ analysis of carboxyl and sulfhydryl groups of extracellular polymeric secretions by confocal laser scanning microscopy. Anal. Biochem. **304**, 266–267 (2002)

Kommedal, R., Bakke, R., Dockery, J., Stoodley, P.: Modelling production of extracellular polymeric substances in a Pseudomonas aeruginosa chemostat culture. Water Sci. Technol. **43**, 129–134 (2001)

Kreft, J.U., Wimpenny, J.W.T.: Effect of EPS on biofilm structure and function as revealed by an individual-based model of biofilm growth. Water Sci. Technol. **43**, 135–141 (2001)

Laspidou, C.D., Rittmann, B.E.: A unified theory for extracellular polymeric substances, soluble microbial products, and active and inert biomass. Water Res. **36**, 2711–2720 (2002a)

Laspidou, C.S., Rittmann, B.E.: Non-steady state modeling of extracellular polymeric substances, soluble microbial products, and active and inert biomass. Water Res. **36**, 1983–1992 (2002b)

Laspidou, C.S., Rittmann, B.E.: Modeling the development of biofilm density including active bacteria, inert biomass, and extracellular polymeric substances. Water Res. **38**, 3349–3361 (2004)

Leterm, B., van Clooster, M., van der Linden, T., Tiktak, A., Rounsevell, M.D.A.: Including spatial variability in Monte Carlo simulations of pesticide leaching. Environ. Sci. Technol. **41**, 7444–7450 (2007)

Leone, L., Loring, J., Sjoberg, S., Persson, P., Shchukarev, A.: Surface characterization of the gram-positive bacteria *Bacillus subtilis* – an XPS study. Surf. Interface Anal. **38**, 202–205 (2006)

Liao, B.Q., Allen, D.G., Droppo, G., Leppard, G.G., Liss, S.N.: Surface properties of sludge and their role in bioflocculation and settleability. Water Res. **35**, 339–350 (2001)

Liao, B.Q., Allen, D.G., Leppard, G.G., Droppo, I.G., Liss, S.N.: Interparticle interactions affecting the stability of sludge flocs. J. Colloid Interface Sci. **249**, 372–380 (2002)

Li, J.K., Asali, E.C., Humphrey, A.E.: Monitoring cell concentration and activity by multiple excitation fluorometry. Biotechnol. Prog. **7**, 21–27 (1991)

Li, W.H., Sheng, G.P., Liu, X.W., Yu, H.Q.: Characterizing the extracellular and intracellular fluorescent products of activated sludge in a sequencing batch reactor. Water Res. **42**, 3173–3181 (2008)

Liu, Y., Yang, C.H., Li, J.: Influence of extracellular polymeric substances on *Pseudomonas aeruginosa* transport and deposition profiles in porous media. Environ. Sci. Technol. **41**, 198–205 (2007)

Lowry, O.H., Farr, A.L., Randall, R.J.: Protein measurement with the folin phenol reagent. J. Biol. Chem. **193**, 265–275 (1951)

Luo, J., Farrell, J.: Examination of hydrophobic contaminant adsorption in mineral micropores with grand canonical Monte Carlo simulations. Environ. Sci. Technol. **37**, 1775–1782 (2003)

Mantoglou, A., Kourakos, G.: Optimal groundwater remediation under uncertainty using multi-objective optimization. Water Res. **21**, 835–847 (2007)

McSwain, B.S., Irvine, R.L., Hausner, M., Wilderer, P.A.: Composition and distribution of extracellular polymeric substances in aerobic flocs and granular sludge. Appl. Environ. Microbiol. **71**, 1051–1057 (2005)

Miller, S.A., Landis, A.E., Theis, T.L.: Use of Monte Carlo analysis to characterize nitrogen fluxes in agroecosystems. Environ. Sci. Technol. **40**, 2324–2332 (2006)

Mikkelsen, L.H., Nielsen, P.H.: Quantification of the bond energy of bacteria attached to activated sludge floc surfaces. Water Sci. Technol. **43**, 67–75 (2001)

Ni, B.J., Yu, H.Q.: An approach for modeling two-step denitrification in activated sludge systems. Chem. Eng. Sci. **63**, 1449–1459 (2008)

Ni, B.J., Zeng, R.J., Fang, F., Xu, J., Sheng, G.P., Yu, H.Q.: A novel approach to evaluate the production kinetics of extracellular polymeric substances (EPS) by activated sludge using weighted nonlinear least-squares analysis. Environ. Sci. Technol. **43**, 3743–3750 (2009)

Pons, M.N., Bonté, S.L., Potier, O.: Spectral analysis and fingerprinting for biomedia characterisation. J. Biotechnol. **113**, 211–230 (2004)

Sakamoto, H., Matsumoto, K., Kuwahara, A., Hayami, Y.: Acceleration and stabilization techniques for the Levenberg–Marquardt method. IEICE Trans. Fund. **E88-A**, 1971–1978 (2005)

Salman, A., Al-Shammiri, M. A.: New computational intelligence model for predicting evaporation rates for saline. Desalination. **214**, 273–286 (2007)

Saloranta, T.M., Armitage, J.M., Haario, H., Nes, K., Cousins, I.T., Barton, D.N.: Modeling the effects and uncertainties of contaminated sediment remediation scenarios in a norwegian fjord by markov chain Monte Carlo simulation. Environ. Sci. Technol. **42**, 200–206 (2008)

Sheintuch, M., Tartakovsky, B.: Activated-sludge system design for species selection: analysis of a detailed multispecies model. Chem. Eng. Sci. **52**, 3033–3046 (1997)

Sheng, G.P., Yu, H.Q.: Characterization of extracellular polymeric substances of aerobic and anaerobic sludge using three-dimensional excitation and emission matrix fluorescence spectroscopy. Water Res. **40**, 1233–1239 (2006)

Sheng, G.P., Yu, H.Q., Li, X.Y.: Stability of sludge flocs under shear conditions: roles of extracellular polymeric substances (EPS). Biotechnol. Bioeng. **6**, 1095–1102 (2006)

Sheng, G.P., Zhang, M.L., Yu, H.Q.: Characterization of adsorption properties of extracellular polymeric substances (EPS) extracted from sludge. Colloids Surf. B **62**, 83–90 (2008)

Sin, G., de Pauw, D.J.W., Weijers, S., Vanrolleghem, P.A.: An efficient approach to automate the manual trial and error calibration of activated sludge models. Biotechnol. Bioeng. **100**, 516–528 (2008)

Sin, G., Guisasola, A., de Pauw, D.J.W., Baeza, J.A., Carrera, J., Vanrolleghem, P.A.: A new approach for modelling simultaneous storage and growth processes for activated sludge systems under aerobic conditions. Biotechnol. Bioeng. **92**, 600–613 (2005)

Smith, L.H., McCarty, P.L., Kitanidis, P.K.: Spreadsheet method for evaluation of biochemical reaction rate coefficients and their uncertainties by weighted nonlinear least-squares analysis of the integrated Monod equation. Appl. Environ. Microbiol. **64**, 2044–2050 (1998)

Turakhia, M.H., Characklis, W.G.: Activity of *Pseudomonas aeruginosa* in biofilms: effect of calcium. Biotechnol. Bioeng. **33**, 406–414 (1988)

Urbain, V., Block, J.C., Manem, A.: Biofloculation in activated sludge: an analytical approach. Water Res. **27**, 829–838 (1993)

Wang, J., Yu, H.Q.: Biosynthesis of polyhydroxybutyrate (PHB) and extracellular polymeric substances (EPS) by Ralstonia eutropha ATCC 17699 in batch cultures. Appl. Microbiol. Biotechnol. **75**, 871-878 (2007)

Zartarian, F., Mustin, C., Villemin, G., Ait-Ettager, T., Thill, A., Bottero, J.Y., Mallet, J.L., Snidaro, D.: Three-dimensional modeling of an activated sludge floc. Langmuir **13**, 35–40 (1997)

Chapter 8
Fractionating and Determination of the Soluble Microbial Products

The subfractions of the soluble microbial products (SMP), i.e., utilization-associated products (UAP), and biomass-associated products (BAP), were characterized in terms of formation sequence, MW and chemical natures, using MW and dissolved organic carbon (DOC) measurements, coupled with oxygen utilization rate determination, polysaccharide and protein measurement, 3-dimensional excitation emission matrix (EEM) fluorescence spectroscopy and Fourier transform infrared spectroscopy (FTIR) analysis. A new approach for determining SMP, UAP and BAP, and their production kinetics was then established. The relationships among the formation of the three subfractions of the SMP and the substrate utilization, as well as the SMP formation mechanisms, were elucidated.

8.1 Fractionating the Soluble Microbial Products

8.1.1 Introduction

SMP, resulting from microbial growth and decay, comprise a wide range of compounds with high or low MWs (Noguera et al. 1994; Barker and Stuckey 1999; Laspidou and Rittmann 2002a). SMP are usually the major component of the soluble organic matter, e.g., soluble COD (SCOD), in effluents from biological wastewater treatment plants (Barker and Stuckey 1999; Aquino and Stuckey 2008; Shon et al. 2004; Jarusutthirak et al. 2005; Rosenberger et al. 2006). Hence, their presence is a matter of great interest not only in terms of achieving current discharge standards, but also because they effectively affect the lower limit for treatment plants (Barker and Stuckey 1999; Gao et al. 2004; Grunheid et al. 2005; Labbs et al. 2006; Holakoo et al. 2006; Ichihashi et al. 2006; Magbanua and Bowers 2006).

SMP have been classified into two groups based on the bacterial phase from which they are derived: the utilization-associated products (UAP) derived from the

B.-J. Ni, *Formation, Characterization and Mathematical Modeling of the Aerobic Granular Sludge*, Springer Theses, DOI: 10.1007/978-3-642-31281-6_8,

original substrate in microbial growth and the BAP generated in the endogenous phase (Namkung and Rittmann 1986; Grady et al. 1999; Laspidou and Rittmann 2002a; Jarusutthirak and Amy 2006). Regarding the characterization, most of the previous related studies have focused on the global characteristics of SMP (Barker and Stuckey 1999). SMP have been found to have a very wide range of MW distribution and different structures/functions (Barker and Stuckey 1999; Grady et al. 1999; Rosenberger et al. 2006; Labbs et al. 2006; Magbanua and Bowers 2006; Jarusutthirak and Amy 2006). The MW distribution of SMP has been extensively examined, because it is useful in assessing the efficiency and suitability of the subsequent treatment facilities (Barker and Stuckey 1999). It is found that a majority of SMP has a MW less than 1 kDa or greater than 10 kDa, and that a minority of SMP has a MW between 1 and 10 kDa (Jarusutthirak and Amy 2006). However, it is also reported that the SMP exhibited a bimodal distribution with 30 % of MW $<$ 1 kDa and 25 % of MW $>$ 100 kDa (Barker and Stuckey 1999). These contradictory findings about the MW of SMP suggest that SMP might be further fractionated for better understand their formation mechanisms and characteristics.

In terms of the measurement methods, many analytical methods have been employed to characterize SMP. GPC has been widely used to determine the MW distribution of SMP from different origins (Barker and Stuckey 1999; Labbs et al. 2006; Jarusutthirak and Amy 2006). Fourier transform infrared spectroscopy (FTIR) and dissolved organic carbon (DOC) have been utilized to analyze the functional group characteristics and total content of SMP, respectively (Grunheid et al. 2005; Jarusutthirak and Amy 2006). Chemical analysis has also been performed for the determination of polysaccharides and proteins present in SMP (Barker and Stuckey 1999; Aquino and Stuckey 2008; Holakoo et al. 2006). However, in these studies the characteristics of global SMP, rather than their subfractions, i.e., UAP and biomass-associated products (BAP), are explored. Furthermore, SMP are a complex mixture of organic compounds with various compositions and characteristics, and thus it is impossible to elucidate the comprehensive characteristics of SMP and quantification of UAP and BAP by using a single chemical approach. If UAP and BAP could be accurately and quantitatively identified, it is possible to examine how refractory the individual compounds are and to determine which type of SMP is the most difficult to remove from biological treatment plant effluents.

Therefore, the main objective of this study is to elucidate SMP formation mechanisms with chemical and mathematical approaches via: (1) conducting an integrated analysis on SMP characteristics and quantitatively identifying UAP and BAP separately; and (2) further fractionating the SMP of activated sludge under aerobic conditions. In this study, we present the integrated attempt to investigate and quantify the subfractionation of SMP of activated sludge through selectively combining MW determination by GPC with chemical analysis by OUR, polysaccharide, and protein determination, 3-dimensional excitation emission matrix (EEM) fluorescence spectroscopy, FTIR, and DOC measurements. Another feature of this work is the development of a mathematical model to further quantitatively and qualitatively describes the production kinetics and subfractionation of SMP.

8.1.2 Materials and Methods

8.1.2.1 Sludge, Reactor, and Wastewater

Sludge, reactor, and wastewater used in Chap. 7 were employed to investigate SMP generation.

8.1.2.2 Experiments

Sludge was sampled from the SBR when no organic substrate was present in the medium. The samples were washed twice with distilled water to remove bulk soluble organic materials. A batch reactor was inoculated with 1 L of diluted activated sludge (65 % inoculum of the sampled sludge from the SBR and aerated continuously to keep the DO concentration above 4 mg L^{-1} in the entire aerobic phase. The sludge concentration in the reactor was 1,600 mg MLVSS L^{-1}. A reactor without the organic substrate dose was also operated as a control. After wastewater was dosed, the OUR was monitored. The feeding wastewater composition of batch experiments was same as the synthetic wastewater of the SBR influent. The experiments were conducted at 20 °C and pH of 7.0. All the experiments were performed repeatedly and the reproducible observations were reported in this chapter.

8.1.2.3 Analysis

DOC was determined using a total organic carbon (TOC) analyzer (V_{CPN}, Shimadzu Co., Japan). The acetate concentration was measured using a gas chromatograph (6890NT, Agilent Inc., USA) equipped with a flame ionization detector and a 30 cm × 0.25 mm × 0.25 mm fused-silica capillary column (DB-FFAP). The total SMP were determined from the difference as follows:

$$\text{Total SMP(as COD)} = \text{Soluble COD} - 1.07 * \text{Acetate} \quad (8.1)$$

$$\text{Total SMP(as TOC)} = \text{Total DOC} - 0.4 * \text{Acetate} \quad (8.2)$$

Fractionated SMP, i.e., UAP and BAP, were obtained from total SMP measurement based on associated phases.

The MW distribution of the organic matters in SMP was determined using a GPC (Waters 1515, Waters Co., USA) with deionized water as eluent at a flow rate of 1.0 ml min^{-1}, as described in Chap. 7. The fractionation of SMP (UAP and BAP) and subfractionation of BAP were performed according to their formation sequence and different compounds sizes as well as their GPC chromatograms. Based on the DOC analysis of each fraction of SMP, their concentrations could then be determined. All EEM spectra were measured using a luminescence spectrometry (LS-55, Perkin-Elmer Co., USA). Parallel factor analysis (PARAFAC) was used to interpret the EEM fluorescence data, as described in Chap. 7. FTIR (Magna-IR 750, Nicolet Instrument Co., USA) was employed to quantify the functional group characteristics of SMP. Measurement of MLSS, MLVSS, and COD was performed according to the standard methods (APHA 1995).

8.1.2.4 Modeling the Fractionated SMP Formation

SMP are a complex mixture of organic compounds with various compositions and characteristics, and thus it is impossible to measure the dynamics of the fractionated SMP formation by using experimental approaches directly. The chemical analysis of UAP and BAP should be integrated with effective mathematical model for quantitatively and qualitatively describes the production kinetics and subfractionation of SMP. In the present work, BAP were further classified into GBAP produced in the microbial growth phase and EBAP generated in the endogenous phase (Fig. 8.1). EBAP had a larger MW than GBAP. Thus, these two fractions of BAP had different formation rates and distinct biodegradation kinetics. In our model, part of the external substrate was used for biomass synthesis, and the other substrate electrons were diverted into the formation of UAP and EPS (Fig. 8.1). UAP were released to the aqueous solution, while EPS were released as a polymer to form the aggregate. In the growth phase, the hydrolysis of EPS produces GBAP, while the hydrolysis of EPS produces EBAP in the endogenous phase. The formation of GBAP occured only when the external substrate was available. Substrate oxidation and respiration of the electrons to reduce O_2 generate the energy required to fuel the formation of active biomass, i.e., EPS and UAP.

The model established in our work describes the relationships among the three solid species: active bacteria (X_H), EPS (X_{EPS}), and residual inert biomass (X_I); four soluble species: external substrate (S_S) and three unique forms of SMP, i.e., UAP (S_{UAP}), and two fractions of BAP (S_{BAP}); and an electron acceptor, i.e., DO (S_O). The hydrolysis rates of GBAP and EBAP are proportional to the EPS concentration, with k_{hyd1} and k_{hyd2} as the two different first-order rate coefficients for GBAP and EBAP, respectively. $\frac{S_S}{K_{S,BAP}+S_S}$ is a switch kinetic function for the GBAP formation controlled by a relatively low half-saturation constant. $\frac{K_{S,BAP}}{K_{S,BAP}+S_S}$ is an inhibition function to ensure that EBAP are produced after the external substrate depletion. In this approach GBAP could be formed when S_S is available, and EBAP could be produced when S_S is utilized. The UAP, GBAP, and EBAP are biodegradable. The kinetic equations for SMP, UAP, BAP, GBAP, and EBAP are expressed as follows:

$$\frac{dS_{SMP}}{dt} = \frac{dS_{UAP}}{dt} + \frac{dS_{BAP}}{dt} \tag{8.3}$$

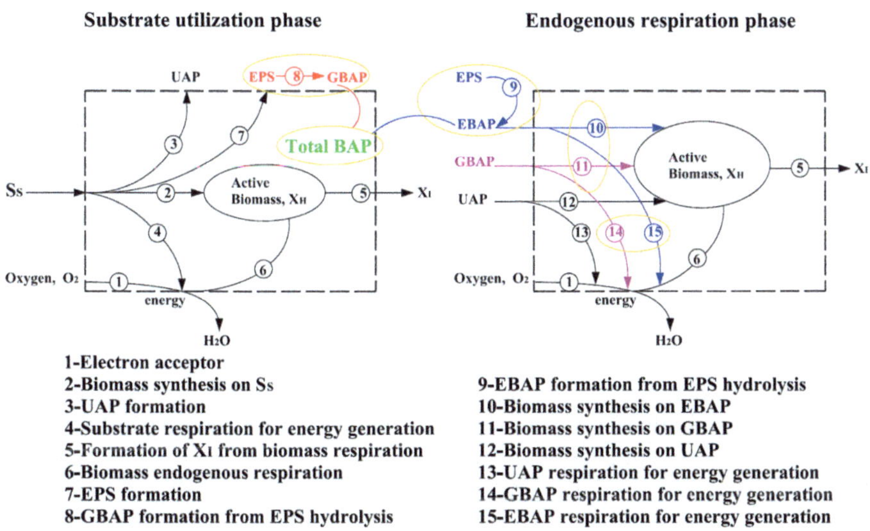

Fig. 8.1 Schematic diagram of electron flows in the active microbial growth phase (*left side*) and endogenous respiration phase (*right side*) for the developed model (Reprinted from Ni et al. (2010) with permission from Elsevier)

$$\frac{\mathrm{d}S_{\mathrm{UAP}}}{\mathrm{d}t} = k_{\mathrm{UAP}}\frac{\mu_{\mathrm{H}}}{Y_{\mathrm{H}}}\frac{S_{\mathrm{S}}}{K_{\mathrm{S}}+S_{\mathrm{S}}}\frac{S_{\mathrm{O}}}{K_{\mathrm{O}}+S_{\mathrm{O}}}X_{\mathrm{H}} - \frac{\mu_{\mathrm{UAP}}}{Y_{\mathrm{UAP}}}\frac{S_{\mathrm{UAP}}}{K_{\mathrm{UAP}}+S_{\mathrm{UAP}}}\frac{S_{\mathrm{O}}}{K_{\mathrm{O}}+S_{\mathrm{O}}}X_{\mathrm{H}} \tag{8.4}$$

$$\frac{\mathrm{d}S_{\mathrm{BAP}}}{\mathrm{d}t} = \frac{\mathrm{d}S_{\mathrm{GBAP}}}{\mathrm{d}t} + \frac{\mathrm{d}S_{\mathrm{EBAP}}}{\mathrm{d}t} \tag{8.5}$$

$$\frac{\mathrm{d}S_{\mathrm{GBAP}}}{\mathrm{d}t} = k_{\mathrm{hyd1}}\frac{S_{\mathrm{S}}}{K_{\mathrm{S,BAP}}+S_{\mathrm{S}}}X_{\mathrm{EPS}} - \frac{\mu_{\mathrm{GBAP}}}{Y_{\mathrm{GBAP}}}\frac{S_{\mathrm{GBAP}}}{K_{\mathrm{GBAP}}+S_{\mathrm{GBAP}}}\frac{S_{\mathrm{O}}}{K_{\mathrm{O}}+S_{\mathrm{O}}}X_{\mathrm{H}} \tag{8.6}$$

$$\frac{\mathrm{d}S_{\mathrm{EBAP}}}{\mathrm{d}t} = k_{\mathrm{hyd2}}\frac{K_{\mathrm{S,BAP}}}{K_{\mathrm{S,BAP}}+S_{\mathrm{S}}}X_{\mathrm{EPS}} - \frac{\mu_{\mathrm{EBAP}}}{Y_{\mathrm{EBAP}}}\frac{S_{\mathrm{EBAP}}}{K_{\mathrm{EBAP}}+S_{\mathrm{EBAP}}}\frac{S_{\mathrm{O}}}{K_{\mathrm{O}}+S_{\mathrm{O}}}X_{\mathrm{H}} \tag{8.7}$$

The EPS formation rate is proportional to the substrate utilization rate (Laspidou and Rittmann 2002a, b). The first term on the right side of Eq. 8.8 is for EPS formation, while the second term is for EPS loss.

$$\frac{\mathrm{d}X_{\mathrm{EPS}}}{\mathrm{d}t} = k_{\mathrm{EPS}}\frac{\mu_{H}}{Y_{\mathrm{H}}}\frac{S_{\mathrm{S}}}{K_{\mathrm{S}}+S_{\mathrm{S}}}\frac{S_{\mathrm{O}}}{K_{\mathrm{O}}+S_{\mathrm{O}}}X_{\mathrm{H}} - \left(k_{\mathrm{hyd1}}+k_{\mathrm{hyd2}}\right)X_{\mathrm{EPS}} \tag{8.8}$$

Equation 8.9 is the kinetic equation for the external substrate, while the kinetics for active biomass is described in Eq. 8.10.

$$\frac{\mathrm{d}S_{\mathrm{S}}}{\mathrm{d}t} = -\frac{\mu_{\mathrm{H}}}{Y_{\mathrm{H}}}\frac{S_{\mathrm{S}}}{K_{\mathrm{S}}+S_{\mathrm{S}}}\frac{S_{\mathrm{O}}}{K_{\mathrm{O}}+S_{\mathrm{O}}}X_{\mathrm{H}} \tag{8.9}$$

$$\frac{dX_H}{dt} = -(1 - k_{EPS} - k_{UAP})Y_H\frac{dS_S}{dt} + \mu_{UAP}\frac{S_{UAP}}{K_{UAP} + S_{UAP}}\frac{S_O}{K_O + S_O}X_H + \left(\frac{\mu_{GBAP}S_{GBAP}}{K_{GBAP} + S_{GBAP}} + \frac{\mu_{EBAP}S_{EBAP}}{K_{EBAP} + S_{EBAP}}\right)\frac{S_O}{K_O + S_O}X_H - b_H\frac{S_O}{K_O + S_O}X_H \tag{8.10}$$

The last term in Eq. 8.10 represents the loss of active biomass attributed to the endogenous respiration, which produces X_I.

$$\frac{dX_I}{dt} = f_I b_H \frac{S_O}{K_O + S_O}X_H \tag{8.11}$$

Equation 8.12 itemizes the respiration processes: utilization of S_S (term 1), endogenous respiration of X_H (term 2), utilization of UAP (term 3), utilization of GBAP (term 4), and utilization of EBAP (term 5).

$$\begin{aligned}\frac{dS_O}{dt} =& [1 - k_{EPS} - k_{UAP} - Y_H(1 - k_{EPS} - k_{UAP})]\frac{dS_S}{dt} - \frac{(1-f_I)}{f_I}\frac{dX_I}{dt}\\ &- \frac{1 - Y_{UAP}}{Y_{UAP}}\frac{\mu_{UAP}S_{UAP}}{K_{UAP} + S_{UAP}}\frac{S_O}{K_O + S_O}X_H\\ &- \frac{1 - Y_{GBAP}}{Y_{GBAP}}\frac{\mu_{GBAP}S_{GBAP}}{K_{GBAP} + S_{GBAP}}\frac{S_O}{K_O + S_O}X_H\\ &- \frac{1 - Y_{EBAP}}{Y_{EBAP}}\frac{\mu_{EBAP}S_{EBAP}}{K_{EBAP} + S_{EBAP}}\frac{S_O}{K_O + S_O}X_H\end{aligned} \tag{8.12}$$

The simulation was performed with the AQUASIM software package. The model parameters definitions are given in Table 8.1.

8.1.3 Results and Discussion

8.1.3.1 SMP Production

The time dependence of acetate consumption and SMP formation in batch experiments is shown in Fig. 8.2. The concentration of SCOD or DOC decreased rapidly and continuously in the initial 50 min, as the external substrate was consumed (see Fig. 8.2a). Because of the SMP production, a complete depletion of the SCOD or DOC pool did not occur despite the complete consumption of acetate. A minimum DOC was reached after approximately 1 h. Later, the DOC slightly increased, presumably as a consequence of the released SMP.

The SMP initially increased to 15 mg C L^{-1} at ~1 h and slightly decreased after 100 min, and later gradually increased to 22 mg C L^{-1} at 6 h (Fig. 8.2c). In the rapid growth period, the acetate consumption resulted in an increase in total

Table 8.1 Kinetic and stoichiometric parameters used in the model simulation

Parameter	Definition	Value	Unit
Y_H	Yield coefficient for growth on S_S	0.34	g COD_X g^{-1} COD_S
f_I	Fraction of X_I in decay	0.20	g COD_X g^{-1} COD_X
Y_{AP}	Yield coefficient for growth on UAP	0.45	g COD_X g^{-1} COD_{UAP}
Y_{BAP}	Yield coefficient for growth on GBAP	0.45	g COD_X g^{-1} COD_{BAP}
Y_{BAP}	Yield coefficient for growth on EBAP	0.45	g COD_X g^{-1} COD_{BAP}
μ_H	Maximum growth rate on S_S	0.40	h^{-1}
K_S	External substrate affinity constant	9.8	g COD_S m^{-3}
$K_{S,BAP}$	Affinity constant for BAP formation	0.02	g COD_S m^{-3}
K_O	Dissolve oxygen affinity constant	0.2	g COD_O m^{-3}
b_H	Decay rate coefficient of X_H	0.021	h^{-1}
k_{PS}	EPS formation coefficient	0.18	g COD_{EPS} g^{-1} COD_S
k_{yd1}	EPS hydrolysis rate for GBAP	0.0095	h^{-1}
k_{yd2}	EPS hydrolysis rate for EBAP	0.0071	h^{-1}
k_{AP}	UAP formation coefficient	0.05	g COD_{UAP} g^{-1} COD_S
μ_{AP}	maximum rate of UAP degradation	0.053	h^{-1}
K_{UAP}	Biomass affinity constant for UAP	100	g COD_{UAP} m^{-3}
μ_{GBAP}	Maximum rate of GBAP degradation	0.0029	h^{-1}
K_{GBAP}	Biomass affinity constant for GBAP	85	g COD_{GBAP} m^{-3}
μ_{EBAP}	Maximum rate of EBAP degradation	0.0021	h^{-1}
K_{EBAP}	Biomass affinity constant for EBAP	106	g COD_{EBAP} m^{-3}

SMP. With the evolution of net microbial decay, the SMP gradually increased again and became the main fraction of DOC. These observations were in agreement with those of Jarusutthirak and Amy (2007). Figure 8.2c shows that SMP built up slowly and continually after 100 min, attributed to the BAP formation and a much slower utilization of BAP. Thus, UAP were the main fraction of the total SMP before 50 min, whereas BAP were the major content of the total SMP thereafter.

The contents of the different components released in SMP were not same. The content of polysaccharides was much higher than that of proteins, and the polysaccharide profiles in SMP generally tracked the SMP and they occupied more than 95 % of the total SMP, as shown in Fig. 8.2b.

8.1.3.2 Transformation of Organic Compounds

Figure 8.3 shows the GPC chromatograms of the SMP samples as a result of the transformation of acetate in the activated sludge. In the initial 50 min, acetate was transformed to intermediates with an MW higher than 10 kDa. As shown in Fig. 8.2c, acetate was consumed, whereas the SMP concentration increased with the acetate utilization. The GPC peaks of MWs of 10, 80, 250, and 1,200 kDa appeared. The GPC peaks of the portion with MWs of 250 and 80 kDa decreased and were transformed into the compounds with a much smaller MW after 50 min.

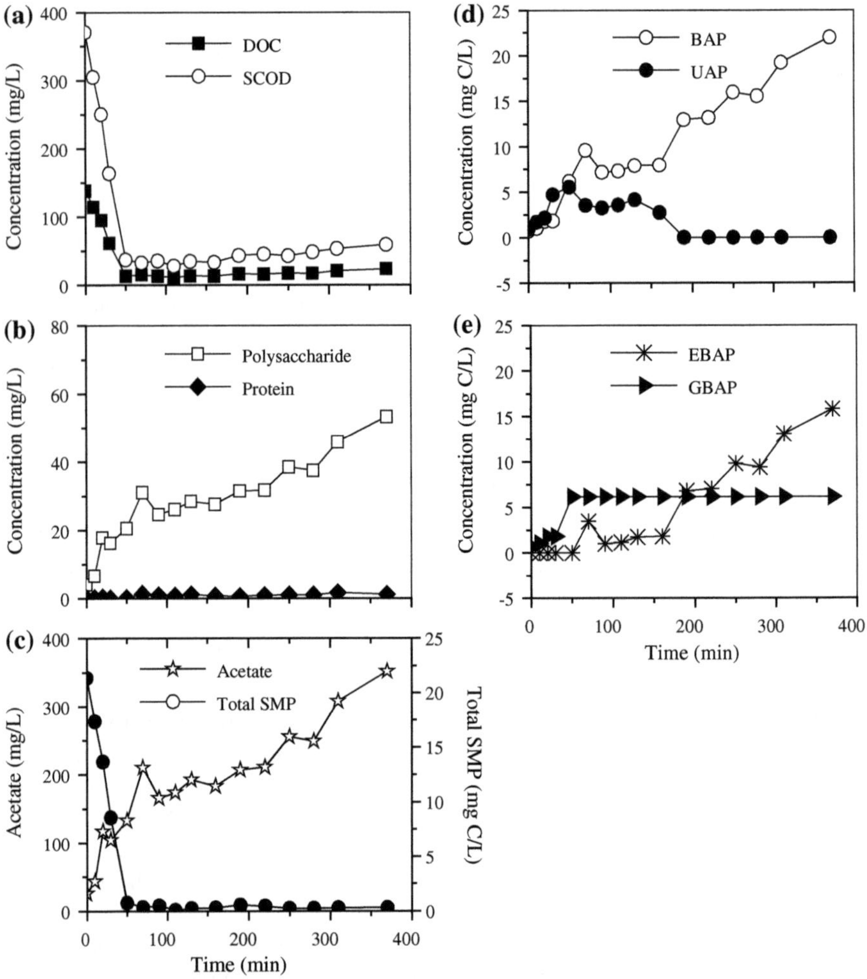

Fig. 8.2 External substrate consumption and SMP formation in activated sludge process: **a** DOC and COD concentrations; **b** polysaccharide and protein profiles; **c** acetate and total SMP profiles; **d** UAP and BAP profiles; and **e** GBAP and EBAP profiles (Reprinted from Ni et al. (2010) with permission from Elsevier)

However, the GPC peaks of the portion with MWs of 250 and 1,200 kDa did not change significantly. These observations indicated that the so-called UAP were produced in the microbial growth and substrate utilization phases. The BAP with an MW of 1,200 kDa were also produced in the growth phase, which might be the hydrolysis products of EPS. The UAP dominated under the substrate-rich conditions but they were likely to be readily biodegraded.

The DOC peaked at 50 min, and then slightly increased from 10 to 20 mg C L^{-1} (Fig. 8.2a). As shown in Fig. 8.3, the GPC peaks of the larger compounds appeared and became more prominent with time. The MW distributions of SMP after 50 min

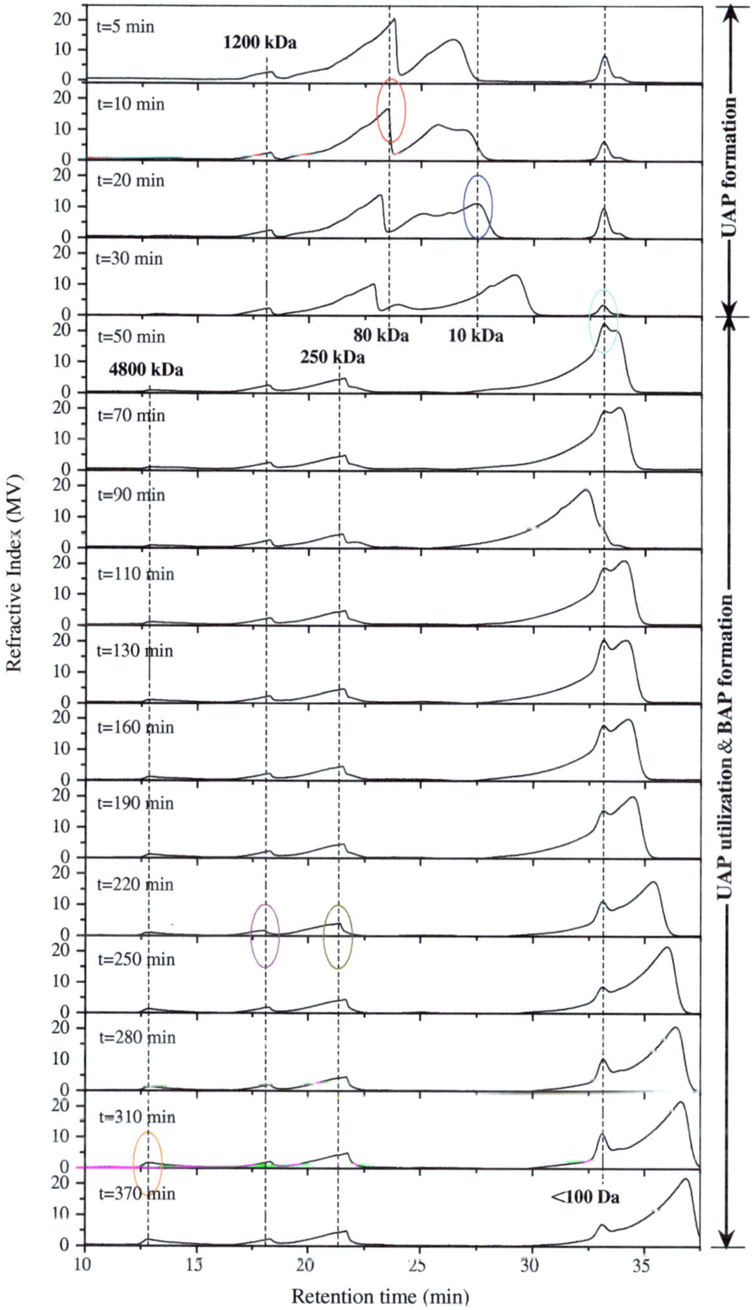

Fig. 8.3 GPC chromatograms of the transformation of organic compounds at different phases of activated sludge process with initial SCOD = 370 mg L^{-1} (Reprinted from Ni et al. (2010) with permission from Elsevier)

were similar and primarily composed of organic compounds with an MW larger than 4,000 kDa. These results suggested the formation of the BAP. The starvation and microbial decay appeared to be major reasons responsible for the formation of SMP in the form of BAP, which were expected to dominate under the substrate-deficient conditions. Our experimental data showed a clear dominance of high-MW compounds in the effluent, and were consistent with other observations (Schiener et al. 1998; Jarusutthirak and Amy 2007).

The BAP produced in the endogenous phase had an MW larger than 4,000 kDa, were cellular macromolecules, whereas the BAP produced in the microbial growth phase had an MW lower than 1,200 kDa. These two different categories of high-MW compounds were expected to have different formation rates, chemical natures, and distinct biodegradation kinetics.

To further explore the different categories of SMP, the control experiment without substrate dose and those with different initial substrate levels was conducted. The GPC chromatograms of these SMP samples illustrated in Fig. 8.4 show that only 4,800 kDa-SMP were produced in the endogenous phase (Fig. 8.4a). The concentration of SMP with an MW lower than 250 kDa increased with the increasing initial substrate concentration (Fig. 8.4b). On the other hand, the 1,200 kDa-SMP did not change substantially with the increasing initial substrate dosage. However, the 1,200 kDa-BAP were not detected when no external substrate was dosed (Fig. 8.4a). These results indicted that the 4,800 kDa-SMP should be the BAP generated in the endogenous phase. The SMP with an MW lower than 250 kDa were substrate associated, i.e., UAP. The 1,200 kDa-SMP were also the biomass-associated SMP (BAP), rather than the accumulating products of UAP. However, they have sufficiently different chemical natures compared with those of the 4,800 kDa-BAP produced in the endogenous phase. The 1,200 kDa-BAP could only be generated in the growth phase in the presence of external substrates (Fig. 8.4b).

Therefore, according to their formation sequence and different compounds sizes, BAP were further classified into GBAP, which were produced in the growth phase, and EBAP, which were generated in the endogenous phase. As shown in the GPC chromatograms, after the depletion of the external substrate, the low-MW BAP fractions did not increase, while the high-MW portions of BAP increased and became more significant.

8.1.3.3 EEM and FTIR Spectra of SMP

The 3-D EEM spectroscopy was also applied to characterize the SMP. Nine SMP–EEM fluorescence spectra were acquired. They were similar in the peak locations, but had different fluorescence intensities. To find out the appropriate number of PARAFAC components, core consistency diagnostic was also used to determine the correct number of components in the PARAFAC solution, and the results are shown in Fig. 8.5. Two components of fluorescence components were identified:

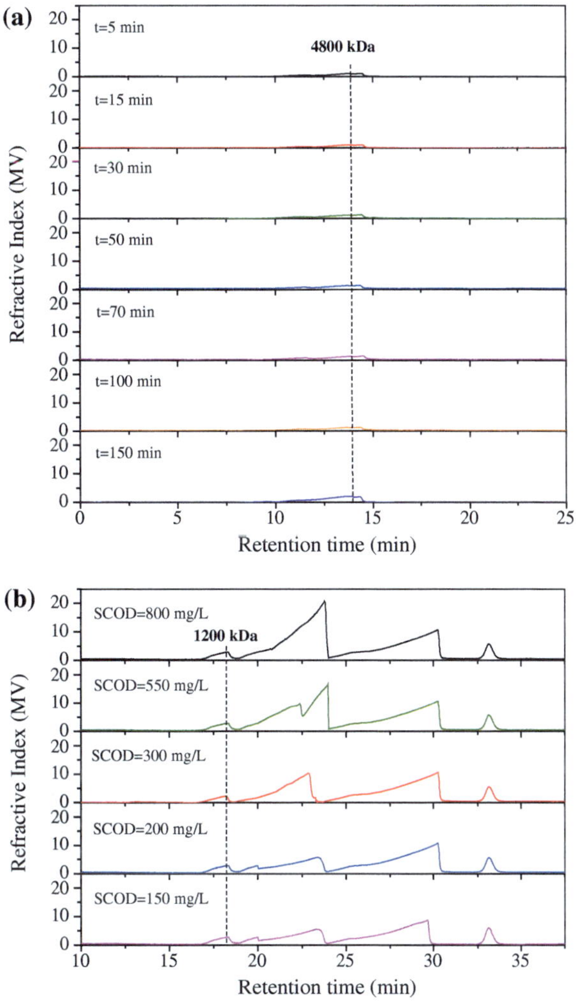

Fig. 8.4 GPC chromatograms of the transformation of organic compounds: **a** SMP samples collected at different times in the control experiment without substrate dose; and **b** SMP samples collected at a series of experiments with different initial substrate dosages at 15 min (Reprinted from Ni et al. (2010) with permission from Elsevier)

fulvic-acid-like substances and humic-like substances, whose fluorescence peaks were at Ex/Em 320/400 nm and 390/460 nm, respectively.

The fluorescence peak locations did not change much in the activated sludge process, while the fluorescence peak intensities changed substantially. EEM results revealed that the relative concentrations of humic-like substances increased in the substrate consumption phase, but decreased in the endogenous phase. The decrease in the humic-like substances concentration might be attributed to their hydrolysis and partial utilization by microorganisms, especially in the endogenous decay phase. The fulvic-acid-like component contained in the polymer matrix might be attributed to either fulvic acids directly or NADH and pyridoxine. In addition, the soluble microbial compounds showing a fluorescent signal maybe excreted by the microorganisms, whereas others were derived from their decay. The humic-like

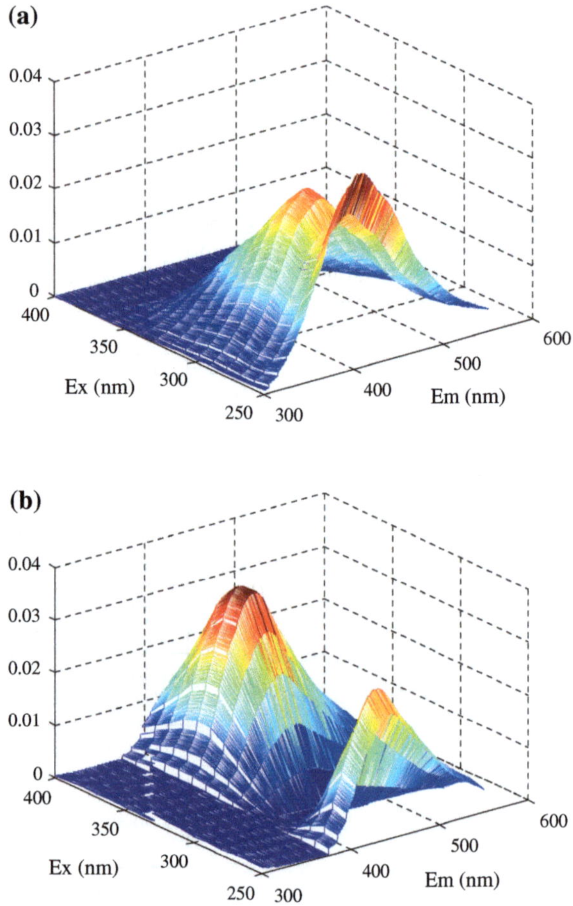

Fig. 8.5 Two components of SMP decomposed by PARAFAC based on EEM spectra of the SMP samples: **a** fulvic-like substance; and **b** humic-like substance (Reprinted from Ni et al. (2010) with permission from Elsevier)

substances were mainly substrate-utilization-associated, while the fulvic-acid-like substances were nongrowth associated.

FTIR spectra of SMP show the functional groups of SMP at different sampling times. Figure 8.6 illustrates the FTIR spectra of the SMP samples. The peaks at 1,040 and 2,900 cm^{-1} indicate the presence of the polysaccharide-like materials. In addition, FTIR spectra of the SMP sampled in the substrate utilization phase (before 45 min), which were mainly associated with UAP, show a significant different peak from those sampled in the endogenous phase (after 45 min), which were mainly attributed to BAP. The SMP sampled in the endogenous phase had a main peak at 2,900 cm^{-1}, whereas the SMP sampled in the substrate utilization phase had a dominant peak at 1,040 cm^{-1}.

These results were consistent with the GPC results, and provided a further evidence for the SMP fractionation. In addition, the relative peaks of 1,450 and 1,640 cm^{-1} increased during the whole activated sludge process. On the contrary, the relative peaks of 1,550 cm^{-1} decreased gradually. Therefore, the

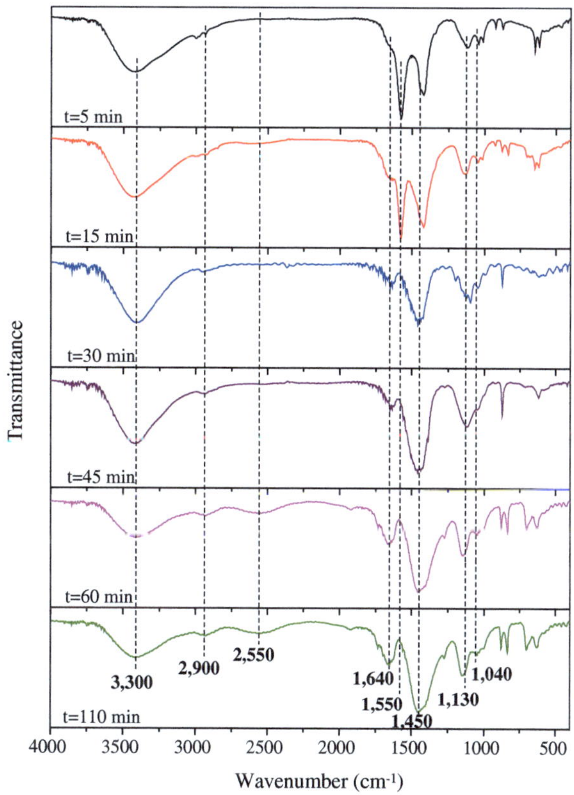

Fig. 8.6 FTIR spectra of the SMP samples collected at different times (Reprinted from Ni et al. (2010) with permission from Elsevier)

polysaccharide-like materials with peaks of 1,450 and 1,640 cm^{-1} might be related to the BAP, while polysaccharide-like materials with peaks of 1,550 cm^{-1} might be related to the UAP. The relative peaks of 2,550 cm^{-1} did not change much during the substrate consumption phase (initial 50 min) but increased during the endogenous phase, suggesting its possible relation to EBAP.

8.1.3.4 MW Fractionation of SMP

The MW fractionation of the accumulated SMP in the reactor for another set of experiment is shown in Fig. 8.7. The fraction of the organics with an MW lower than 1,200 kD increased from ~1.0 mg C L^{-1} at 5 min to 3.5 mg C L^{-1} at 50 min, then gradually decreased. The fraction of the organics with an MW of 4,500 kDa increased from 4.0 mg C L^{-1} at 50 min to 16 mg C L^{-1} at 370 min. Meanwhile, it is interesting to note that, despite an increase in the percentage of SMP with an MW of 4,500 kDa from 50 to 370 min, the concentration of the SMP subfraction with an MW of 1,200 kDa did not change significantly in this phase.

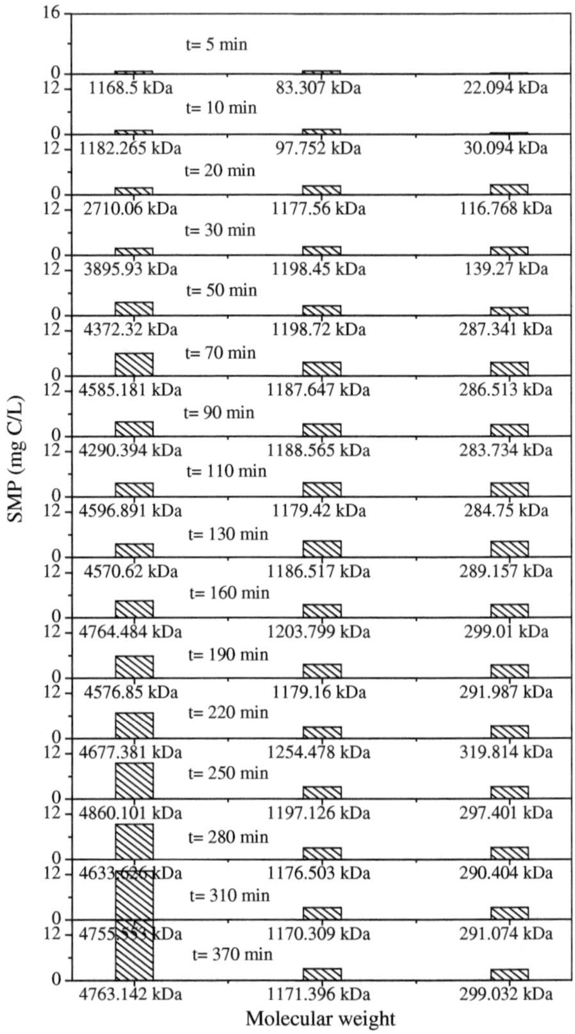

Fig. 8.7 MW fractionation of SMP in the activated sludge under aerobic conditions (Reprinted from Ni et al. (2010) with permission from Elsevier)

These observations also suggest that SMP had a broad spectrum of MW, especially a greater amount of high-MW compounds. The UAP were mainly small carbonaceous compounds derived from the original substrate (i.e., acetate in this work), while the BAP were cellular macromolecules including GBAP and EBAP. Compared to the other studies (Jarusutthirak and Amy 2006), a higher percentage of SMP with an MW of 4,500 kDa was observed. This is likely to be attributed to the relatively higher biomass concentration and the lower ratio of substrate concentration to sludge concentration. In this case, the specific growth rate was much lower than the maximum growth rate. This might cause the more production of BAP over UAP, and result in the accumulation of higher MW fractions.

Jarusutthirak and Amy (2007) found that the MW distribution patterns of all SMP–BAP samples were similar and primarily composed of one category of organics with an MW larger than 10,000 Da. Furthermore, such organics accumulated and were produced continuously in the entire bioreaction process: both in the microbial growth and the endogenous phases (Jarusutthirak and Amy 2007). However, our work showed that two different categories of BAP with significantly different MWs were mainly produced in the microbial growth phase.

8.1.3.5 Formation and Utilization Dynamics of the Fractionated SMP

The UAP and BAP concentrations in the activated sludge process under aerobic conditions are shown in Fig. 8.2. The UAP profiles exhibited a peak followed by a declining curve (Fig. 8.2d), reflecting a decomposition of the biodegradable UAP. The UAP concentration increased rapidly over the initial 50 min. The substrate was then taken up completely, and the UAP concentration decreased gradually and became depleted after approximately 200 min, indicating that the UAP might be utilized to provide for new synthesis with a relatively low rate in the subsequent endogenous phase.

The UAP were formed when substrate was utilized, while the BAP built up slowly in the entire process (Fig. 8.2d). The BAP accumulated slowly and continually because of the slower production of BAP as well as a much slower utilization of BAP. UAP accounted for over half of the total SMP at 50 min, BAP were about 95 % of the SMP at 6 h. For one of the two subfractions of BAP, EBAP is the BAP formed prior to 50 min, whereas the GBAP is the BAP formed prior to that time based on the above MW analysis. GBAP were mainly produced in the growth phase (0–50 min) at a linear increasing rate. In the subsequent endogenous phase (50–370 min), the GBAP content did not change significantly, whereas the EBAP concentration increased gradually (Fig. 8.2e).

8.1.3.6 Modeling Fractionation of SMP

To further quantitatively and qualitatively describe the production kinetics and subfractionation of SMP, the experimental data about fractionation of SMP are simulated using our model. The SCOD, OUR, acetate, total SMP, UAP, BAP, GBAP, and EBAP profiles simulated with the established model are illustrated in Fig. 8.8. Model parameters used in the simulation and their values are shown in Table 8.1. The simulation results matched the measured data well, suggesting that the model established in this work properly captures the connection among the three types of SMP and substrate consumption in activated sludge under aerobic conditions.

The decay rate for active biomass (b_H) was 0.021 h^{-1}, which was appropriately three times larger than the two hydrolysis rates for EPS of 0.0095 h^{-1} (k_{hyd1}) and 0.0071 h^{-1} (k_{hyd2}). Such a difference means that the EPS were not used as an energy- and electron-storage material by microorganisms, as they were slowly hydrolyzed to GBAP or EBAP. Furthermore, the EPS hydrolysis rate for GBAP (k_{hyd1}) was higher

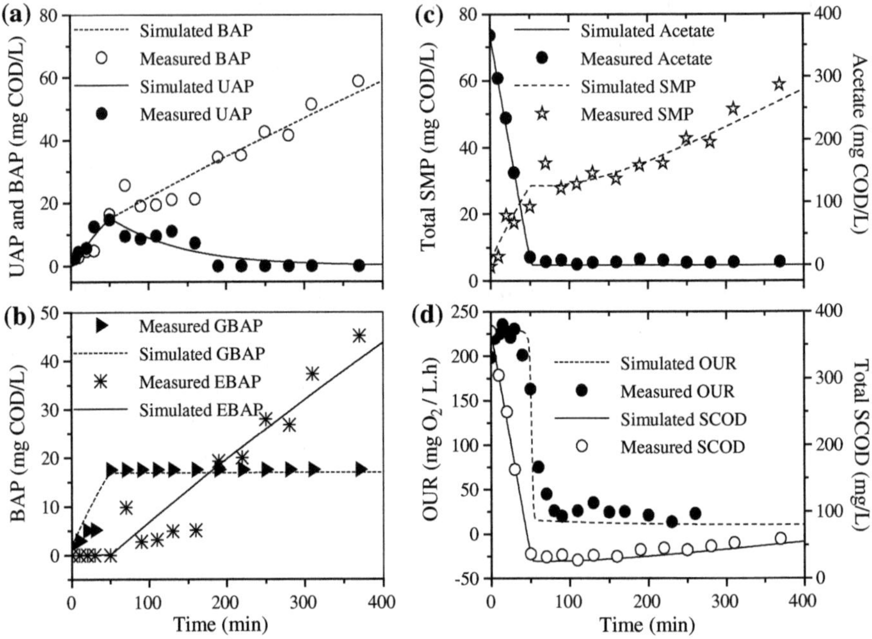

Fig. 8.8 Modeling of the SMP formation in the activated sludge process: **a** OUR and SCOD profiles; **b** acetate and total SMP profiles; **c** UAP and BAP profiles; and **d** GBAP and EBAP profiles (Reprinted from Ni et al. (2010) with permission from Elsevier)

than that for EBAP (k_{hyd2}). This is in accordance with the experimental observations. The half-saturation coefficients for the UAP, GBAP, and EBAP were much larger than K_S, indicating the slow biodegradation kinetics in most cases.

The modeling results show that the microorganisms in activated sludge consumed the electron-donor substrate to build active biomass. At the same time, they produced EPS and UAP. EPS were hydrolyzed to GBAP and EBAP, while active biomass undergone endogenous decay to form residual dead cells. The formation rate of GBAP was faster than that for EBAP. The UAP were mainly small carbonaceous compounds derived from the original substrate, whereas GBAP and EBAP were cellular macromolecules. The SCOD in activated sludge plant effluents consists of residual substrate, UAP and GBAP, and EBAP. The established model is useful in quantifying the relationships among the different soluble species in the effluent, i.e., original substrate, UAP, GBAP, and EBAP. In addition, it might serve as a basis for the further development of more comprehensive models.

8.1.3.7 Implications of a Further Fractionation of SMP

The presence and characteristics of SMP in the effluent of wastewater treatment plant are of a great importance with respect to discharge quality and efficiency of the subsequent treatment facilities. In the present work, a combination of modeling approach with chemical analysis by MW, DOC, OUR, EEM, and FTIR was successfully used to explore the formation kinetics and further fractionate SMP in activated sludge. An integration of MW, DOC, and OUR was applied to characterize the SMP formation process, while the polysaccharide and protein determination as well as EEM and FTIR analysis results were used to reveal the chemical characteristics of SMP. With such a combination of chemical analysis and mathematical modeling, unique, and useful information on the characteristics, the quantification, the formation kinetics, and mechanisms of SMP of activated sludge was obtained.

The UAP, produced in the substrate utilization process, were found to be carbonaceous compounds with an MW lower than 290 kDa. The BAP were mainly cellular macromolecules with an MW in a range of 290–5,000 kDa, and could be further classified into the GBAP with an MW of 1,000 kDa, which were produced in the microbial growth phase, and the EBAP with an MW of 4,500 kDa, which were generated in the endogenous phase. Dynamic quantification data were obtained for UAP, GBAP, and EBAP separately and used for parameter estimation, allowing more trust in parameter estimation and resulting in reasonable kinetic parameter values for SMP production. The GBAP and EBAP had different formation rates from the hydrolysis of EPS and distinct biodegradation kinetics. These results suggest that the major contributions of organic compound in the secondary effluent of activated sludge plants might come from EBAP.

The experimental and modeling approaches in this work could be useful to select, design, and optimize the post-biological wastewater treatment facilities. In addition, with this information, engineers might be able to evaluate which SMP fractions can be preferentially removed by individual or integrated unit processes, e.g., coagulation, granular activated carbon, ozonation, biological treatment methods, and membrane filtration. Also, our results could be used to identify which SMP components are problematical for a selected wastewater treatment process. Furthermore, this work implies the possible important roles of EBAP in membrane fouling and flux decline in wastewater reclamation/reuse applications. But this warrants further investigations.

8.1.4 Conclusions

SMP in activated sludge are characterized, quantified, and fractionized using integrated chemical analysis and mathematical approach. The UAP, produced in the substrate utilization process, are found to be carbonaceous compounds with an MW lower than 290 kDa and are quantified separately from BAP. The BAP are mainly cellular macromolecules with an MW in a range of 290–5,000 kDa, and could be further classified into the GBAP with an MW of 1,000 kDa, which are produced in the microbial growth phase, and the EBAP with an MW of 4,500 kDa, which are generated in the endogenous phase. The GBAP and EBAP have different formation rates from the hydrolysis of EPS and distinct biodegradation kinetics. The major contributions of organic compound in the secondary effluent of activated sludge plants might come from EBAP.

8.2 Determinating Utilization-Associated and Biomass-Associated Products

8.2.1 Introduction

SMP are a myriad of soluble organic matters produced by microbial populations in bioreactors (Noguera et al. 1994; Barker and Stuckey 1999; Laspidou and Rittmann 2002a, b). SMP have been found to comprise the majority of soluble organic materials in the effluents from biological treatment systems (Barker and Stuckey 1999; Aqino 2004; Rosenberger et al. 2006). Therefore, their presence is of a particular interest in terms of achieving discharge consent levels for wastewater treatment plants (Shon et al. 2004; Grunheid et al. 2005; Jarusutthirak and Amy 2006; Labbs et al. 2006; Gray et al. 2007, 2008).

SMP are found to mainly contain small carbonaceous compounds derived from the original substrate in microbial growth, as the UAP, and cellular macromolecules generated in the endogenous phase, as the BAP (Grady et al. 1999; Laspidou and Rittmann 2002a; Jarusutthirak and Amy 2006). If UAP and BAP could be accurately and quantitatively identified, it is possible to examine how refractory the individual compounds are and determine which type of SMP is the most difficult to remove from wastewater treatment plant effluents (Smith et al. 2006; Ciston et al. 2009). Therefore, the quantitative determination of the formation of UAP and BAP in activated sludge processes should be explored.

However, since SMP are a complex mixture of organic compounds with various compositions and characteristics (Al-Halbouni et al. 2009; Dong and Jiang 2009; Kimura et al. 2009; Okamura et al. 2009), experimental approaches to directly measure the UAP and BAP formation are not available. In our previous study (Ni et al. 2010), SMP in activated sludge were characterized and fractionized using chemical analysis and mathematical modeling, but the quantification could be only achieved through selectively combining chemical analysis such as MW determination, OUR, polysaccharide, and protein determination, 3-D EEM fluorescence spectroscopy, FTIR, and DOC measurements with a mathematical model. Such a fractionation and characterization are a complicated and time consuming, but the results have shown combining experimental and modeling approach is an applicable way to determine UAP and BAP quantitatively (Ni et al. 2010).

Several approaches have been used to model SMP production. Laspidou and Rittmann (2002a, b) formulated a model about biofilm density and assumed that UAP production was proportional to the substrate utilization rate, while BAP were generated from EPS hydrolysis. Aquino and Stuckey (2008) considered that both soluble EPS and cell lysis products were the sources of BAP. The biochemical reactions in these models are expressed using nonlinear equations, resulting in complexity of their models and difficulty in estimating kinetic parameters for SMP production. In addition, the predictions of UAP and BAP using these models cannot be verified.

Therefore, an effective and convenient approach, other than the complicated chemical methods or complex models, to quantitatively and qualitatively determine the production of UAP and BAP, is highly desirable. This chapter describes an effective and convenient approach to determine the UAP and BAP by activated sludge using weighted nonlinear least squares analysis (WNLSA) based on their production kinetics. The integrated form of the Monod-based equations for substrate utilization production and BAP formation is formulated, and the minimization of the objective function of the sum of the squared weighted errors is achieved with the iterative search routine in a spreadsheet program. The experimental and literature data are used to validate this approach. Moreover, the comparison between this approach and others is performed to obtain better understanding of UAP and BAP production.

8.2.2 Materials and Methods

8.2.2.1 Experimental Setup and Analysis

Three different sets of experiments (I-III) were conducted at three different substrate levels (400, 800, 1,600 mg COD L^{-1}, respectively) to evaluate the SMP production by activated sludge. Each experiment was carried out in triplicates. The seeding sludge was sampled from the SBR when no substrate was present in the medium, and was washed twice with distilled water to remove external soluble organic material. A reactor was inoculated with 1 L of diluted activated sludge and aerated it continuously to keep DO over 2 mg L^{-1}. Then, the wastewater with predetermined concentrations was added. The feeding wastewater composition was same as the synthetic wastewater from the SBR. Samples were taken at given intervals for the analysis of COD, SMP, and MLVSS. The experiments were conducted at 20 °C and pH of 7.0 ± 0.05.

Mixed liquid samples were centrifuged for 15 min at 12,000 rpm, and then were prefiltered through 0.45-μm acetate cellulose membranes to represent the SMP, which were used for COD analyses. The measurement of MLVSS and COD was performed according to the standard methods (APHA 1995).

8.2.2.2 Kinetics for UAP and BAP Production

Table 8.2 lists the kinetic equations for UAP and BAP production. The total SMP are divided into UAP and BAP, as shown in Eq. 1. A Monod based kinetics for UAP formation is evaluated in this work. For the flow of electrons from the external substrate, the external substrate is partially used for biomass synthesis, and part of the external electrons is diverted to the formation of UAP. UAP are released to the aqueous solution. Substrate oxidation and respiration of the electrons to reduce oxygen and generate the energy needed to fuel for formation of

Table 8.2 Kinetic equations for UAP and BAP production by activated sludge

Component	Equation	No.
SMP	$\frac{ds_{SMP}}{dt} = \frac{ds_{UAP}}{dt} + \frac{ds_{BAP}}{dt}$	(1)
UAP	$\frac{ds_{UAP}}{dt} = k_{UAP}\frac{\mu_H}{Y_H}\frac{S_S}{K_S+S_S}X_H$	(2)
X_H	$\frac{dx_H}{dt} = (1-k_{UAP})\mu_H\frac{S_S}{K_S+S_S}X_H$	(3)
S_S	$\frac{ds_S}{dt} = -\frac{\mu_H}{Y_H}\frac{S_S}{K_S+S_S}X_H$	(4)
BAP	$\frac{ds_{BAP}}{dt} = k_{BAP}X_H$	(5)

active biomass and UAP (Laspidou and Rittmann 2002a). The rate of UAP formation is proportional to the substrate utilization rate (Laspidou and Rittmann 2002a), and the kinetics is expressed as Eq. 2, where X_H is the active bacteria (mg COD L^{-1}), S_{UAP} is the produced UAP (mg COD L^{-1}), S_S is the rate-limiting external substrate (mg COD L^{-1}), t is time (h), k_{UAP} is the UAP formation coefficient (g COD_{UAP} g^{-1} COD_S), μ_H is the maximum specific growth rate of active biomass (h^{-1}), Y_H is its yield for active cell synthesis (g COD_X g^{-1} COD_S), and K_S are the half-maximum-rate concentration for S_S (g COD_S m^{-3}). Equation 3 is the kinetic equation for active biomass, while the kinetics for external substrate is described with Eq. 4. The biomass, in turn, undergoes endogenous decay and releases cell lysis products, which are the source of BAP. The BAP formation rate is proportional to the biomass concentration, with k_{BAP} as the first-order rate coefficient for BAP production, as shown in Eq. 5, where X_H is the active bacteria (mg COD L^{-1}), S_{BAP} is the produced BAP (mg COD L^{-1}), t is time (h), *and* k_{BAP} is the BAP formation coefficient (h^{-1}).

8.2.2.3 Integrated UAP and BAP Production Equation

Table 8.3 defines all the integrated UAP and BAP production equations by activated sludge. According to Eqs. 2 and 3, the active cell and UAP production associated with growth can be described by Eqs. 6 and 7, where the subscript 0 denotes time zero. Equations 4 and 6 are combined into the Eq. 8. Then, Eq. 8 can be integrated to calculate the integrated substrate utilization by activated sludge, as expressed by Eq. 9. Therefore, the best estimates of the rate coefficients which are found out in Eq. 10, such as k_{UAP} and Y_H, can be determined by comparing the model predictions with the observed values of S_S and t.

Thereafter, UAP can be calculated from Eq. 7. With the total SMP measurement, BAP then can be determined from their difference. In order to determine the BAP formation kinetics, the integrated BAP production equation should be formulated. According to Eq. 5, the active cell associated with cell lysis for BAP production can be described by Eq. 11. Equations 5 and 11 are combined into Eq. 12. In a similar way, the BAP production kinetics can be determined from Eqs. 12–14.

Table 8.3 Integrated UAP and BAP production equations by activated sludge

Equation	No.
$X_H = X_{Ho} + Y_H(1 - k_{UAP})(S_{S0} - S_S)$	(6)
$S_{UAP} = S_{UAPo} + k_{UAP}(S_{S0} - S_S)$	(7)
$\frac{K_S + S_S}{(X_{Ho} + Y_H(1-k_{UAP})(S_{S0}-S_S))S_S} dS_S = -\frac{\mu_H}{Y_H} dt$	(8)
$\int_{S_{S0}}^{S_S} \frac{K_S+S_S}{(X_{Ho}+Y_H(1-k_{UAP})S_{S0}-Y_H(1-k_{UAP})S_S)S_S} dS_S = \int_{t0}^{t} -\frac{\mu_H}{Y_H} dt$	(9)
$t = \frac{Y_H}{\mu_H} \frac{K_S}{X_{H0} + Y_H(1-k_{UAP})S_{S0}} \ln\left(\frac{(X_{H0} + Y(1-k_{UAP})S_{S0} - Y(1-k_{UAP})S_S)}{K_S S_S}\right) + \frac{1}{\mu_H(1-k_{UAP})} \ln((X_{H0} + Y_H(1-k_{UAP})S_{S0} - Y_H(1-k_{UAP})S_S)) - \frac{Y_H}{\mu_H} \frac{K_S}{X_{H0} + Y_H(1-k_{UAP})S_{S0}} \ln \frac{X_{H0}}{K_S S_{S0}} - \frac{1}{Y_H(1-k_{UAP})} \ln X_{H0}$	(10)
$X_H = X_{H0} e^{-k_{BAP} t}$	(11)
$dS_{BAP} = k_{BAP} X_{H0} e^{-k_{BAP} t} dt$	(12)
$\int_{S_{BAP0}}^{S_{BAP}} dS_{BAP} = \int_{t0}^{t} k_{BAP} X_{H0} e^{-k_{BAP} t} dt$	(13)
$S_{BAP} = S_{BAP0} + X_{H0}\left(1 - e^{-k_{BAP} t}\right)$	(14)
$SSWE_{UAP} = \sum_{i=1}^{n} \left[w_i \left(t_i^{obs} - t_i^{pred}\right)\right]^2$	(15)
$SSWE_{UAP} = \sum_{i=1}^{n} \left[\frac{\Delta S_S}{\Delta t} \left(t_i^{obs} - t_i^{pred}\right)\right]^2 \approx \sum_{i=1}^{n} \left(S_{Si}^{obs} - S_{Si}^{pred}\right)^2$	(16)
$SSWE_{BAP} = \sum_{i=1}^{n} \left(S_{BAPi}^{obs} - S_{BAPi}^{pred}\right)^2$	(17)

For the UAP production, the WNLSA is used to construct the objective function and to determine the kinetic parameters by comparing the model predictions with the observed substrate concentrations and t by using the known values. Thus, the estimation of the kinetic parameters is successively performed by minimizing the sum of the squared weighted errors ($SSWE_{UAP}$) based on n observations using the spreadsheet method, as shown in Eq. 15, where w_i is an appropriate weighting factor; $t_i^{obs}$is the time of the ith observation; t_i^{pred} is the t value predicted by the model for the measured S_S value. The differences between the predicted and the observed S_S values can be estimated by multiplying $t_i^{obs} - t_i^{pred}$ by the local slope of the substrate utilization curve, $\Delta S_S/\Delta t$. Therefore, the logical weighting factor is the local slope of the substrate utilization curve (Eq. 16). S_{Si}^{obs} and S_{Si}^{pre} are the ith measured and predicted substrate concentrations, respectively. Similarly, the kinetic BAP parameters can be estimated from Eq. 17.

All of the calculations were done on a personal computer by using a spreadsheet program (Microsoft Excel 2003).

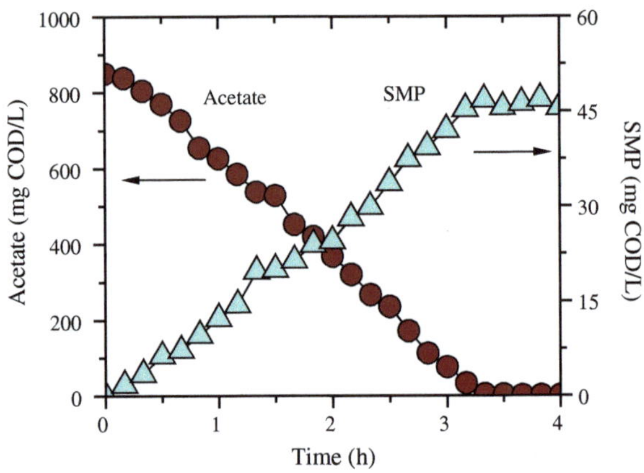

Fig. 8.9 External substrate consumption (acetate) and SMP formation by activated sludge in batch test II (Reprinted from Ni et al. (2012) with permission of American Chemical Society)

8.2.3 *Results and Discussion*

8.2.3.1 SMP Formation in Activated Sludge

The profiles of measured substrate consumption and SMP formation in the three sets of batch tests were similar. The time dependence of acetate consumption and SMP formation in batch test II is shown in Fig. 8.9 for representation. The acetate concentration decreased rapidly and continuously within the initial 3 h, as the external substrate was consumed. Because of the SMP production, a complete depletion of the SCOD pool in the mixed liquid would not occur despite the complete consumption of acetate after 3 h.

External substrate consumption resulted in an increase in the amount of total SMP. The SMP concentrations increased to 45 mg COD L^{-1} at ~3 h gradually (Fig. 8.9). In the rapid growth period, the acetate consumption resulted in an increase in total SMP. In addition, with the evolution of net microbial decay, the SMP also gradually increased (Jiang et al. 2008). These two processes might lead to an increase in total SMP. Our observations are in agreement with those reported by Jarusutthirak and Amy (2007). They are also in accordance with the unified theory for SMP production proposed by Laspidou and Rittmann (2002a, b), which couple the production and degradation of SMP with the formation of extracellular polymeric substances (EPS) for aerobic systems.

8.2.3.2 Appling the Approach for UAP and BAP Determination

The experimental data of batch test II for substrate utilization by activated sludge were analyzed by using the new approach to determine UAP formation based on the estimation of rate coefficients μ_H and K_S and the two yield coefficients for active bacteria (Y_H) and UAP (k_{UAP}), by fitting Eq. 10 to the substrate utilization data using the spreadsheet method with WNLSA (Eqs. 15 and 16). The input value for X_{H0} was estimated based on the results of several experiments. With the calculated UAP data with Eq. 7, the BAP production as well as the rate coefficients k_{BAP} could then be determined by fitting Eq. 14 to the BAP production data by using the spreadsheet method with the WNLSA (Eq. 17).

The spreadsheet for fitting the integrated substrate utilization equation to the data is shown in Fig. 8.10. The values of the rate coefficients and yield constants, which were used as fitting parameters, are given in rows 1–4. The values of the fitting parameters shown were initial guesses that were subsequently changed by the program as the model was fitted to the experimental data. The initial guess of the parameters obtained from the literature was used for the initial estimation procedure. The experimental data are listed in columns A and B. Column C contains the calculated t value for each observed S_S value (Eq. 10). The calculated values are shown in Fig. 8.10. Column D contained the S_S values used to calculate the values in column C. These S_S values were the same as the values in column B. The UAP concentrations were calculated (column E) which were required for the BAP determination. Additional model predictions used to calculate the weighting factors are shown in columns F and G; the S_S values in column G were slightly lower (0.1 mg L^{-1}) than the values in column D, and the t values in column F were calculated from the values in column G using Eq. 10. Column H shows the local slope of the model curve, which was calculated for each observation by using the two sets of model predictions. For example, the value at H11 (column H, row 11) equaled (D11–G11)/(C11–F11). The differences between model and predicted t values (column A–column C) were calculated (column I), and the results were multiplied by the weighting factors (column H) to give the weighted errors (column J). The weighted errors were squared (column K), and the squared weighted errors were summed (cell K35).

The kinetic model was fitted to the data by using the Solver function under the Formula menu in Excel to adjust the parameter estimates to minimize the $SSWE_{UAP}$ (cell K35). The best fit was obtained with Solver, which uses an iterative search for the parameter values that yield the minimum $SSWE_{UAP}$, and the initial estimates of the parameters were automatically replaced by the best estimates in rows 1–4. Excel solver can rapidly search for a maximum, minimum, or specified value for any selected cell by varying the values for one or more other selected cells in a spreadsheet. The best estimates of the rate coefficients μ_H (0.268 h^{-1}), K_S (30.590 g COD_S m^{-3}), k_{BAP} (0.0041 h^{-1}) obtained along with the two yield coefficients of Y_H (0.531 g COD_X g^{-1} COD_S) and k_{UAP} (0.042 g COD_{UAP} g^{-1} COD_S), yielded the simulation curves that matched all the four parameters data (acetate, total SMP, UAP, and BAP) in batch test II shown in Fig. 8.11.

8.2.3.3 Verification of the Approach

Verification was performed through comparing the measured and simulated results. In our work, one data set (set II) was used to determine UAP and BAP, and

	A	B	C	D	E	F	G	H	I	J	K
1	μ_H = 0.268 h^{-1} (fitting parameters-varies)										
2	K_S = 30.590 g COD_S m^{-3} (fitting parameters-varies)										
3	k_{UAP} = 0.042 g COD_{UAP} g^{-1} COD_S (fitting parameters-varies)										
4	Y_H = 0.531 g COD_X g^{-1} COD_S (fitting parameters-varies)										
5						Extra data for slope					
6	Observed data		Predicted data (model)			calculations		Slope	Error	Weighted	W. Error
7	t	S_S	t	S_S	UAP	t	S_S	dS_S/dt	t_{obs}-t_{pre}	error	squared
8	(h)	(mg L^{-1})	(h)	(mg L^{-1})	(mg L^{-1})	(h)	(mg L^{-1})	(mg L^{-1} h^{-1})	(h)	(mg L^{-1})	(mg L^{-1})^2
9	0.000	850.040	0.000	850.040	0.000	0.000	849.940		0.000		
10	0.167	817.840	0.166	817.840	1.320	0.167	817.740	-197.436	0.000	-0.066	0.004
11	0.333	784.290	0.333	784.290	2.696	0.333	784.190	-205.418	0.000	-0.071	0.005
12	0.500	749.350	0.500	749.350	4.128	0.500	749.250	-213.674	0.000	-0.059	0.003
13	0.667	713.040	0.666	713.040	5.617	0.667	712.940	-222.182	0.000	-0.075	0.006
14	0.833	674.350	0.837	674.350	7.203	0.838	674.250	-231.156	-0.004	0.873	0.763
15	1.000	636.240	0.999	636.240	8.766	0.999	636.140	-239.887	0.001	-0.262	0.069
16	1.167	594.570	1.169	594.570	10.474	1.170	594.470	-249.288	-0.002	0.572	0.328
17	1.333	553.210	1.332	553.210	12.170	1.333	553.110	-258.437	0.001	-0.198	0.039
18	1.500	508.490	1.502	508.490	14.004	1.502	508.390	-268.075	-0.002	0.567	0.322
19	1.667	464.180	1.665	464.180	15.820	1.665	464.080	-277.298	0.002	-0.662	0.438
20	1.833	416.180	1.835	416.180	17.788	1.835	416.080	-286.808	-0.002	0.513	0.263
21	2.000	367.640	2.001	367.640	19.778	2.002	367.540	-295.737	-0.001	0.419	0.175
22	2.167	319.710	2.161	319.710	21.744	2.162	319.610	-303.594	0.006	-1.723	2.970
23	2.333	266.430	2.335	266.430	23.928	2.335	266.330	-310.646	-0.002	0.533	0.284
24	2.500	215.170	2.499	215.170	26.030	2.499	215.070	-314.720	0.001	-0.470	0.221
25	2.667	161.460	2.669	161.460	28.232	2.669	161.360	-313.787	-0.002	0.656	0.430
26	2.833	112.450	2.827	112.450	30.241	2.828	112.350	-303.291	0.006	-1.709	2.920
27	3.000	61.690	3.003	61.690	32.322	3.004	61.590	-266.568	-0.003	0.932	0.869
28	3.167	22.720	3.176	22.720	33.920	3.176	22.620	-174.029	-0.009	1.481	2.194
29	3.333	3.880	3.353	3.880	34.693	3.355	3.780	-46.030	-0.020	0.901	0.811
30	3.500	1.377	3.435	1.377	34.795	3.441	1.277	-17.219	0.065	-1.117	1.247
31	3.667	0.930	3.465	0.930	34.814	3.474	0.830	-11.577	0.202	-2.336	5.458
32	3.833	0.580	3.501	0.580	34.828	3.515	0.480	-7.039	0.332	-2.338	5.465
33	4.000	0.440	3.522	0.440	34.834	3.541	0.340	-5.191	0.478	-2.483	6.166
34											
35									weighted SSE		31.451

Fig. 8.10 Spreadsheet for fitting the integrated substrate utilization equation to the data and the UAP determination process using the WNLSA (Reprinted from Ni et al. (2012) with permission of American Chemical Society)

the resulting values for the coefficients were used to generate the two sets of additional curves including acetate, SMP, UAP, and BAP for comparison with the other two data sets (sets I and III with initial 400 and 1,600 mg COD L^{-1} in Fig. 8.11). The maximum difference between the measured and calculated values was 15 %, and about 65 % of the results had a difference of less than 5 %. Furthermore, the simulation shows no systematic deviations, suggesting the validity of our approach for UAP and BAP determination.

To further validate our approach, the predicted SMP and external substrate profiles in Fig. 8.12a were compared with the corresponding experimental data from Hsieh et al. (1994) for a transient batch experiment. Overall, the model outputs captured all the experimental trends for this experiment in terms of SMP production. The external substrate declined at an increasing rate over the initial 10 h. During the period of rapid microbial growth, the SMP increased gradually. Also, the UAP and BAP concentrations were determined as shown in Fig. 8.12b.

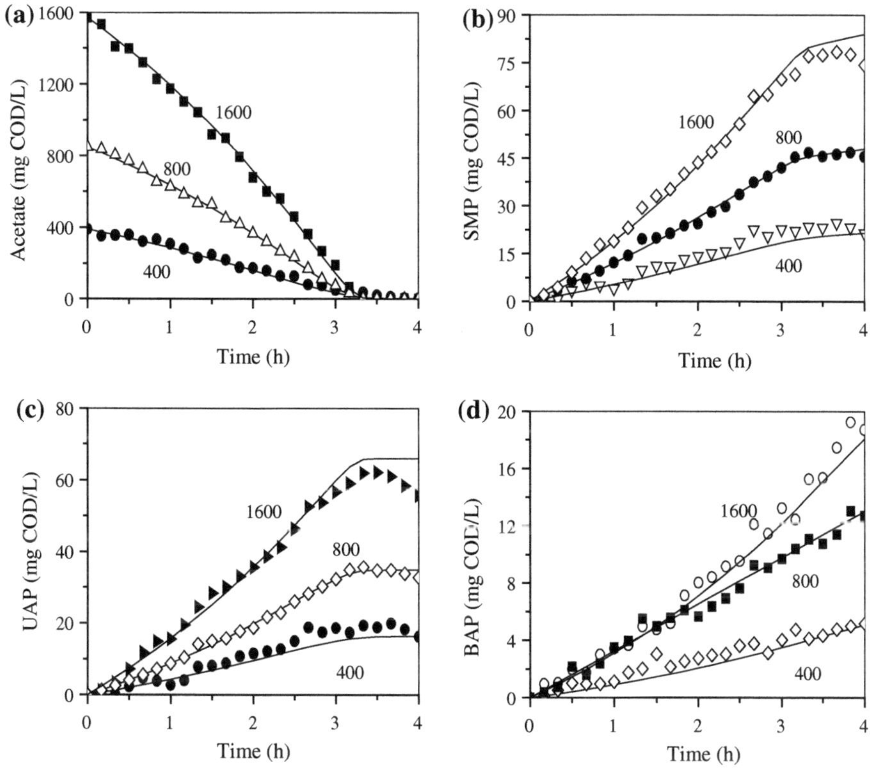

Fig. 8.11 Results of fitting the model to experimental data. The model (*solid line*) is fitted to one data set of substrate and SMP concentrations (scatter, 800 mg COD L^{-1} initial acetate concentration), which resulted in the parameter values shown in Fig. 8.10. Model curves obtained with the same parameter values are shown for other two data sets (400 and 1,600 mg COD L^{-1} initial acetate concentration) for comparison: **a** substrate utilization; **b** total SMP formation; **c** UAP production; and **d** BAP formation (Reprinted from Ni et al. (2012) with permission of American Chemical Society)

The two most appropriate estimates of the k_{UAP} and k_{BAP} were 0.03 g COD_{UAP} g^{-1} COD_S and 0.002 h^{-1}, respectively, which were comparable to the values obtained in the present work. In our approach, only one data set was needed to obtain UAP and BAP concentrations by using a convenient kinetic approach, as demonstrated in this work.

In addition, our approach was further evaluated by using the experimental results of Jiang et al. (2008). In their work, dedicated batch experiments were designed to produce BAP and UAP separately. The BAP experiment was conducted under starvation conditions without substrate addition, while the UAP test was performed with substrate dose. As shown in Fig. 8.12c, d, the model predictions matched the experimental data, suggesting that our approach was appropriate to determine the UAP and BAP formation. However, the estimated k_{BAP} value, 0.021 h^{-1}, was much higher than that calculated using our approach

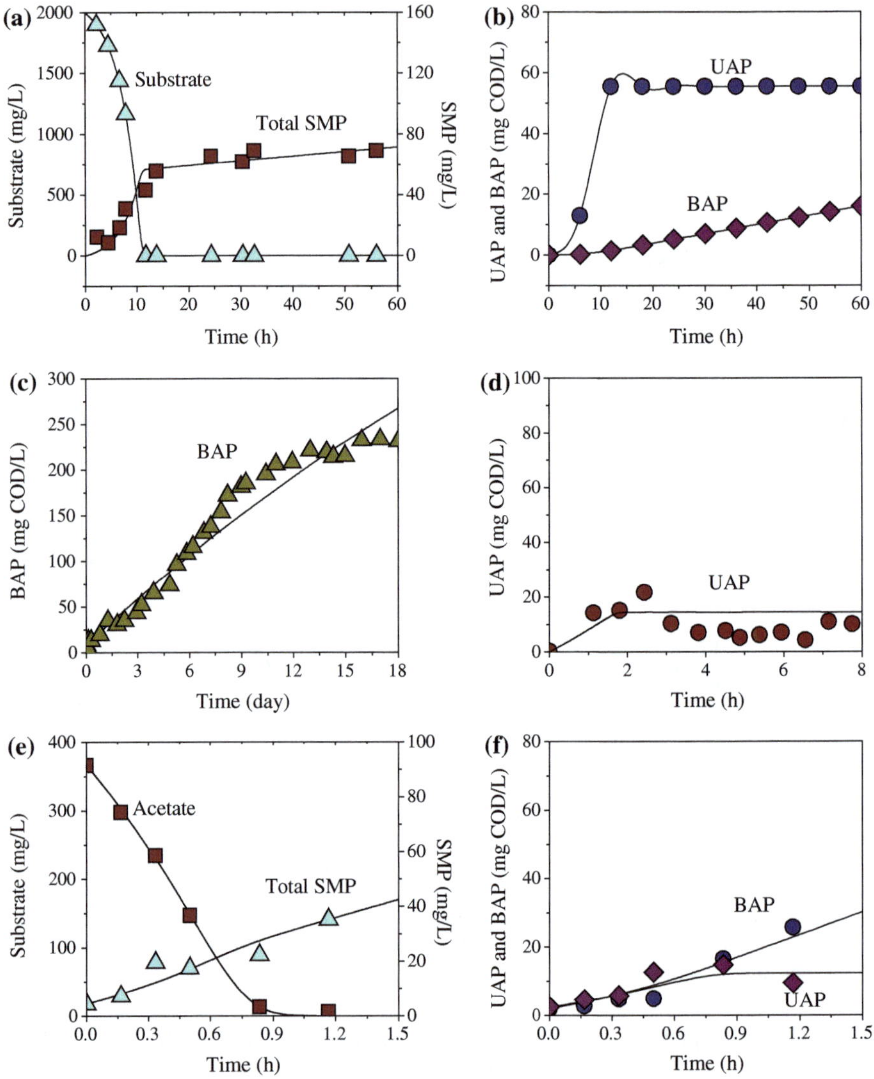

Fig. 8.12 Comparisons between the model simulations and the experimental data for the verification of the approach: **a** SMP data set of Hsieh et al. (1994); **b** UAP and BAP determinations corresponding to **a**; **c** BAP data set of Jiang et al. (2008); **d** UAP data set of Jiang et al. (2008); **e** SMP data of Ni et al. (2010); and **f** UAP and BAP data of Ni et al. (2010). (Reprinted from Ni et al. (2012) with permission of American Chemical Society)

(0.0041 h^{-1}). In their work the total SCOD was the estimate of the BAP concentration under starvation condition, which might be over estimation of real BAP concentration, resulting in the higher k_{BAP} value. On the contrary, their k_{UAP} value of 0.035 COD_{UAP} g^{-1} COD_S was slightly lower than that of our work (0.042 g COD_{UAP} g^{-1} COD_S). Jiang et al. (2008) also obtained the SMP production kinetic parameters using a complex mathematical model. Their estimated k_{UAP} and k_{BAP} values, 0.0963 g COD_{UAP} g^{-1} COD_S and 0.0215 h^{-1}, were comparable to those estimated using our approach. However, their model structure included many complex mass balance equations, and had to be calibrated using extensive experimental data and complex model solution method. Again, only one data set was needed to obtain UAP and BAP concentrations in our approach.

Finally, we also employed SMP data reported in our previous work (Ni et al. 2010) to validate our approach. In Ni et al. (2010), a complicated methods combining chemical analysis and modeling approach are applied to determine UAP and BAP quantitatively. As shown in Fig. 8.12e, f, the model predictions matched the experimental data, suggesting that our approach was appropriate to determine the UAP and BAP formation again. Ni et al. (2010) also obtained the SMP production kinetic parameters and the UAP and BAP formation data. However, their model structure included 12 nonsteady-state kinetic equations, and had to be combined with extensive chemical analysis method, such as molecular weight (MW) determination and DOC measurements, to determine the UAP and BAP formation.

8.2.3.4 Comparison with Other Approaches

Rittmann and coworkers have developed a series of SMP models, which are summarized as a unified SMP and EPS theory (Laspidou and Rittmann 2002a, b). Accordingly to their theory, UAP and EPS are produced in proportional to the substrate utilization rate and BAP are described as hydrolysis products of EPS. In this unified theory eight SMP-associated model parameters are introduced. Later, a very complex SMP model has been incorporated into ASM1 and ASM3 in MBR studies (Jiang et al. 2008). These existing SMP models exhibit very heterogeneous model structures. A common problem of these models is that the models are too complex and over-parameterized with strong parameter correlations. Indeed, the available measurements for model calibration are usually limited. Thus, the validity of the SMP model structure and the obtained parameter values is questionable. Moreover, in most cases, few SMP data (and mostly only steady-state SCOD data) are available for parameter estimation. Thus, model parameters are often estimated using trial and error methods, no parameter confidence interval is given and no independent model validations are conducted.

On the contrary, in this work a simple kinetic model including only five SMP-related parameters was introduced and integrated to determine the SMP production. Dynamic batch data were determined for BAP and UAP separately. Such an approach was validated using independent SMP production tests and results reported in the literature. Moreover, with the integrated Monod-based equation in this work UAP and BAP could be determined through progress curves from a few batch experiments or even one batch experiment. It becomes attractive when data are difficult to obtain, such as UAP and BAP production in activated sludge.

Another motivation for developing this method is the complexity and time consuming of the existing SMP production methods. This work just uses a computer spreadsheet program to determine UAP and BAP accurately from experimental data. It provides a flexible simulation tool capable of accounting in a straightforward manner.

Smith et al. (1998) proposed a computer spreadsheet program to estimate the kinetic parameters for biochemical reaction, based on the conventional Monod equation for utilization of a substrate in a batch reactor. In the conventional Monod equation, the SMP production is not considered and has been lumped into the yields for cell synthesis. Thus, their methods cannot be used for the determination of UAP and BAP because of no incorporation of SMP production. In addition, the substrate electrons for the synthesis of SMP which are accounted for the biomass yield in Smith et al. (1998) would result in a higher biomass yield value. Actually, SMP are important sinks for electrons and carbon derived from the original substrate. The diversion of electrons and carbon for SMP production could otherwise be invested in the cell yield and the growth rate. An ignoring of SMP formation by the method of Smith et al. (1998) could lead to an overestimation of true cellular growth rates.

In our approach, the SMP formation in biological wastewater treatment processes has been incorporated into the integrated Monod equation (not the conventional Monod equation) for UAP and BAP determination. For better comparison, the experimental data in our work are also used to estimate the bioreaction kinetics using Smith et al.'s method. The best-fit biomass yield on substrate for Smith et al.'s method is 0.60 g COD_X g^{-1} COD_S, which is higher than the corresponding value of biomass yield on substrate consumed by using our approach (i.e., 0.531 g COD_X g^{-1} COD_S). Obviously, Smith et al.'s method results in an overestimation of true cellular growth rates. The electron distribution for active biomass and UAP production should be 4.2 and 53.1 %, respectively. Thus, a significant amount of substrate electrons is distributed to produce SMP in addition to the active cells.

8.2.4 Conclusions

The integrated SMP production method presented in this work is a convenient procedure to determine the UAP and BAP production of activated sludge. This approach is very attractive for the processes, because the effective methods for separately measuring UAP and BAP are not defined yet. Dynamic quantification data are obtained for UAP and BAP separately in our work. Changes in SMP production attributed to toxicity or other conditions do not limit the applicability of this method, as they can be included in the integrated form of the equation. The weighting method applied in this work has a rational statistical basis. The difference between the observed and predicted substrate concentrations is estimated by multiplying the slope of the substrate utilization curve by the difference between the measured and predicted sample times. This approach is appropriate because it approximates the error in the measurement of substrate concentration and such errors are usually larger than those in time measurement in biological wastewater treatment experiments.

Another important feature of our approach is the economy of obtaining determination of UAP and BAP from few batch experiments or even one single batch experiment, rather than collecting large numbers of initial rate measurements. Statistical software packages for a personal computer may also allow convenient application of the approach if they provide for weighting in the nonlinear least squares fitting. The approach established in this work can also be applied for other biological processes with similar difficulty for quantitative product determination. In this work, the determination of UAP and BAP is selected as an example for the application of this approach and it appears successfully. If this methodology is used to determine complex components for other biological processes, the results can be obtained after changing the objective function, the optimized variables, and the ranges of the variables.

References

Al-Halbouni, D., Dott, W., Hollender, J.: Occurrence and composition of extracellular lipids and polysaccharides in a full-scale membrane bioreactor. Water Res. **43**, 97–106 (2009)

APHA: Standard Methods for the Examination of Water and Wastewater, 19th edn. American Public Health Association, New York (1995)

Aqino, S.F.D.: Formation of soluble microbial products (SMP) in anaerobic digesters during stress conditions. Ph.D. thesis, Imperial College, London (2004)

Aquino, S.F., Stuckey, D.C.: Integrated model of the production of soluble microbial products (SMP) and extracellular polymeric substances (EPS) in anaerobic chemostats during transient conditions. Biochem. Eng. J. **38**, 138–146 (2008)

Barker, D.J., Stuckey, D.C.: A review of soluble microbial products (SMP) in wastewater treatment systems. Water Res. **33**, 3063–3082 (1999)

Ciston, S., Lueptow, R.M., Gray, K.A.: Controlling biofilm growth using reactive ceramic ultrafiltration membranes. J. Membr. Sci. **342**, 263–268 (2009)

Dong, B., Jiang, S.: Characteristics and behaviors of soluble microbial products in sequencing batch membrane bioreactors at various sludge retention times. Desalination **243**, 240–250 (2009)

Gao, M., Yang, M., Li, H., Yang, Q., Zhang, Y.: Comparison between a submerged membrane bioreactor and a conventional activated sludge system on treating ammonia-bearing inorganic wastewater. J. Biotechnol. **108**, 265–269 (2004)

Grady Jr, C.P.L., Daigger, G.T., Lim, H.C.: Biological Wastewater Treatment, 2nd edn, pp. 282–284. Marcel Dekker, New York (1999)

Gray, S.R., Ritchie, C.B., Tran, T., Bolto, B.A.: Effect of NOM characteristics and membrane type on microfiltration performance. Water Res. **41**, 3833–3841 (2007)

Gray, S.R., Ritchie, C.B., Tran, T., Bolto, B.A., Greenwood, P., Busetti, F., Allpike, B.: Effect of membrane character and solution chemistry on microfiltration performance. Water Res. **42**, 743–753 (2008)

Grunheid, S., Amy, G., Jekela, M.: Removal of bulk dissolved organic carbon (DOC) and trace organic compounds by bank filtration and artificial recharge. Water Res. **39**, 3219–3228 (2005)

Holakoo, L., Nakhla, G., Yanful, E.K., Bassi, A.S.: Chelating properties and molecular weight distribution of soluble microbial products from an aerobic membrane bioreactor. Water Res. **40**, 1531–1538 (2006)

Hsieh, K.M., Murgel, G.A., Lion, L.W., Schuller, M.L.: Interactions of microbial biofilms with toxic trace metals 1. Observation and modeling of cell growth, attachment, and production of extracellular polymer. Biotechnol. Bioeng. **44**, 219–231 (1994)

Ichihashi, O., Satoh, H., Mino, T.: Effect of soluble microbial products on microbial metabolisms related to nutrient removal. Water Res. **40**, 1627–1633 (2006)

Jarusutthirak, C., Amy, G., Croue, J.P.: Fouling characteristics of wastewater effluent organic matter (EfOM) isolates on NF and UF membranes. Desalination **145**, 247–255 (2005)

Jarusutthirak, C., Amy, G.: Role of soluble microbial products (SMP) in membrane fouling and flux decline. Environ. Sci. Technol. **40**, 969–974 (2006)

Jarusutthirak, C., Amy, G: Understanding soluble microbial products (SMP) as a component of effluent organic matter (EfOM). Water Res. **41**, 2787–2793 (2007)

Jiang, T., Myngheer, S., De Pauw, D.J.W., Spanjers, H., Nopens, I., Kennedy, M.D., Amy, G., Vanrolleghem, P.A.: Modelling the production and degradation of soluble microbial products (SMP) in membrane bioreactors (MBR). Water Res. **42**, 4955–4964 (2008)

Kimura, K., Naruse, T., Watanabe, Y.: Changes in characteristics of soluble microbial products in membrane bioreactors associated with different solid retention times: relation to membrane fouling. Water Res. **43**, 1033–1039 (2009)

Labbs, C., Amy, G., Jekel, M.: Understanding the size and character of fouling-causing substances from effluent organic matter (EfOM) in low-pressure membrane filtration. Environ. Sci. Technol. **40**, 4495–4499 (2006)

Laspidou, C.D., Rittmann, B.E.: A unified theory for extracellular polymeric substances, soluble microbial products, and active and inert biomass. Water Res. **36**, 2711–2720 (2002a)

Laspidou, C.S., Rittmann, B.E.: Non-steady state modeling of extracellular polymeric substances, soluble microbial products, and active and inert biomass. Water Res. **36**, 1983–1992 (2002b)

Magbanua Jr, B.S., Bowers, A.R.: Characterization of soluble microbial products (SMP) derived from glucose and phenol in dual substrate activated sludge bioreactors. Biotechnol. Bioeng. **93**, 862–870 (2006)

Namkung, E., Rittmann, B.E.: Soluble microbial products (SMP) formation kinetics by biofilms. Water Res. **20**, 795–806 (1986)

Ni, B.J., Fang, F., Xie, W.M., Xu, J., Yu, H.Q.: Formation of distinct soluble microbial products by activated sludge: kinetic analysis and quantitative determination. Environ. Sci. Technol. **46**, 1667–1674 (2012)

Ni, B.J., Zeng, R.J., Fang, F., Xie, W.M., Sheng, G.P., Yu, H.Q.: Fractionating soluble microbial products in the activated sludge process. Water Res. **44**, 2292–2302 (2010)

Noguera, D.R., Araki, N., Rittmann, B.E.: Soluble microbial products in anaerobic chemostates. Biotechnol. Bioeng. **44**, 1040–1047 (1994)

Okamura, D., Mori, Y., Hashimoto, T., Hori, K.: Identification of biofoulant of membrane bioreactors in soluble microbial products. Water Res. **43**, 4356–4362 (2009)

Rosenberger, S., Laabs, C., Lesjean, B., Gnirss, R., Amy, G., Jekel, M., Schrotter, J.C.: Impact of colloidal and soluble organic material on membrane performance in membrane bioreactors for municipal wastewater treatment. Water Res. **40**, 710–719 (2006)

Schiener, P., Nachaiyasit, S., Stuckey, D.C.: Production of soluble microbial products (SMP) in an anaerobic baffled reactor, composition, biodegradability and the effect of process parameters. Environ. Sci. Technol. **19**, 391–400 (1998)

Shon, H.K., Vigneswaran, S., Kim, I.S., Cho, J., Ngo, H.H.: The effect of pretreatment to ultrafiltration of biologically treated sewage effluent: a detailed effluent organic matter (EfOM) characterization. Water Res. **38**, 1933–1939 (2004)

Smith, L.H., McCarty, P.L., Kitanidis, P.K.: Spreadsheet method for evaluation of biochemical reaction rate coefficients and their uncertainties by weighted nonlinear least-squares analysis of the integrated Monod equation. Appl. Environ. Microbiol. **64**, 2044–2050 (1998)

Smith, P.J., Vigneswaran, S., Ngo, H.H., Ben-Aim, R., Nguyen, H.: A new approach to backwash initiation in membrane systems. J. Membr. Sci. **278**, 381–389 (2006)

Chapter 9
Fate of the Microbial Products in Aerobic Granular Sludge

The formation of EPS, SMP, and internal storage products (X_{STO}) in aerobic granular sludge was investigated using experimental and modeling approaches. An expanded unified model describing the production and the consumption of EPS, SMP, and X_{STO} was formulated after integrating the electron flows from the external substrate to EPS, SMP, and X_{STO}. The effect of the sludge retention time (SRT) on the formation of EPS, SMP, and X_{STO} in the aerobic granular sludge was analyzed. The new model could be used for process control and thus for the optimization of aerobic-granule-based reactors.

9.1 Modeling Microbial Products Under Feast-Famine Conditions

9.1.1 Introduction

Activated sludge exhibits a matrix structure that includes biomass components other than the metabolically active bacteria. For example, extensive research has shown that EPS are a major component of the matrix material (McSwain et al. 2005; Su and Yu 2005; Sheng et al. 2006). EPS are sticky solid materials secreted by cells, and they are involved in adhesion phenomena, formation of the matrix structure, controlling the microbial physiology, and the long-term stability of the sludge (Sheng et al. 2006). EPS in activated sludge are hypothesized to bridge two neighboring bacterial cells physically to each other, as well as with inert particulate matter (Liu et al. 2004). This cohesive binding creates large, dense aggregates that settle rapidly. It is hypothesized that EPS production in sludge is promoted by stress conditions (Nichols et al. 2004), such as high hydrodynamic shear force and a feast-and-famine regime.

B.-J. Ni, *Formation, Characterization and Mathematical Modeling of the Aerobic Granular Sludge*, Springer Theses, DOI: 10.1007/978-3-642-31281-6_9,

For some bacteria, intracellular storage products (X_{STO})—such as polyhydroxyalkanoates, lipids, and polysaccharides—are significant sinks for the carbon and electrons removed from the substrate (van Loosdrecht et al. 1997; Majone et al. 1999; Pratt et al. 2004; Ni and Yu 2007). Activated sludge processes can be highly dynamic with respect to the feed regime, especially for SBR (van Loosdrecht et al. 1997) and enhanced biological phosphorus removal processes (Oehmen et al. 2005). In these processes, the active microorganisms are exposed to significant concentrations of the external substrate only for relatively short periods of time. Internally stored products allow them to take advantage of the dynamic feast-and-famine periods. Microorganisms capable of quickly storing substrate as internal storage products during feast conditions and then consuming the stored substrate during famine conditions have a strong competitive advantage over microorganisms without such a capacity.

In addition to making different biomass components, bacteria convert a fraction of the organic substrate into SMP, which account for the bulk of the soluble organic carbon in reactor effluents (de Silva and Rittmann 2000a; b; Benjamin et al. 2006). Models of biological wastewater treatment systems based on Monod kinetics predict that the steady-state effluent concentration of the rate-limiting substrate should be independent of the influent substrate concentration. However, this is not in accord with the actual operating experience for aggregate measures of effluent organic matter, such as soluble biochemical oxygen demand or chemical oxygen demand. The incorporation of SMP brings model predictions into accord with empirical experience in terms of the soluble organic material in treatment effluents (Benjamin et al. 2006; Picioreanu et al. 1998).

EPS, X_{STO}, and SMP are important sinks for electrons and carbon derived from the original substrate. Their production and consumption phenomena are not the same as each other or for the active biomass. Phenomena involved with multiple carbon and electron sinks should have significant implications on the performance of activated sludge systems (van Loosdrecht et al. 1997), particularly when the process experiences feast-and-famine conditions, which could promote the accumulation and subsequent utilization of X_{STO}, EPS, or SMP (Pratt et al. 2004).

Mathematical modeling has proven to be essential for understanding complex microbial systems, such as in biological wastewater treatment. The popular ASMs produced by the IWA have been focused on the heterogeneity of substrate and biomass types (Pratt et al. 2004; Picioreanu et al. 1998). Comparatively little attention has been paid to EPS, X_{STO}, and SMP, although these components are of high importance to aggregation, effluent quality, and process dynamics. Therefore, this paper aims to overcome the limitation by incorporating the latest representations for the production and consumption of EPS, $X_{STO,}$ and SMP in a dynamic biological-treatment model. We developed an expanded model that integrates production and consumption of X_{STO} with the unified model for EPS, SMP, and active and inert biomass in activated sludge. We conducted batch and SBR tests to evaluate the model's ability to represent all the microbial products and to obtain values of model parameters for our expanded model.

9.1.2 Model Development

9.1.2.1 General Model Description

One partial foundation of the model is the unified theory for EPS and SMP developed by Laspidou and Rittmann (2002a). The second partial foundation is a submodel for storage and utilization of X_{STO} (Krishna and van Loosdrecht 1999; Sin et al. 2005). The submodel for storage and utilization of X_{STO} is based on the approach of Ni and Yu (2007), who explored the dynamic kinetics of the one microbial product, X_{STO}. Combining the two parts establishes the expanded unified model, which is applicable for modeling all three microbial products (EPS, SMP, and X_{STO}) in activated sludge under feast-famine conditions. The established model here describes the relationships among four solid (particulate) components: active heterotrophic bacteria, EPS, X_{STO}, and residual inert biomass; three soluble components: external substrate and two unique forms of SMP; and an electron acceptor, which is dissolved oxygen. The model employs the following symbols for concentrations: external substrate (S), active heterotrophic biomass (X_H), internal storage products (X_{STO}), residual inert biomass (X_I), utilization-associated products (S_{UAP}), biomass-associated products (S_{BAP}), soluble microbial products ($S_{SMP} = S_{UAP} + S_{BAP}$), extracellular polymeric substances (X_{EPS}), and dissolved oxygen (S_O). The units for all components are oxygen demand or oxygen (for DO), which is directly proportional to electron equivalents (i.e., 8 g O_2 per e^- equivalent).

Figure 9.1 illustrates schematically all of the electron flows in the expanded unified model. The left side of the figure describes the flow of electrons from the external substrate. According to the unified theory (Laspidou and Rittmann 2002a, b), part of external substrate is used for biomass synthesis (Path 2). Other substrate electrons are diverted to the formation of UAP and EPS. UAP (Path 4) are released to the aqueous solution, while EPS (Path 8) are released as a solid to form the aggregate matrix. Hydrolysis of EPS (Path 9) produces BAP, which are soluble. Substrate oxidation and respiration of the electrons to reduce O_2 (Path 5) generate the energy needed to fuel the formation of active biomass, EPS, and UAP. Expanding the unified model involves adding X_{STO} formation and utilization. During a feast period, some external substrate is converted into X_{STO} (Path 3). Because formation of X_{STO} competes with normal biomass synthesis, formation of X_{STO} occurs only when the substrate concentration is high. The substrate concentration should be much higher than the substrate's half-maximum-rate concentration (K_S). Simultaneously, UAP and EPS also compete with biomass synthesis.

The right side of Fig. 9.1 illustrates non-substrate phenomena represented in the expanded unified model. Active biomass decays simultaneously through endogenous respiration that consumes oxygen (Path 7) and by generating residual inert biomass (Path 6), which is not biodegradable. Since UAP and BAP are biodegradable, some of their electrons can also be used by the bacteria as ''recycled'' substrate (Paths 11 and 12), while the remaining electrons also go to the electron

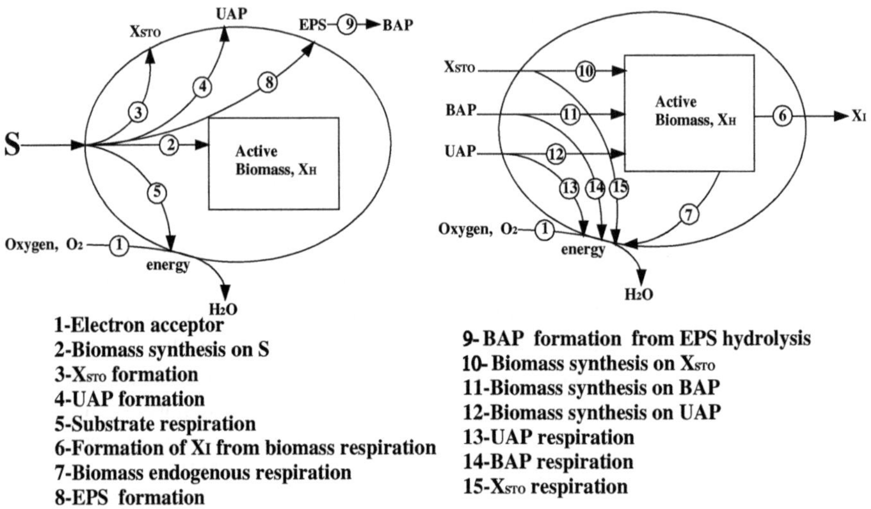

Fig. 9.1 Schematic diagram of electron flows from the external substrate (*left* side), and regarding active and inert biomass, EPS, SMP, and X_{STO} (*right* side) for the expanded unified model (Reprinted from Ni et al. (2009) with permission from American Chemical Society)

acceptor for energy generation (Paths 13 and 14). X_{STO} also is biodegradable and is utilized by the active biomass when it enters a famine period. Some of the electrons in X_{STO} are used for biomass synthesis (Path 10), while the other electrons are respired for energy generation (Path 15). By convention (Laspidou and Rittmann 2002b), utilizations of UAP, BAP, and X_{STO} do not result in formation of new UAP, EPS, or X_{STO}.

Table 9.1 lists the equations for all components in the expanded unified model. The following sections explain the bases for the equations involving the microbial products. Table 9.2 defines all the parameters used in the expanded unified model, their symbols, and their units.

9.1.2.2 Kinetics of the Formation and Hydrolysis of EPS

Most bacteria produce EPS, which is critical for the formation of microbial aggregates (McSwain et al. 2005; Liu et al. 2004). Equation 9.1 is the non-steady-state mass balance for EPS in a batch system. The first term on the right side is for EPS formation, while the second term is EPS loss. Laspidou and Rittmann (2002b) established that the rate of EPS formation (r_{EPS}) is proportional to the rate of substrate utilization ($r_{EPS} = k_{EPS} r_S X_H$), and k_{EPS} is the fraction of substrate electrons diverted to EPS formation. Hydrolysis of solid-phase EPS converts it into soluble BAP, with EPS hydrolysis being the only source of the BAP. The hydrolysis rate is proportional to the EPS concentration ($r_{hyd} = k_{hyd} X_{EPS}$), with k_{hyd} being the first-order rate coefficient (Laspidou and Rittmann 2002b). $M_S(t) = \frac{S(t)}{K_S + S(t)}$ and $M_O(t) = \frac{S_O(t)}{K_{O_2} + S_O(t)}$

Table 9.1 Nonsteady-state mass-balance equations for all components in the model

	Equation	No.
EPS	$\frac{dX_{EPS}(t)}{dt} = k_{EPS}\frac{\mu_{H,S}}{Y_{H,S}}M_S(t)M_O(t)X_H(t) - k_{hyd}X_{EPS}(t)$	(9.1)
UAP	$\frac{dS_{UAP}(t)}{dt} = k_{UAP}\frac{\mu_{H,S}}{Y_{H,S}}M_S(t)M_O(t)X_H(t) - \frac{\mu_{UAP}}{Y_{UAP}}M_{UAP}(t)M_O(t)X_H(t)$	(9.2)
BAP	$\frac{dS_{BAP}(t)}{dt} = k_{hyd}X_{EPS}(t) - \frac{\mu_{BAP}}{Y_{BAP}}M_{BAP}(t)M_O(t)X_H(t)$	(9.3)
X_{STO}	$\frac{dX_{STO}(t)}{dt} = k_{STO}\frac{\mu_{H,S}}{Y_{H,S}}M_S(t)M_O(t)X_H(t) - \frac{\mu_{H,STO}}{Y_{H,STO}}M_{STO}(t)I_S(t)M_O(t)X_H(t)$	(9.4)
S	$\frac{dS(t)}{dt} = -\frac{\mu_{H,S}}{Y_{H,S}}M_S(t)M_O(t)X_H(t)$	(9.5)
X_H	$\frac{dX_H(t)}{dt} = (1 - k_{EPS} - k_{UAP} - k_{STO})\mu_{H,S}M_S(t)M_O(t)X_H(t) + \mu_{H,STO}M_{STO}(t)I_S(t)M_O(t)X_H(t) + \mu_{UAP}M_{UAP}(t)M_O(t)X_H(t) + \mu_{BAP}M_{BAP}(t)M_O(t)X_H(t) - b_HM_O(t)X_H(t)$	(9.6)
X_I	$\frac{dX_I(t)}{dt} = -f_Ib_HM_O(t)X_H(t)$	(9.7)
DO	$\frac{dS_O(t)}{dt} = k_La(S_O^* - S_O(t)) - \frac{[1 - k_{EPS} - k_{UAP} - k_{STO} - Y_{H,S}(1 - k_{EPS} - k_{UAP} - k_{STO})]}{Y_{H,S}}\mu_{H,S}M_S(t)M_O(t)X_H(t) - \frac{1 - Y_{H,STO}}{Y_{H,STO}}\mu_{H,STO}M_{STO}(t)I_S(t)M_O(t)X_H(t) - (1 - f_I)b_HM_O(t)X_H(t) - \frac{1 - Y_{UAP}}{Y_{UAP}}\mu_{UAP}M_{UAP}(t)M_O(t)X_H(t) - \frac{1 - Y_{BAP}}{Y_{BAP}}\mu_{BAP}M_{BAP}(t)M_O(t)X_H(t)$	(9.8)

Table 9.2 Stoichiometric and kinetic coefficients (20 °C) for the model

Parameter	Definition	Value	Units
$Y_{H,S}$	Yield coefficient for growth on S_S	0.34	g COD_X g^{-1} COD_S
$Y_{H,STO}$	Yield coefficient for growth on X_{STO}	0.53	g COD_X g^{-1} COD_{STO}
f_I	Fraction of X_I in decay	0.20	g COD_X g^{-1} COD_X
Y_{UAP}	Yield for growth on UAP	0.45	g COD_X g^{-1} COD_{UAP}
Y_{BAP}	Yield for growth on BAP	0.45	g COD_X g^{-1} COD_{BAP}
k_{STO}	X_{STO} formation coefficient	0.23	g COD_{STO} g^{-1} COD_S
$\mu_{H,STO}$	Maximum growth rate on X_{STO}	0.31	h^{-1}
$\mu_{H,S}$	Maximum specific growth rate	0.40	h^{-1}
K_S	Substrate affinity constant	11.38	g COD_S m^{-3}
K_{STO}	Biomass affinity constant for X_{STO}	1.0	g COD_{STO} g^{-1} COD_X
K_O	Dissolved oxygen affinity constant	0.2	g O_2 m^{-3}
b_H	Decay rate coefficient of X_H	0.021	h^{-1}
k_{EPS}	EPS formation coefficient	0.18	g COD_{EPS} g^{-1} COD_S
k_{hyd}	EPS hydrolysis rate coefficient	0.0071	h^{-1}
k_{UAP}	UAP formation coefficient	0.05	g COD_{UAP} g^{-1} COD_S
μ_{UAP}	Maximum rate of UAP degradation	0.053	h^{-1}
K_{UAP}	Biomass affinity constant for UAP	100	g COD_{UAP} m^{-3}
μ_{BAP}	Maximum rate of BAP degradation	0.0029	h^{-1}
K_{BAP}	Biomass affinity constant for BAP	85	g COD_{BAP} m^{-3}

are Monod kinetic functions for the external substrate and DO, respectively. The ratio $\mu_{H,S}/Y_{H,\,S}$ equals the maximum specific utilization rate of the external substrate.

9.1.2.3 Kinetics for SMP Formation and Utilization

SMP are divided into two categories (Laspidou and Rittmann 2002a): UAP produced as a direct result of substrate utilization, and BAP is formed from hydrolysis of EPS. Equations 9.2 and 9.3 are the non-steady-state mass balances for UAP and BAP. Laspidou and Rittmann (2002b) represented the rate UAP-formation as $r_{UAP} = k_{UAP} r_S X_H$, where k_{UAP} is the proportion of electrons from the external substrate that are diverted to UAP formation. The rate of BAP formation from EPS hydrolysis is $r_{BAP} = k_{hyd} X_{EPS}$. UAP and BAP are biodegradable, and their degradation kinetics can be described with Monod-type expressions for which BAP and UAP have distinct biodegradation kinetic parameters. $M_{UAP}(t) = \frac{S_{UAP}(t)}{K_{UAP}+S_{UAP}(t)}$ and $M_{BAP}(t) = \frac{S_{BAP}(t)}{K_{BAP}+S_{BAP}(t)}$ stand for Monod kinetic functions of UAP and BAP, respectively.

9.1.2.4 Kinetics for X_{STO} Formation and Utilization

The simultaneous storage and growth concept (Karahan et al. 2006) describes the kinetics of X_{STO} formation and degradation under feast-and-famine conditions for the external substrate. Equation 9.4 integrates X_{STO} formation and utilization kinetics into the structure of the unified model. $M_{STO}(t) = \frac{X_{STO}(t)/X_H(t)}{K_{STO}+X_{STO}(t)/X_H(t)}$ stands for a saturation-kinetic function for utilization of X_{STO}; and $I_S(t) = \frac{K_S}{K_S+S(t)}$ is a Monod-type inhibition function for the effect of high S on utilization of X_{STO}. X_{STO} formation is correlated directly with growth, as defined by the Monod expression of Eq. 9.4. Our approach allows a high rate of X_{STO} formation when S is large (i.e., feast) and utilization of X_{STO} to support synthesis of active biomass when normal synthesis is precluded by low S (i.e., famine).

9.1.2.5 Kinetics of Substrate Consumption and Active Biomass and X_I Accumulations

Equation 9.5 is the non-steady-state mass balance for the external substrate. Loss of external substrate is coupled to synthesis of active biomass, EPS, UAP, and X_{STO}. The non-steady-state mass balance on active biomass is Eq .9.6. Aerobic heterotrophic growth takes place through utilization of S (first term), X_{STO} (second term), UAP (third term), and BAP (fourth term). The fifth term represents the loss of active biomass due to endogenous respiration. Equation 9.7 describes the accumulation of X_I as a result of endogenous respiration of active biomass.

9.1.2.6 Kinetics for Oxygen Transfer and Consumption

Equation 9.8 gives the non-steady-state mass balance for dissolved oxygen. Equation 9.8 shows that DO is added to the system by gas–liquid mass transfer that has a mass-transfer rate of k_La (Nicolella et al. 1998). S_O^* is the maximum oxygen solubility in liquid phase, and S_O is the oxygen concentration in the bulk liquid. The latter five terms in Eq. 9.8 represents oxygen uptake by all respiration processes: utilization of S (term 2), utilization of X_{STO} (term 3), endogenous respiration of X_H (term 4), utilization of UAP (term 5), and utilization of BAP (term 6).

9.1.2.7 Aggregate Parameters

The measurable aggregate outputs—total soluble COD (SCOD) and MLVSS, both expressed in mg COD L^{-1}—are calculated from:

$$SCOD(t) = S(t) + S_{UAP}(t) + S_{BAP}(t) \tag{9.9}$$

$$\text{MLVSS}(t) = X_{\text{H}}(t) + X_{\text{I}}(t) + X_{\text{EPS}}(t) + X_{\text{STO}}(t) \tag{9.10}$$

9.1.2.8 Model Simulation

We performed model simulations using the software package AQUASIM (Reichert 1998). This program offers a flexible definition of the kinetic model, flow scheme, and process control strategies; it also provides support for graphic display of the support of the simulation results, corresponding experimental data, and communication with spreadsheet programs (Reichert 1998).

9.1.3 Experimental Materials and Methods

9.1.3.1 Reactor Setup and Operation

The SBR setup in this chapter is similar to that reported in Chap. 3. The SBR was fed with filtered and diluted (tenfold) soybean-processing wastewater having a COD of around 800 mg L^{-1}. The experimental setup and feeding solutions used in this work are similar to those reported in Ni and Yu (2007), but distinctly different and designed to evaluate the expanded unified model. First, the characteristics of the two feeding wastewaters are different. Ni and Yu (2007) fed a fatty-acids wastewater, while we fed a carbohydrate and protein wastewater. Second, the experimental strategies and results are substantially different. Whereas Ni and Yu (2007) measured only OUR, X_{STO}, and original substrate, we measured OUR, SCOD, VSS, EPS, SMP, and X_{STO}, which are the parameters needed to evaluate the expanded unified model. Virtually all of the COD was soluble and biodegradable, which corresponds to the description of the substrate in the model.

We set the SRT at 20 days by controlling the amount of sludge wasted from the reactor in each cycle. The reactor was extensively monitored for DO and sampled for measurements of COD, storage polymers, and oxygen uptake rate. The criteria for reaching steady state were a constant storage-polymer concentration and consistent patterns of DO, OUR, and COD in repetitive cycles. The SBR operated over 100 d in order to reach steady state. The constant storage-polymer concentration refers to the value after one full cycle of 4 h under global steady-state conditions. The deviation for this storage-polymer concentration was approximately 2 mg L^{-1}.

9.1.3.2 Parameter Determination

We estimated the substrate half-saturation coefficient (K_{S}) and endogenous-decay coefficient (b_{H}) as described by Henze et al. (1987). We fit the increase in μ_{H} with increasing S concentration in independent batch experiments to find the best-fit K_S value for the diluted soybean-processing wastewater. To estimate b_H, we removed sludge from a completely mixed reactor and put it into an aerated and non-fed batch reactor. We measured the OUR approximately thrice over a period of 10 days. A plot of the logarithm of the respiration rate versus time gave a straight line with slope b_H.

We took all parameters for the EPS, UAP, and BAP, as well as the heterotrophic biomass yield on substrate (Y_H), directly from Laspidou and Rittmann (2002b). Because the utilization and formation rates that resulted from using these parameters matched the rates found in the experiments, we considered them to be sufficient and did not calibrate the model further. We estimated the storage-process parameters ($Y_{H,\ STO}$, k_{STO}, and $\mu_{H,\ STO}$) with batch respirometry experiments. Prior to a respirometry test, we washed the sludge twice with distilled water to remove carry-over material. Then, we added diluted soybean-processing wastewater as a pulse and measured the DO concentration over time. We used the calculated OUR for parameter estimation with the methods of Avcioglu et al. (2003).

9.1.3.3 Experimentally Testing the Model

We conducted independent sets of batch experiments at known initial conditions to evaluate the model and its parameter values, and we also conducted intensive one-cycle measurements of the SBR. For the batch experiments, we sampled sludge from the SBR when no substrate was present in the medium. The sludge was kept in the famine condition to deplete it of internal storage products (X_{STO}) and washed twice with distilled water to remove external soluble organic material (SCOD). We inoculated a batch reactor with 1 L of diluted activated sludge and aerated it continuously so that the DO concentration was above 4 mg L^{-1}. We measured the endogenous OUR over 10 min. Then, we added external substrate at a predetermined concentration and monitored the OUR until the substrate became depleted and the endogenous activity resumed. In the first batch experiment (Set I), the initial MLVSS and substrate concentrations were 500 mg L^{-1} and 320 mg L^{-1}, respectively. We designed much different initial conditions for a second batch experiment (Set II). The initial MLVSS and substrate concentrations for the second batch test were 3,400 mg L^{-1} and 700 mg L^{-1}, respectively. The experimental conditions were $T = 20 \pm 1$ °C, pH $= 7.0 \pm 0.05$. We also took samples every 5–10 min for the analysis of SCOD, MLVSS, EPS, SMP, and X_{STO}.

To further explore the universality of this expanded unified model, we operated a conventional CSTR system with the same soybean-processing wastewater (Set III). The initial operating conditions for the CSTR were: $S_S = 800$ mg L^{-1}, $X_{STO} = 30$ mg L^{-1}, $X_{EPS} = 300$ mg L^{-1}, and MLVSS $= 1{,}150$ mg L^{-1}. We took samples for the analysis of EPS, SMP, and X_{STO} at given time intervals.

9.1.3.4 Analytical Procedures

The analysis of DO and OUR was described in Chap. 3. We followed Standard Methods (APHA 1995) for measurements of COD, MLSS, and MLVSS. In this work, the MLVSS reported in terms of COD was obtained by conversion of the measured VSS value using the coefficient 1.42 mg COD mg^{-1} VSS.

The EPS of the sludge samples were extracted using the CER technique (Dowex Marathon C, 20–50 mesh, sodium form, Fluka 91973), as described in Chap. 7. Mixed liquid samples were centrifuged at 12,000 rpm for 15 min and then were pre-filtered through 0.45-μm acetate cellulose membranes to represent the SMP. The internal storage polymers (X_{STO}) were measured according to Pratt et al. (2004). The storage polymer measured is poly-hydroxyalkanoate (PHA) in this work.

9.1.4 Results and Discussion

9.1.4.1 Stoichiometry and Kinetic Parameters

Table 9.2 summarizes the values for all the stoichiometric and kinetic parameters used in the model. The obtained stoichiometric and kinetic parameter values are within the scope of those reported in the literature (Henze et al. 1987; Avcioglu et al. 2003) for the same parameter and units. Comparison of model parameter values provides insight into the roles and fates of X_{STO}, EPS, UAP, and BAP.

The heterotrophic maximum specific growth rate through utilization of storage polymers ($\mu_{H,\ STO}$) is 0.31 h^{-1}, which is only slightly slower than the maximum specific growth rate for the external substrate ($\mu_{H,\ S}$) of 0.4 h^{-1}. Thus, growth through utilization of X_{STO} is commensurate with growth on external substrate. On the other hand, the maximum specific growth rates through utilization of UAP and BAP are much smaller: 0.053 and 0.0029 h^{-1}, respectively. The half-maximum-rate concentration for X_{STO} (K_{STO}) has a value of 1.0 g COD_{STO} g^{-1} COD_X. This value cannot be compared directly to K_S, K_{UAP}, or K_{BAP}, because the units are different. However, the K_{STO} value suggests that the kinetics are usually first order, as the actual X_{STO}/X_H ratio is much lower than 1. The half-saturation coefficients for UAP and BAP are much larger than K_S, ensuring slow biodegradation kinetics under most conditions.

Electrons from the external substrate are distributed in the following order when X_{STO} is produced (feast): new biomass synthesis 34 %, X_{STO} 23 %, O_2 for respiration 20 %, EPS 18 %, and UAP 5 %. Thus, most substrate electrons are diverted to biomass and storage polymers, and fewer are distributed to produce EPS and UAP when ample substrate is available. Taken together with the much faster utilization rate for X_{STO} than for UAP and BAP, the relatively high distribution of electrons to X_{STO} quantifies why X_{STO} is the primary route by which the heterotrophs store and later utilize excess electrons and energy. In short, X_{STO} can have a high turnover rate when the biomass is in a feast-and-famine regime.

The decay rate for active biomass (b_H) is 0.021 h^{-1}, which is about three times larger than the hydrolysis rate for EPS (k_{hyd}) of 0.0071 h^{-1}. This difference means that EPS are not used by the bacteria as energy- and electron-storage materials, since they are hydrolyzed too slowly to BAP. Furthermore, the utilization kinetics of BAP are even slower. Thus, formation of EPS seems to be focused on its relationship to aggregation of sludge, and not on being a storage material for supporting biomass growth during feast and famine.

In addition, our expanded unified model, which incorporates the production and consumption of EPS, X_{STO}, and SMP, has different parameter values from those reported in Ni and Yu (2007). The lower true microbial yield in this work [0.34 vs. 0.52 in Ni and Yu (2007)] is probably attributable to the fact that the formation of SMP and EPS does not subtract from the net yield of X_H in Ni and Yu (2007). In addition, the soybean-processing wastewater is a complex mixture of carbohydrates and proteins, which are macromolecules compared with fatty acids. The complex composition of the wastewater used in this work is likely to be associated with a lower true microbial yield. The different kinetic parameters of X_{STO} formation and consumption (k_{STO}, $\mu_{H,\ STO}$, $Y_{H,\ STO}$) are related to the different feeding wastewater characteristics, which lead to different microbial communities. One parameter is the same: the biomass-decay coefficient, b_H, which does not depend on the different forms of products in the two models or the different substrates.

9.1.4.2 Model Evaluation

We solved the model for the set of parameters shown in Table 9.2 in order to evaluate how well the model can describe the experimental results from independent batch experiments. The evaluation also helps identify relationships among the different components of the expanded unified model. We estimated the initial concentration of active biomass ($X_H(0)$) using the baseline endogenous *OUR* prior to substrate addition:

$$\mathrm{OUR}_{\mathrm{end}}(0) = (1 - f_{\mathrm{I}}) \cdot b_{\mathrm{H}} \cdot X_{\mathrm{H}}(0) \tag{9.11}$$

where f_I is 0.2 mg COD mg^{-1} COD, as used in ASM3. Thus, for a given f_I and measured b_H, $X_H(0)$ could be calculated from the $OUR_{end}(0)$ data. In the first batch experiment, the initial concentration of active biomass was determined as 480 mg COD L^{-1}. Since the biomass was washed, the initial concentrations for BAP and UAP were set to zero. Likewise, we made the initial X_{STO} concentration zero, because we starved the cells to deplete internal storage products before the experiment. The initial value of X_I was set as the difference between MLVSS and $X_H(0)$.

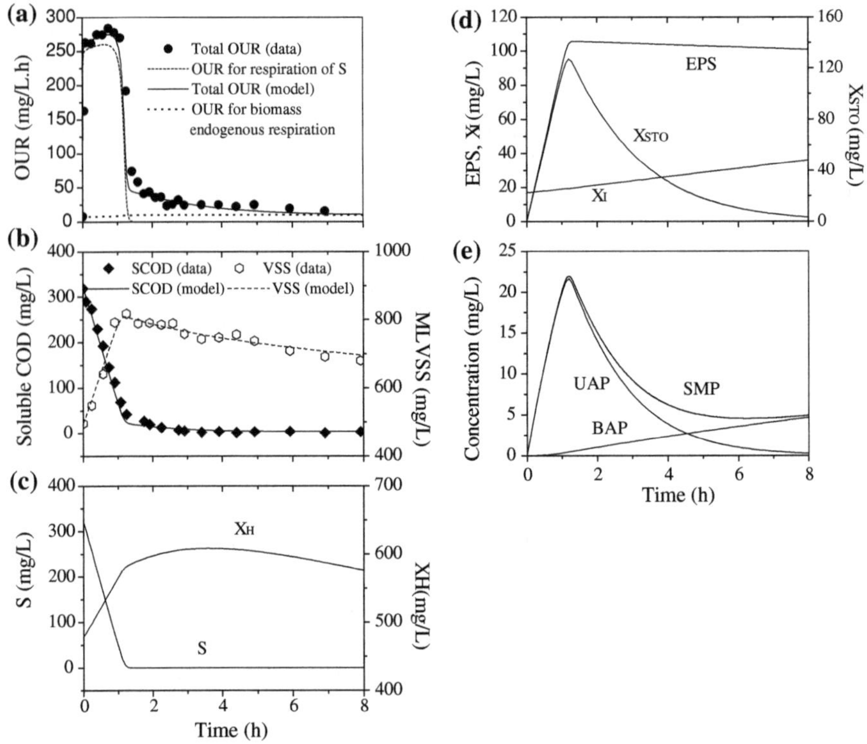

Fig. 9.2 Model evaluation of **a** OUR profiles, **b** SCOD data (SCOD = $S + S_{UAP} + S_{BAP}$) and MLVSS profiles (MLVSS = $X_H + X_I + X_{EPS} + X_{STO}$), and model simulation of **c** S and biomass profiles, **d** formation and hydrolysis of EPS, formation of X_{STO} and X_I, and **e** production and degradation of SMP during the first batch experiment (Set I). Initial conditions: $S_S = 320$ mg L^{-1} and MLVSS = 500 mg L^{-1} (Reprinted from Ni et al. (2009) with permission from American Chemical Society)

Figure 9.2 compares the simulation and the first batch experimental results for OUR, MLVSS, and SCOD, as well as the model-predicted concentrations for all the soluble and solid components. Model and experimental data agree very well for the three measured parameters in Fig. 9.2a and b. After the addition of substrate, OUR increases due to the rapid oxidation of external substrate, with about 20 % of the electron flow going directly to respiration. The sharp inflection in the OUR profile corresponds with the complete removal of the external substrate added (Fig. 9.2c). The change of OUR is a direct indicator of the transition from feast to famine, which occurs when the external substrate is consumed completely. However, OUR does not immediately return to the endogenous respiration level because of the oxidation of X_{STO} and UAP, which are formed when the external substrate is being utilized, but are net utilized only after S becomes depleted (Fig. 9.2d and e).

SCOD decreases rapidly and continuously for the first ~1 h (Fig. 9.2b), as the external substrate is consumed. However, a complete depletion of the soluble COD pool does not occur despite the complete consumption of S, because UAP forms when S is being utilized, while BAP builds up slowly in the entire process (Fig. 9.2e). The net effect is that SCOD is almost constant from 2 to 8 h.

MLVSS initially increases from 500 to 800 mg L^{-1} at approximately 1 h and later gradually decreases to 700 mg L^{-1} at 8 h (Fig. 9.2b). Figure 9.2d shows that, in the period of rapid biomass growth, X_{STO} and X_{EPS} increase rapidly and occupy significant fractions of MLVSS. However, once S is consumed and rapid growth ceases, only X_{STO} is utilized fast enough to provide for new synthesis with a relatively high rate under famine conditions. Thus, in the period of net biomass decay, X_{EPS} and X_H become the main fractions of MLVSS, while X_{STO} declines. These observations are in qualitative agreement with those of Beun et al. (2000), who used acetate as the sole substrate.

Figure 9.2d and e also shows that X_I and X_{BAP} build up slowly and continually. X_I accumulates steadily due to active-biomass endogenous decay. X_{BAP} build up slowly and continually because of the slow hydrolysis of EPS to BAP, but even slower utilization of BAP. Whereas UAP is about 95 % of the total SMP content at 1 h, BAP is nearly 100 % of the SMP at 8 h.

Figure 9.3 compares the simulation and experimental results for OUR, MLVSS, SCOD, EPS, SMP, and X_{STO} during the second batch experiment. Using Eq. (9.11) and the endogenous respiration data, we estimated the initial concentration of active biomass in this second batch test as 540 mg COD L^{-1}. Having experimental measurements of EPS, SMP, and X_{STO} in this second batch experiment allows us to test directly the models' ability to represent those microbial products. Furthermore, the initial conditions for this experiment are substantially different from those in Fig. 9.2. The added external substrate (700 mg L^{-1}) is more than twice that of the first batch test (320 mg L^{-1}). Also, the initial sludge content (3,400 mg L^{-1}) is much higher than the first one (500 mg L^{-1}). Correspondingly, the initial active biomass and EPS concentrations are different, and the EPS and X_{STO} contributions to total MLVSS are much larger. The experimental trends and absolute concentrations for all components are described well by the expanded unified model. These trends include the sharp drop-off of OUR, the peaking of X_{STO}, and the leveling-off of EPS as the original substrate is depleted at approximately 2 h. The agreement between model outputs and the experimental data for the two independent experiments strongly supports that this expanded unified model properly captures the relationship among the different types of microbial products.

We also evaluated the model with the experimental results of one cycle of the SBR operation at global steady state and with an SRT of 20 days and MLVSS of 4,700 mg L^{-1}. Again, we first used Eq. (9.11) to calculate the initial concentration of active biomass. The determined initial concentration is 530 mg COD L^{-1}.

Figure 9.4 compares the measured and simulated profiles of the soluble COD, X_{STO}, and OUR through a feast-and-famine cycle of the SBR. X_{STO} (Fig. 9.4a) increases to maximum of 200 mg L^{-1} during the period of rapid microbial growth,

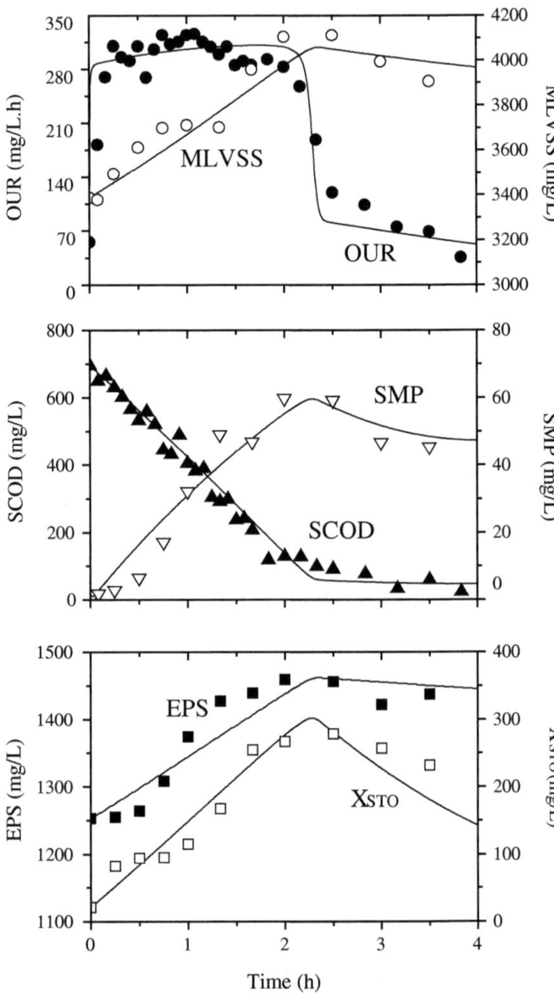

Fig. 9.3 Model evaluation of OUR profiles, SCOD data, MLVSS profiles, and the EPS, SMP, X_{STO} data during the second batch experiment (Set II). Initial conditions: $S_S = 700$ mg L^{-1}, $X_{STO} = 20$ mg L^{-1}, $X_{EPS} = 1{,}100$ mg L^{-1}, and MLVSS $= 3{,}400$ mg L^{-1} (Reprinted from Ni et al. (2009) with permission from American Chemical Society)

when the external substrate (S) is consumed at a rapid rate, as is evidenced by the sharp decline in OUR at about 1 h (Fig. 9.4c). During the famine phase (after about 1 h), X_{STO} declines due to its rapid utilization. It should be noted that sludge has to be partially discharged according to the predetermined SRT value after one cycle. Thus, residual storage polymers should also be reduced to some extent after sludge discharge. In the steady-state cycle shown in Fig. 9.4, the initial concentration of X_{STO} is 30 mg COD L^{-1}, which is lower than its end point value of 45 mg COD L^{-1} after one full cycle of 4 h. However, after sludge discharge, the storage polymers decrease by 15 mg COD L^{-1} (according to the discharged sludge content) and then have the same level as the initial value. X_{STO} never builds up to a large fraction of MLVSS, because it is rapidly consumed during the famine phase in a repeated feast-and-famine cycle. When feast and famine are repeated, as

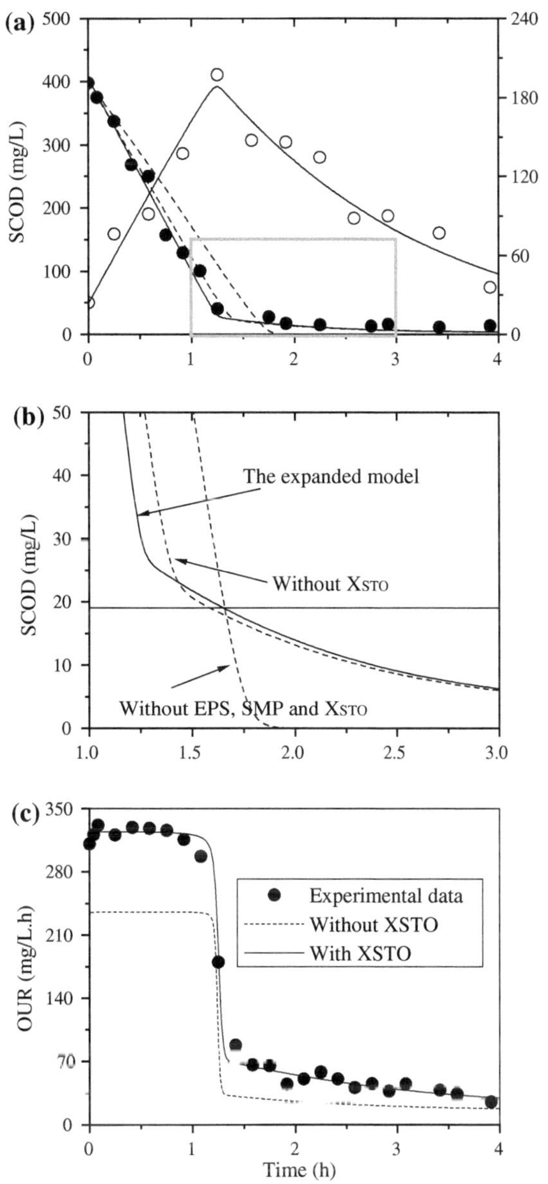

Fig. 9.4 Experimental (*symbol*) and model (*line*) values of **a** SCOD and X_{STO}, and comparing experimental SCOD results with model simulations without X_{STO} or EPS, SMP, and X_{STO}, **b** an expanded scale for famine phase, where SMP becomes an important part of effluent SCOD, and **c** OUR data during one steady-state cycle of the SBR operation at SRT of 20 days and comparing experimental OUR results with model simulations without X_{STO}. The MLVSS was 4,700 mg L^{-1} (Reprinted from Ni et al. (2009) with permission from American Chemical Society)

In an SBR, rapid uptake and utilization of X_{STO} is a strong tool that the bacteria can use for balancing the growth rate in the face of periodic famine conditions. The correspondence of experimental data and model results for OUR, soluble COD, and X_{STO} concentration in an SBR cycle further supports the validity of the expanded unified model.

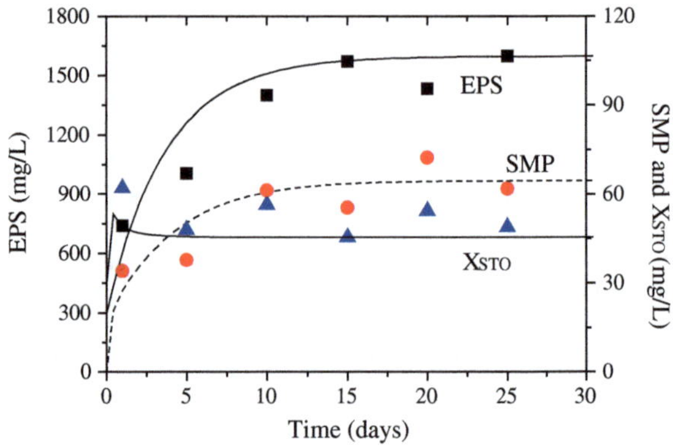

Fig. 9.5 Experimental (*symbol*) and model (*line*) values of EPS, SMP, and X_{STO} data for a CSTR operated at an SRT of 10 days. Initial conditions: $S_S = 800$ mg L^{-1}, $X_{STO} = 30$ mg L^{-1}, $X_{EPS} = 300$ mg L^{-1}, and MLVSS = 1,150 mg L^{-1} [Reprinted from Ni et al. (2009) with permission of American Chemical Society]

The important role of X_{STO} in the dynamic feast-and-famine conditions is demonstrated by eliminating it from the model. The SCOD and OUR data and model simulations with and without the X_{STO} component for the 4 h of one cycle of the SBR operation are presented in Fig. 9.4. In Fig. 9.4c, the model absent of the X_{STO} component cannot simulate the initial high OUR value, the sharp transition toward the second OUR level, or the long-term OUR. In all cases, the model predicts OUR values substantially too low when X_{STO} is removed. In Fig. 9.4b , the model that has no X_{STO}, EPS, and SMP components seriously under-simulates the effluent SCOD, showing the practical importance of including the microbial products.

To further explore the universality of this expanded unified model, we also apply the model to simulate the data from the conventional CSTR system. For the mass balance equations for the influent of the CSTR, each mass balance has an advective term, $QS^0/V - QS/V$; where V is the liquid volume, Q is the flow rate, and S^0 and S represent influent and reactor concentrations, respectively. Rate terms in each mass balance are explained in the equations shown in Table 9.1.

The simulation and experimental results for EPS, SMP, and X_{STO} of the CSTR system are compared in Fig. 9.5. The experimental trends for the three microbial products are described well by the expanded unified model. SMP and EPS take about a bit more than one SRT (10 d) to approach plateaus with the model and the experiments. X_{STO} stabilizes after about 5 days of operation, since it does not change much, and the sharp increase at about one day is present for experimental and model results. The good agreement between the model outputs and the experimental data for this flow-through system further supports that the expanded unified model has good universality in terms of representing the components and their formation/consumption, as long as the feed material is the same.

9.1.5 Conclusions

In summary, the expanded unified model successfully integrates production and consumption of internal storage products (X_{STO}) with unified model for EPS, SMP, and active and inert biomass in activated sludge. The good agreement between model simulations and experimental EPS, SMP, and X_{STO} data from four distinctly different experiments supports that the expanded unified model properly captures the relationships among the forms of microbial products. In particular, the model illustrates how X_{STO} cycles up and down rapidly during feast and famine periods, EPS and biomass components are relatively stable despite feast and famine, and SMP controls effluent SCOD.

9.2 Formation and Utilization of Microbial Products in Aerobic Granular Sludge

9.2.1 Introduction

Recently, aerobic granule has been extensively investigated (Peng et al. 1999; Beun et al. 2002; de Kreuk et al. 2005; Su and Yu 2005). Compared with the conventional sludge flocs, the aerobic granules have a more compact structure, better settleability, and greater biomass retention. Studies have shown that the aerobic granule could be applied for high-strength organic wastewater treatment, simultaneous removal of organics matter, nitrogen and phosphorus (de Kreuk et al. 2005), and toxic wastewater treatment (Jiang et al. 2002). SBR has been used for the granulation of activated sludge (Su and Yu 2005; de Kreuk and van Loosdrecht 2006). Aerobic-granule-based SBR has been proven to be applicable for the treatment of wastewaters from various industries, such as malting (Schwarzenbeck et al. 2004), dairy (Arrojo et al. 2004), and soybean-processing (Su and Yu 2005), as well as municipal wastewater (de Kreuk and van Loosdrecht 2006). These results demonstrate that the aerobic granule has a great potential in municipal and industrial wastewater treatment.

EPS are a major component of the matrix material in granules (McSwain et al. 2005; Sheng et al. 2006). In addition to EPS, all bacteria convert a fraction of the organic substrate into SMP, which account for the bulk of the soluble organic carbon in reactor effluents (de Silva and Rittmann 2000a, b; Benjamin et al. 2006). The incorporation of SMP formation has paved the way for more accurate modeling of biological wastewater treatment processes. On the other hand, the storage of the carbon sources as intracellular polymers, such as polyhydroxyalkanoates, lipids, and polysaccharides, is likely to play a significant role in the carbon turnover (van Loosdrecht et al. 1997; Majone et al. 1999; Pratt et al. 2004). Aerobic granule SBR reactors are typically operated under highly dynamic conditions, attributed to the great substrate concentration gradients versus time (van Loosdrecht et al. 1997). As a result, microorganisms in granules experience alternative feast and famine conditions, which promote the occurrence of carbon storage. Microorganisms, which are capable of quickly storing substrate and consume this stored substrate in a more balanced way, have a strong competitive advantage over the microorganisms without such a capacity. As an essential intermediate, storage polymers play an important role in the overall substrate removal.

The EPS, SMP, and X_{STO} are important sinks for electrons and carbon derived from the original substrate. The microorganisms in aerobic granules may promote the accumulation of X_{STO}, EPS, or both. For example, some microorganisms rapidly take up the external substrate and store it as X_{STO} under feast conditions. The stored carbon is used for growth and maintenance in the subsequent famine period (Pratt et al. 2004). The carbon and electron sinking is important for the aerobic-granule-based reactors (van Loosdrecht et al. 1997).

Mathematical models are essential for optimizing design and improving operation of the complex biological processes (Lee et al. 1999; Ajbar and Alhumazi 2000; Nopens et al. 2005). Therefore, this paper aims to elucidate the production and utilization of EPS, SMP, and X_{STO} with a mathematical model. Furthermore, the work describes the mechanisms for the substrate removal of granules that have diffusion resistance for soluble components (Beyenal et al. 1997). The sensitivity of these microbial products' concentrations toward the key model parameters is analyzed using independent experiments. The model is evaluated using one-cycle operation experimental results of a laboratory-scale aerobic granule-based SBR. Then, it is used to generate simulations about the effects of substrate and biomass concentrations on the dynamics of EPS, SMP, and X_{STO} in aerobic-granule-based reactors.

9.2.2 Materials and Methods

9.2.2.1 Reactor Setup and Operation

Aerobic granules were cultivated in a laboratory-scale SBR fed with a fatty-acids-rich wastewater at a COD of approximately 800 mg L^{-1}, as described in Chap. 3.

9.2.2.2 Parameter Determination

The substrate half-saturation coefficient (K_S) and endogenous-decay coefficient (b_H) were estimated as described by Henze et al. (1987). The increase in maximum growth rate ($\mu_{H,S}$) with the increasing S_S concentration in independent batch experiments was fitted to find the best-fit K_S value. To estimate b_H, granules were removed from the SBR and put into an aerated batch reactor without feed. The O_2 respiration rate was measured at a given time interval over a period of several days. A plot of the logarithm of the respiration rate versus time gave a straight line with a slope b_H. The storage process parameters ($Y_{H, STO}$, k_{STO} and $\mu_{H, STO}$) were estimated with the batch respirometry experiments. Prior to a respirometry test, the sludge was washed twice with distilled water to remove carry-over materials. Then, fatty-acids-rich wastewater was added as a pulse and the DO concentration was measured over time. The experiments were repeated with three different food-to-microorganism ratios for the parameters' estimation. The initial substrate concentrations were set at 200, 300, and 400 mg COD L^{-1}, resulting in food-to-microorganism ratios of 0.2, 0.3, and 0.4 mg COD mg^{-1} VSS, respectively. The calculated OUR and measured COD variation were used for parameter estimation with the methods of Avcioglu et al.(2003). The EPS and SMP formation parameters (k_{EPS}, $k_{UAP,}$ and k_{hyd}) were estimated from the batch experimental results using aerobic granular sludge with the measurements of EPS and SMP under known initial conditions.

9.2.2.3 Batch Experiments for Model Evaluation

Independent sets of batch experiments under known initial conditions were conducted to evaluate the model. The aerobic granules were sampled from the SBR when substrate was depleted in the reactor. The sludge was kept under famine conditions and no internal storage products (X_{STO}) was formed inside it. After washed twice with distilled water to remove the external soluble organic material (SCOD), the granules were transferred to beakers with a working volume of 1 L. After that, the external substrate at predetermined concentrations was dosed, and the OUR was monitored until the substrate became depleted and the endogenous activity was resumed. Much different initial conditions from cycle operation were designed for model evaluation. Samples were taken every 10–20 min for the analysis of SCOD, EPS, SMP, and X_{STO} (PHB).

9.2.2.4 Analytical Procedures

The analysis of DO, OUR, and sludge image was described in Chap. 3 Determinations of COD, MLSS, and MLVSS followed the Standard Methods (APHA 1995). The internal storage polymers, i.e., PHB, SMP, and EPS were measured using the same methods described in the previous section. Transmission electron microscopy (TEM) analysis of PHB in the granules was performed according to Jendrossek et al. (2007). The EPS distribution within the granules was examined using confocal laser scanning microscopy (CLSM) (LSM 5 Pascal, Zeiss, Jena, Germany). For florescent staining of both cells and EPS, two probes were applied collectively: SYTO9 (25 μm, Molecular Probe, Eugene, OR) to target all microbes, and ConA-TRITC lectin (250 mg L^{-1}, Sigma) to target the polysaccharides with D-glucose or D-mannose.

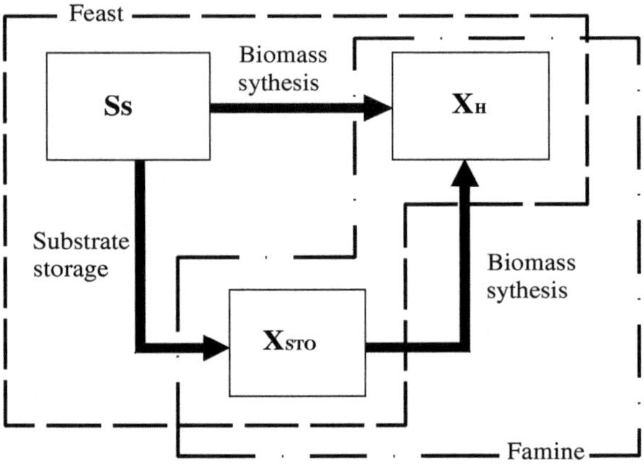

Fig. 9.6 Schematic diagram of the substrate utilization mechanisms for simultaneous growth and storage [From Ni and Yu (2010), reprinted with permission from Wiley-Blackwell]

9.2.3 Mathematical Modeling

Modeling of a granule-based SBR involves oxygen transfer, diffusion within granules, and biological reactions. In this work, aerobic granules with a wide size of 0.41–3.5 mm are modeled with a consideration of substrate diffusion. The first step is its diffusion into the granules. Then, biological reactions involving each given model components occur within the granules.

A unified theory for EPS and SMP has been developed by Laspidou and Rittmann (2002a, b). In the present work, based on the unified model, a new model is established through introduction of simultaneous storage and growth concepts. Our model describes the relationships among the four solid species: active heterotrophic bacteria, EPS, X_{STO}, and residual inert biomass; three soluble species: external substrate and two unique forms of SMP; and an electron acceptor, which is dissolved oxygen. The model employs the following symbols for concentrations: external substrate (S_S), active heterotrophic biomass (X_H), internal storage products (X_{STO}), residual inert biomass (X_I), utilization-associated products (S_{UAP}), biomass-associated products (S_{BAP}), SMP ($S_{SMP} = S_{UAP} + S_{BAP}$), EPS ($X_{EPS}$), and DO ($S_O$). The units for all species are oxygen demand or oxygen (for DO), which is directly proportional to electron equivalents (8 g O_2 per e^- equivalent).

The simultaneous storage and growth concepts are schematically shown in Fig. 9.6. The external substrate is primarily used by active biomass X_H for the generation of new biomass X_H. Excess amount of S_S available is converted into X_{STO}. After the consumption of the primary external substrate, the secondary growth process occurs on the stored X_{STO} in the famine phase.

Figure 9.7 illustrates the consistent approach for the fate of the substrate electrons, which directly enter a cell (i.e., active biomass primarily growth on S_S in Fig. 9.6). There are four possible ways. Part of the external substrate is used for biomass synthesis. Other substrate electrons are diverted to the formation of UAP and EPS. UAP are released to the aqueous solution, while EPS are released as a solid to form the aggregate matrix. The hydrolysis of EPS produces BAP, which

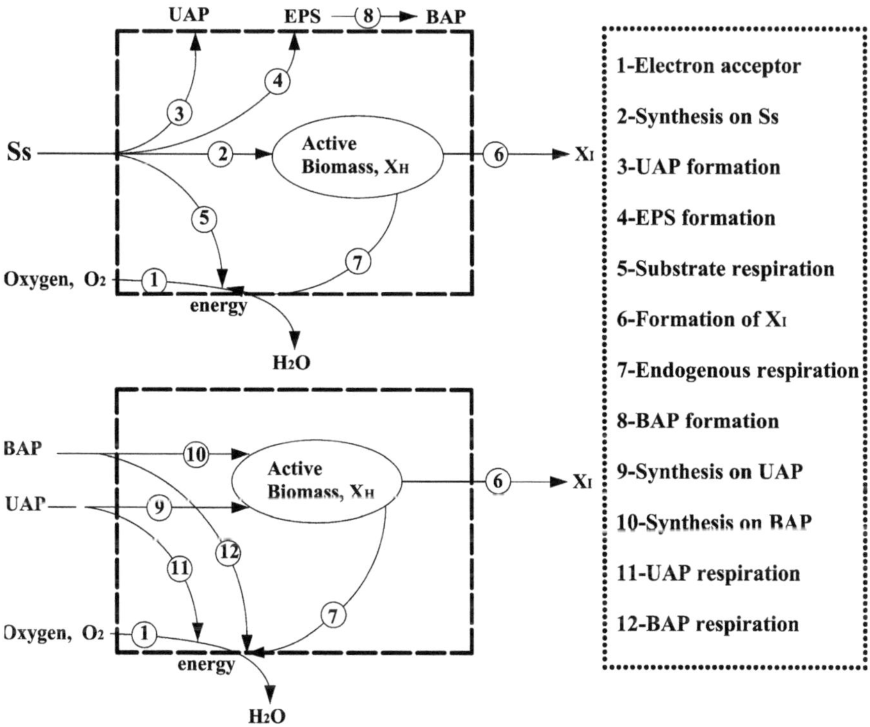

Fig. 9.7 Schematic diagram of electron flows regarding original substrate, active biomass, EPS, and SMP adapted from Laspidou and Rittmann (2002b) [From Ni and Yu (2010), reprinted with permission from Wiley-Blackwell]

are soluble. Substrate oxidation and respiration of the electrons to reduce O_2 and generate the energy needed to fuel for formation of active biomass, EPS, and UAP.

In our approach the unified model is extended through the introduction of X_{STO} formation and utilization. In a feast period, some of the external substrate is converted into X_{STO}. Because the formation of X_{STO} competes with the usual biomass synthesis, the formation of X_{STO} occurs only when the substrate concentration is high. The active biomass decays in two ways. First, the active biomass is oxidized through the endogenous respiration to yield energy for maintenance. Second, the decay also produces residual inert biomass, which is not biodegradable. Since both UAP and BAP are biodegradable, some of their electrons can also be used by the microorganisms as the ''recycled'' substrate, while the remaining electrons also go to the electron acceptor for energy generation. X_{STO} is also biodegradable and can be utilized by the active biomass under the famine conditions. Some of the electrons in X_{STO} are used for the biomass synthesis, while the other electrons are respired for energy generation. By convention (de Silva and Rittmann 2000a, b; Laspidou and Rittmann 2002a, b), the utilization of UAP, BAP, and X_{STO} does not result in the formation of new UAP, EPS, or X_{STO}.

Related process kinetics and stoichiometry describing the interactions and transformations among model components are expressed in a way to be compatible with previous mathematical models, which have been proposed to formulate biochemical reactions of microbial populations fed with different types of substrate (Henze et al. 1987; Gujer et al. 1999). The structure of our model is presented in a matrix format reflecting the basic stoichiometric relationships constituting the backbone of the model (Henze et al. 1987; Gujer et al. 1999). The matrix format is outlined in Tables 9.3 and 9.4, where model components are listed in the upper row; the rightmost column in Table 9.4 gives the process rate expressions; the relevant stoichiometric coefficients are incorporated in appropriate matrix cells. In this way, the generation or utilization rate in a model component for a given biochemical process is obtained by multiplication of related process stoichiometry and kinetics. Table 9.5 defines all the parameters, their units, and the values used in modeling and simulation results.

Modeling and simulation are performed using a software package AQUASIM (Reichert 1998). In order to simulate an aerobic-granule-based bioreaction process, the method of de Kreuk et al. (2007) is adopted in this work. The model is integrated into the biofilm compartment of AQUASIM. The reactor in the model is described with a biofilm compartment connected to a mixed compartment with an advective link, and a high recirculation flow rate is incorporated from the biofilm compartment to the completely mixed one. The high recirculation rate is chosen to ensure the same concentrations in the liquor of biofilm compartment and the mixed compartment (Beun et al. 2001). The biofilm compartment with a volume of 1 L contains the biomass granules and bulk liquid volume. The mixed compartment of 1 L includes the remaining liquor volume (Beun et al. 2001).

In modeling, the aerobic granules are assumed to be spherical in shape and their size distribution to be constant in one cycle of operation. A biofilm model implemented in AQUASIM is used to model the aerobic granules in this work (Wanner and Reichert 1996). One-dimensional conservation laws are formulated as a balance between the mass conserved and utilized in this granule model.

$$\frac{\partial \hat{\rho}}{\partial t} + \frac{\partial \hat{j}}{\partial z} = \hat{r} \tag{9.12}$$

where $\hat{\rho}$ is the one-dimensional density (amount of conserved quantity per unit compartment length), $\hat{j}$ is the one-dimensional flux (amount of the conserved quantity transported per unit time), $\hat{r}$ is the one-dimensional source term (amount produced per unit compartment length and per unit time), t is time, and z is the space coordinate.

The granules cultivated in this study had a size distribution between 0.41 mm and 3.5 mm with an average of 1.2 mm. It is true that granules with a wide range in size were formed in the reactor and that their size distribution also changed in time. However, the granule size distribution was not taken into account in this

Table 9.3 Stoichiometric matrix for the established model

Component	S_O	S_S	S_{UAP}	S_{BAP}	X_I	X_{EPS}	X_H	X_{STO}
Process	O_2	COD	COD	COD	COD	COD	COD	COD
Growth on S_S	$-\frac{[1-k_{EPS}-k_{UAP}-k_{STO}-Y_{H,S}(1-k_{EPS}-k_{UAP}-k_{STO})]}{Y_{H,S}}$	$-\frac{1}{Y_{H,S}}$	$\frac{k_{UAP}}{Y_{H,S}}$			$\frac{k_{EPS}}{Y_{H,S}}$	$1-k_{UAP}-k_{EPS}-k_{STO}$	$\frac{k_{STO}}{Y_{H,S}}$
Growth on X_{STO}	$-\frac{1-Y_{H,STO}}{Y_{H,STO}}$						1	$-\frac{1}{Y_{H,STO}}$
Growth on S_{UAP}	$-\frac{1-Y_{H,UAP}}{Y_{H,UAP}}$		$-\frac{1}{Y_{H,UAP}}$				1	
Growth on S_{BAP}	$-\frac{1-Y_{H,BAP}}{Y_{H,BAP}}$			$-\frac{1}{Y_{H,BAP}}$			1	
Release of X_{EPS}				1		-1		
Endogenous respiration	$-(1-f_I)$				f_I		-1	

Table 9.4 Kinetics rate expressions for the reaction processes

Process	Kinetics rate expressions
Growth on S_S	$\mu_{H,S}\frac{S_S}{K_S+S_S}\frac{S_O}{K_O+S_O}X_H$
Growth on X_{STO}	$\mu_{H,STO}\frac{K_S}{K_S+S_S}\frac{S_O}{K_O+S_O}\frac{X_{STO}/X_H}{K_{STO}+X_{STO}/X_H}X_H$
Growth on S_{UAP}	$\mu_{H,UAP}\frac{S_{UAP}}{K_{UAP}+S_{UAP}}\frac{S_O}{K_O+S_O}X_H$
Growth on S_{BAP}	$\mu_{H,BAP}\frac{S_{BAP}}{K_{BAP}+S_{BAP}}\frac{S_O}{K_O+S_O}X_H$
Release of X_{EPS}	$k_{hyd}X_{EPS}$
Endogenous respiration	$b_H\frac{S_O}{K_O+S_O}X_H$

Table 9.5 Kinetic and stoichiometric coefficients (20 °C) for the model

Parameter	Definition	Values	Unit	Source
Stoichiometry				
$Y_{H,S}$	Yield for growth on S_S	0.39	g COD_X g^{-1} COD_S	(1)
$Y_{H,STO}$	Yield c for growth on X_{STO}	0.59	g COD_X g^{-1} COD_{STO}	(1)
f_I	Fraction of X_I in decay	0.20	g COD_X g^{-1} COD_X	(2)
$Y_{H,UAP}$	Yield for growth on UAP	0.45	g COD_X g^{-1} COD_{UAP}	(3)
$Y_{H,BAP}$	Yield for growth on BAP	0.45	g COD_X g^{-1} COD_{BAP}	(3)
Kinetics				
k_{STO}	X_{STO} formation coefficient	0.26	g COD_{STO} g^{-1} COD_S	(1)
$\mu_{H,STO}$	Growth rate on X_{STO}	0.51	h^{-1}	(1)
$\mu_{H,S}$	Maximum growth rate on S_S	0.40	h^{-1}	(3)
K_S	Substrate affinity constant	39.8	g COD_S m^{-3}	(4)
K_{STO}	Affinity constant for X_{STO}	1.0	g COD_{STO} g^{-1} COD_X	(2)
K_O	DO affinity constant	0.2	g COD_O m^{-3}	(2)
b_H	Decay rate coefficient of X_H	0.035	h^{-1}	(4)
k_{EPS}	EPS formation coefficient	0.23	g COD_{EPS} g^{-1} COD_S	(1)
k_{hyd}	EPS hydrolysis rate	0.0071	h^{-1}	(3)
k_{UAP}	UAP formation coefficient	0.06	g COD_{UAP} g^{-1} COD_S	(1)
$\mu_{H,UAP}$	Growth rate on UAP	0.053	h^{-1}	(3)
K_{UAP}	Affinity constant for UAP	100	g COD_{UAP} m^{-3}	(3)
$\mu_{H,BAP}$	Growth rate on BAP	0.0029	h^{-1}	(3)
K_{BAP}	Affinity constant for BAP	85	g COD_{BAP} m^{-3}	(3)

Notes (1) This work, estimated, (2) Gujer et al. (1999), (3) Laspidou and Rittmann (2002b), and (4) this work, measured

model, as it would significantly increase the complexity of numerical computation and it is not expected to contribute to a better understanding of the system. Therefore, in the simulation the diameter was chosen to be 1.2 mm, which was the most representative for the aerobic granules in this SBR. Since the experimental data measured in one cycle were different in time, depending on granule size, morphology, and reactor operation, a representative cycle measurement was selected to compare to a standard simulation case (de Kreuk et al. 2007). This simulation was performed under the same operational conditions as applied in the

Table 9.6 Parameters for the aerobic granules in model simulation

Parameter	Description	Value	Source
Granules			
R_{min}	Minimum radius, mm	0.41	Measured
R_{max}	Maximum radius, mm	3.50	Measured
R_{mean}	Mean radius, mm	1.20	Measured
Mass transport			
D^{Ss}	Effective diffusivity of Ss, $dm^2\ h^{-1}$	0.000576	Beun et al. (2001)
D^{UAP}	Effective diffusivity of UAP, $dm^2\ h^{-1}$	0.000576	Beun et al. (2001)
D^{BAP}	Effective diffusivity of BAP, $dm^2\ h^{-1}$	0.000576	Beun et al. (2001)
D_{O2}	Effective diffusivity of S_O, $dm^2\ h^{-1}$	0.0023	Horn et al. (2001)
Density of solid phase			
ρ_{XH}	Density of X_H, g COD m^{-3}	35,000	de Kreuk et al. (2007)
ρ_{XI}	Density of X_I, g COD m^{-3}	35,000	de Kreuk et al. (2007)
ρ_{XSTO}	Density of X_{STO}, g COD m^{-3}	350,000	de Kreuk et al. (2007)
ρ_{XEPS}	Density of X_{EPS}, g COD m^{-3}	4,000	Horn et al. (2001)

standard operation of the SBR. Parameters for the aerobic granules are shown in Table 9.6.

9.2.4 Results and Discussion

9.2.4.1 Sensitivity to Key Model Parameters

Prior to the model calibration, sensitivity analysis should be conducted to evaluate the key parameters, which should be strictly kept under control in the experimental calibration and model validation. In the sensitivity analysis, the behavior of the model is evaluated as a consequence of a variation of the input parameters. The sensitivity coefficient represents the change in the output variable, resulting from a change in the input variable. Since the main purpose of this work is to simulate the dynamics of microbial products in the granular sludge, the output values of EPS, SMP, and X_{STO} productions are analyzed. The sensitivity analysis results of key parameters are shown in Figs. 9.8.

According to the simulation results in Fig. 9.8, variation of k_{EPS} (0.03–0.43 g COD_{EPS} g^{-1} COD_S) affects the EPS outcome significantly, while the change of k_{STO} value from 0.03 to 0.43 g COD_{STO} g^{-1} COD_S does affect the X_{STO} outcome substantially. The parameter k_{EPS} responsible for determining the EPS formation affects the outcome of SMP slightly, presumably attributed to the comparatively small amount of BAP, which are directly hydrolyzed from EPS.

The k_{STO} does not considerably influence the dynamics of SMP, because the utilizations of X_{STO} does not result in the formation of new UAP. As shown in Fig. 9.8, after an increase in the k_{UAP} value from 0.02 to 0.10 g COD_{UAP} g^{-1} COD_S, the model predicts a much higher fraction of SMP, suggesting that k_{UAP}

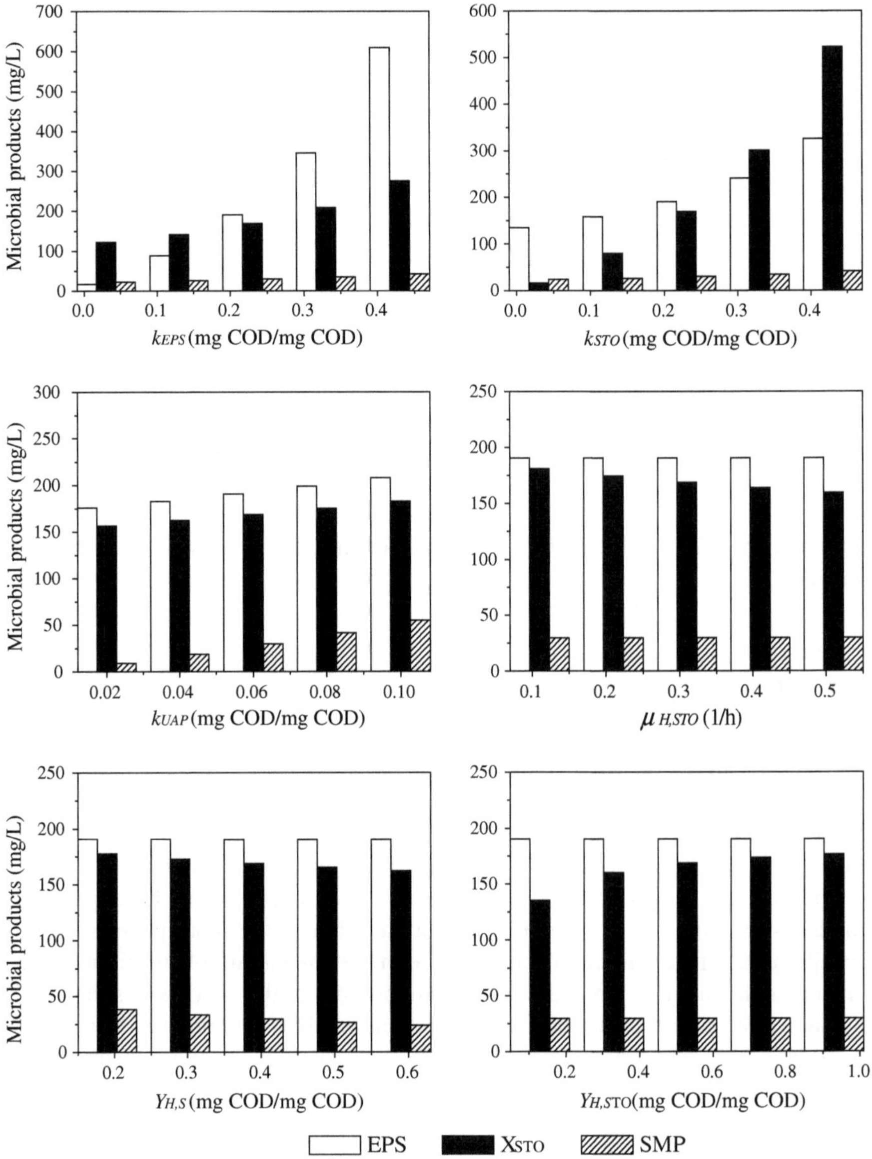

Fig. 9.8 Sensitivity analysis for the key model parameters: variations of EPS, SMP, and X_{STO} productions with varied k_{EPS}, k_{STO}, k_{UAP}, $\mu_{H,\ STO}$, $Y_{H,\ S}$, and $Y_{H,\ STO}$ values [From Ni and Yu (2010), reprinted with permission from Wiley-Blackwell]

should be a key parameter in the SMP formation. It also reveals that the UAP are the main SMP in the substrate utilization. Compared to k_{EPS} and k_{STO}, the kinetic parameters $\mu_{H,\ STO}$ and $Y_{H,\ S}$ affect X_{STO} slightly. With an increase in the $Y_{H,\ STO}$

value from 0.13 to 0.93 h^{-1}, the model predicts an increase in X_{STO}. The sensitivity analysis demonstrates that these key parameters have a great influence on the microbial product formation in the aerobic granular sludge.

9.2.4.2 Model Calibration: Parameter Estimation with Batch Test Results

The parameter values are estimated by minimizing the sum of squares of the deviations between the measured data and the model predictions with the objective function given as below:

$$F(p) = \left(\sum_{i=1}^{n} \left(y_{\exp,i} - y(p)_i \right)^2 \right)^{1/2} \tag{9.13}$$

where $y_{\exp}$ and $y(p)$ are vectors of n measured values and model predictions at times t_i (i from 1 to n), and p is the vector of the model parameters.

To initiate the calibration procedure, an initial guess of the parameters is necessary. Such initial values are obtained with the experimental results and data in the literature. As shown in Table 9.5, parameters b_H and K_S were determined experimentally and independently, and were later used in the model. The initial concentration of active biomass was estimated using the baseline endogenous OUR level prior to substrate addition:

$$\mathrm{OUR}_{\mathrm{end}}(0) = (1 - f_I) \cdot b_H \cdot X_H(0) \tag{9.14}$$

In this equation, f_I is fixed at 0.2 mg COD mg^{-1}COD, as used in ASM3. Hence, for a given f_I and measured b_H, $X_H(0)$ can be calculated from the $\mathrm{OUR}_{\mathrm{end}}(0)$ data. The model was first calibrated for the COD and OUR concentrations in the batch respirometric experiments. $Y_{H,\ STO}$, k_{STO}, and $\mu_{H,\ STO}$ were changed based on the causality of the parameters on the COD and OUR concentrations. The parameter values estimated are summarized in Table 9.5, while the simulation results are shown in Fig. 9.9. The average absolute errors obtained are 9.5 mg L^{-1} and 4.9 mg $L^{-1}h^{-1}$ for COD and OUR concentrations, respectively. There is a good agreement between the model predictions and experimental data. Thereafter, the established model is also calibrated for the measured EPS and SMP concentrations in the batch experiments to determine an appropriate kinetics for simulating the dynamics of EPS and SMP in the aerobic granules. k_{EPS}, k_{UAP}, and $Y_{H,\ S}$ are changed based on the causality of the parameters on the EPS and SMP concentrations. The parameter values estimated are also listed in Table 9.5 with the simulation results shown in Fig. 9.10.

The calibrated heterotrophic maximum specific growth rate through the utilization of storage polymers ($\mu_{H,\ STO}$) is 0.51 h^{-1}, which is slightly higher than the maximum specific growth rate for the external substrate ($\mu_{H,\ S}$) of 0.4 h^{-1}. Thus, growth through the utilization of X_{STO} is commensurate with growth on the external substrate. The X_{STO} formation coefficient k_{STO} is estimated to be 0.26 g

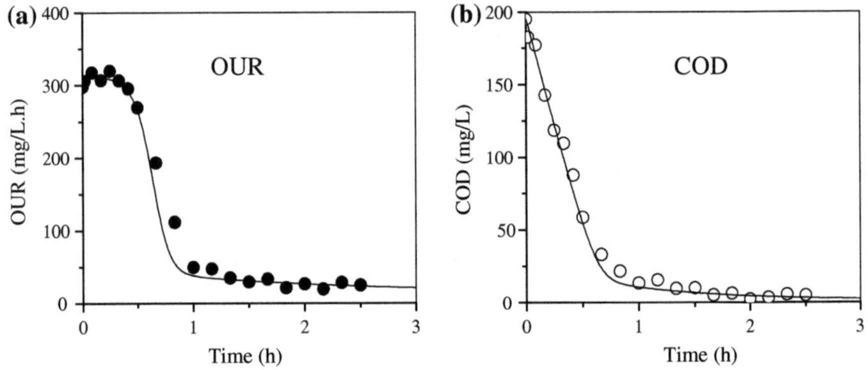

Fig. 9.9 Model calibration for the batch respirometric experimental data: **a** OUR profiles and **b** COD measurements (From Ni and Yu (2010), reprinted with permission from Wiley-Blackwell)

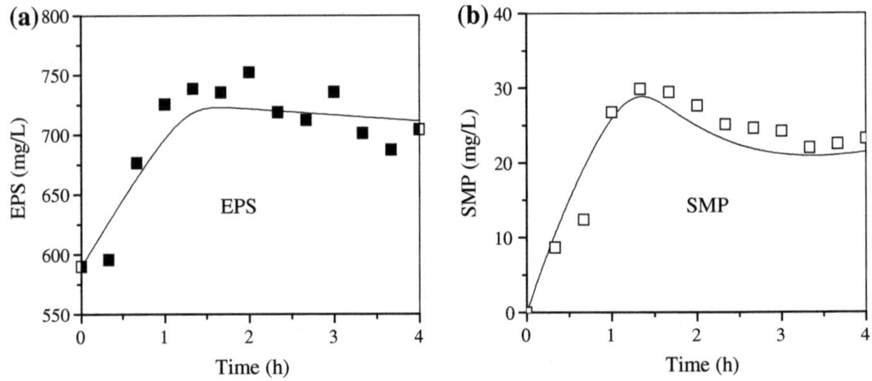

Fig. 9.10 Model calibration for the batch respirometric experimental data: **a** EPS data and **b** SMP concentrations (From Ni and Yu (2010), reprinted with permission from Wiley-Blackwell)

COD_{STO} g^{-1} COD_S, while the EPS formation coefficient k_{EPS} and the UAP formation coefficient k_{UAP} are determined to be 0.23 g COD_{EPS} g^{-1} COD_S and 0.06 g COD_{UAP} g^{-1} COD_S, respectively. Therefore, electrons from the external substrate are distributed in this order when X_{STO} is produced: new biomass synthesis 39 %, X_{STO} 26 %, EPS 23 %, O_2 for respiration 6 %, and UAP 6 %. More substrate electrons are diverted to the storage polymers, but less is distributed to produce EPS and UAP when sufficient substrate is available. Considering the much higher utilization rate for X_{STO} than that for UAP and BAP, the relatively higher distribution of electrons to X_{STO} quantifies why X_{STO} is the primary route by which the heterotrophs store the excess electrons and energy and later utilize them.

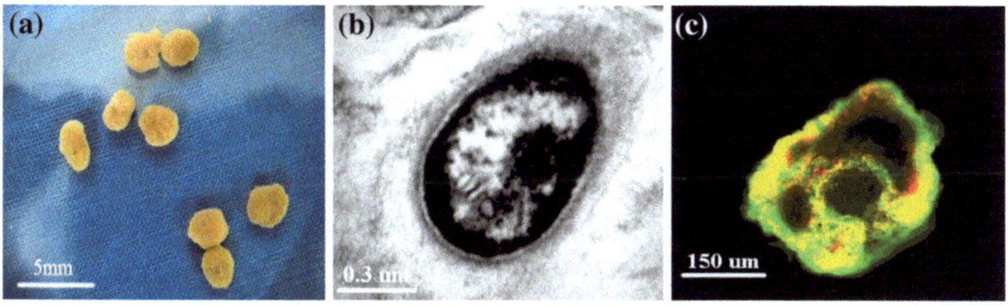

Fig. 9.11 **a** The image of the cultivated aerobic granules, **b** TEM analysis of the storage polymer granule produced within aerobic granule, and **c** CLSM images of the 50-μm cryosections through the center of the aerobic granule. Cells are stained with SYTO9 (*green*) and polysaccharides are stained with ConA (*red*) [From Ni and Yu (2010), reprinted with permission from Wiley-Blackwell]

The yield coefficient for the growth on X_{STO} is determined to be 0.59 g COD_X g^{-1} COD_{STO}, which is much lower than the value suggested by Gujer et al. (1999), partially attributed to the heterotrophic growth on the storage polymers rather than directly on the soluble substrates. A different model structure might be another reason for such a difference. The yield value for the utilization of the external substrate for microbial synthesis is relatively lower (0.39 g COD_X g^{-1} COD_S), compared to the values reported in the literature (Henze et al. 1987; Gujer et al. 1999; Pratt et al. 2004; Sin et al. 2006; Karahan et al. 2006), because a significant amount of substrate electrons is diverted to the microbial storage products, rather than the active biomass. Furthermore, the EPS formation coefficient k_{EPS} is higher than the value suggested by Laspidou and Rittman (2002b), who estimated the parameters from a pure culture with chemostat experiment. Granular biomass is expected to have a significantly different physiology from this pure culture. The enhanced production of EPS observed for the aerobic granules is induced by some so-called stressful culture conditions (Nichols et al. 2004; Qin et al. 2004).

9.2.4.3 Model Evaluation with the Experimental Data

After a one-month operation, aerobic granules became mature. They were yellow in color and spherical or ellipsoid in shape as shown in Fig. 9.11a. The storage polymer production by the aerobic granules and the localization of the newly synthesized storage polymer were observed with TEM (Fig. 9.11b). The TEM images reveal that the storage polymers with an electron-translucent globular structure are non-randomly distributed in the cell lumen, and that they are frequently found on or close to the cytoplasmic membrane. These internally accumulated storage polymers can then be utilized by microorganisms after the depletion of the added substrate.

The CLSM examination after fluorescent staining indicates the EPS distributions in the aerobic granules. The EPS are nearly uniformly distributed throughout

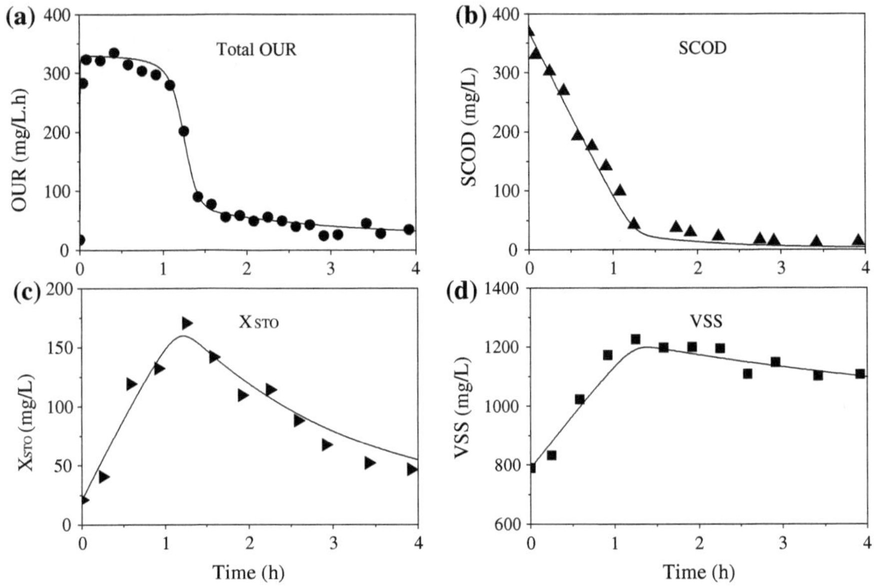

Fig. 9.12 Model evaluation of OUR profiles, X_{STO} data, soluble COD data, and VSS profiles during one-cycle operation of the aerobic granule-based SBR (*Line* for model simulation and *dot* for measured data) [From Ni and Yu (2010), reprinted with permission from Wiley-Blackwell]

the whole granules (Fig. 9.11c). The formation of aerobic granules is mainly attributable to the growth of cell together with the conglutination of EPS under hydrodynamic conditions. The substrate electrons are shuttled toward EPS production by the active microorganisms when they consume substrate electrons for their growth.

The model is solved for the set of parameters shown in Tables 9.5 and 9.6 in order to evaluate how well the model can describe the experimental results of the aerobic granule-based SBR. First, the independent data set for one-cycle operation of the SBR are simulated. The simulation results are shown in Fig. 9.12, together with the measured OUR, X_{STO}, soluble COD (SCOD), and VSS concentrations in the laboratory-scale SBR during a cycle with the same operational conditions. The simulated OUR, X_{STO}, SCOD, and VSS profiles do not differ with values found in the experiments by more than 20 %. Given all the uncertain parameters (e.g., exact granule size distribution and surface area), it is concluded that the model describes the experimental data well.

Second, Fig. 9.13 shows the simulation and experimental results of OUR, SCOD, EPS, SMP, and X_{STO} in a batch experiment, which are not used for model calibration. With the experimental measurements of EPS, SMP, and X_{STO} in this batch experiment, the model's ability to describe those microbial products can be tested directly. Furthermore, the initial conditions for this experiment are substantially different from those in Fig. 9.12. The dosed external substrate (690 mg L^{-1}) is much higher than that of the cycling test (380 mg L^{-1}). The

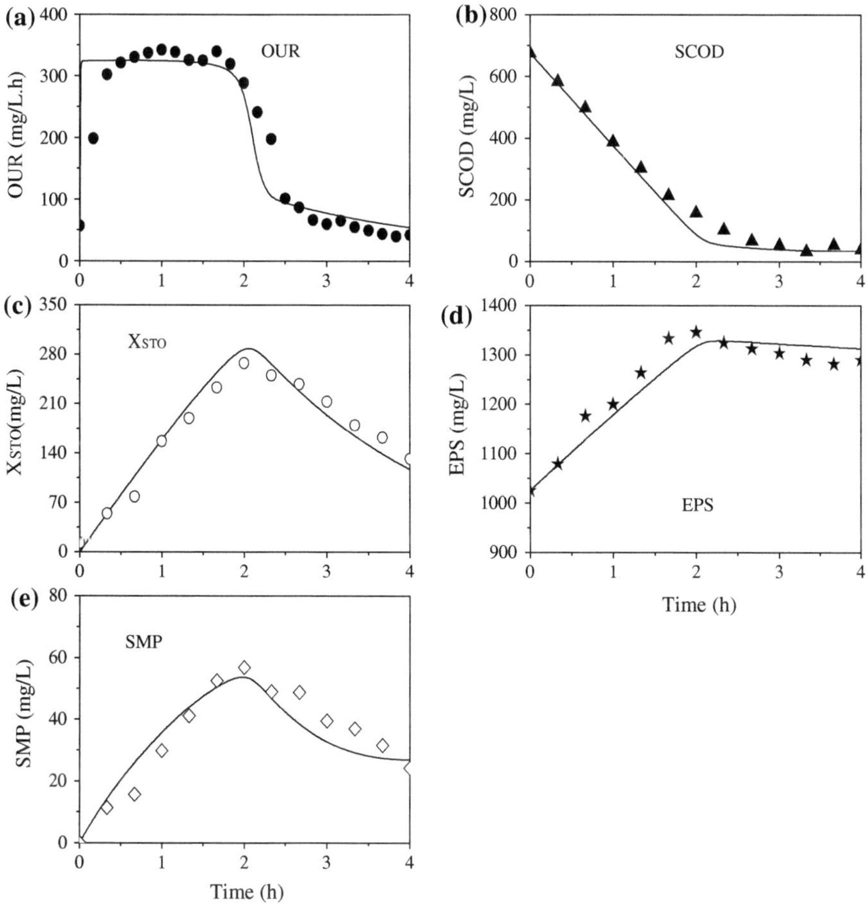

Fig. 9.13 Model evaluations of OUR profiles, X_{STO} data, soluble COD concentrations, EPS variations, and SMP dynamics in batch experiments (*Line* for model simulation and *dot* for measured dada) [From Ni and Yu (2010), reprinted with permission from Wiley-Blackwell]

experimental trends, e.g., the sharp decline of OUR, the peak of X_{STO}, and the leveling off of EPS after the depletion of the original substrate, and their absolute concentrations are described well by our model. The agreement between the model outputs and the measured data from different experiments supports that this model is able to properly describe the formation and utilization of the different types of microbial products in the aerobic granular sludge.

A tailing-off of the DO-consumption curve, which is observed in one cycle, can be appropriately described by the established model. Figures 9.12 and 9.13a show the simulation results for the predicted OUR profiles. The OUR initially increases, attributed to the rapid oxidation of the external substrate for growth and microbial product formation. The sharp bending point in the OUR corresponds with the complete removal of the dosed external substrate. The change in OUR is a direct indication of the transition from the feast phase to the famine phase, which occurs when the external substrate becomes depleted. The established model is able to predict both initial OUR response and smooth transition to the famine phase. However, the OUR does not immediately return to the endogenous respiration level because of the oxidation of X_{STO} and SMP, which are formed when the external substrate is being utilized.

The model is also verified with the SCOD and storage polymer data (Figs. 9.12 and 9.13b). Attributed to the initial rapid storage and consumption of substrate by the heterotrophs, the SCOD is reduced sharply at the initial stage. However, a complete depletion of the soluble COD does not occur because of the SMP formation (Fig. 9.13e). Furthermore, the measured biomass (VSS) variations match with the model predictions. The VSS initially increases from 800 to 1,200 mg L^{-1}, and later gradually decreases to 1,100 mg L^{-1} (Fig. 9.12d). In the period of rapid growth, both X_{STO} and EPS (Fig. 9.13f) increase rapidly and occupy a significant fraction of the total VSS. However, once S_S is consumed and the rapid growth ceases, only X_{STO} is rapidly utilized to provide for new synthesis with a relatively high rate under famine conditions. Thus, in the period of net biomass decay, the EPS and X_H become the main fractions of total VSS, while the X_{STO} fraction declines (Fig. 9.13). There is a good agreement between the model outputs and experimental data of the aerobic granule SBR, demonstrating the validity of the model established in this paper.

The concepts behind the developed model and the modeling approach shown here quantify the interconnections of EPS, SMP, and X_{STO} and offer a new integrated framework to describe these concepts with mathematical expressions and complete mass balances. Ahn et al. (2005) established a model by incorporating the unified EPS and SMP concept into the membrane bioreactor process based on ASM1 in order to predict their fates under various SRT conditions. With their model, EPS and SMP production could be simulated well and the results could be applied for the membrane bio-fouling control. In the unified theory of Laspidou and Rittmann (2002a; b), the hydrolysis of EPS was assumed to be the sole source of BAP. However, there are some disagreements in the literature regarding the unified theory. In addition to the soluble EPS, there are other BAP that might be released into the bulk solution as a result of lysis (i.e., endogenous organics and cell debris) and cell maintenance (i.e., turnover of intracellular components). Aquino and Stuckey (2004) proposed that both soluble EPS and cell lysis products are the sources of BAP. In addition, soluble intracellular components may end up in the bulk solution because of the renewal and turnover of the internal structures or as a result of survival strategies adopted by microorganisms. Ramesh et al. compared the physicochemical characteristics of the SMP and soluble EPS from different sludge samples and concluded that SMP are not identical to the soluble EPS. Because of the complexity of the microbial processes and the difficulty in modeling, information about the BAP formation is still sparse. There are also controversial reports about the BAP mechanisms in previous studies. Since no experimental method is available to accurately quantify the real source of BAP, the model of Laspidou and Rittmann is adopted in our work. Furthermore, compared to the approach of Beun et al.(2001), in our work, the biomass growth and X_{STO} formation kinetics are integrated into one overall reaction, in which biomass and storage product are produced. In this way, the formation of X_{STO} is coupled with the energy generation and respiration, and the X_{STO} formation rate is proportional to the substrate utilization rate. For aerobic granule systems, the model incorporating EPS, SMP, and X_{STO} formation and degradation may offer a rational approach to describe the bioreactor process. The new model provided here would be useful for the design, development, and application of the aerobic granular sludge process.

9.2.4.4 Effect of Substrate Concentration on EPS, SMP, and X_{STO}

Figure 9.14 shows the model predicted profiles of S_S, soluble COD, X_H, MLVSS, X_I, EPS, X_{STO}, SMP, UAP, and BAP in an aerobic-granule-based reactor. A higher substrate concentration results in a greater concentration of EPS, X_{STO}, and X_H.

The EPS and X_{STO} dependence on the substrate concentration is attributed to the fact that S_S governs the substrate utilization rate ($r_S X_H = \frac{\mu_{H,S}}{Y_{H,S}} \frac{S}{K_S+S} \frac{S_O}{K_{O_2}+S_O} X_H$), and accordingly the EPS and X_{STO} formation rate. UAP increases with the increasing substrate concentration, attributed to the same mechanisms for the EPS formation. Both BAP and X_I increase with an increase in substrate concentration, because the "parent" compounds, i.e., EPS and active biomass, increase with the increasing S_S. Thus, a higher initial S_S generates a greater level of total SMP (UAP + BAP). With the increasing X_H, X_I, EPS, and X_{STO}, the VSS also increases with the increasing substrate concentration (Fig. 9.14). These results clearly demonstrate the role of influent substrate concentration in the formation of EPS, SMP, and X_{STO} in the aerobic granular sludge. At a higher influent substrate concentration all reactions in a bioreactor will last longer. This has a significant influence on the SBR performance. In the operating of an SBR, the UAP rapidly decreases in a long cycle, with a low production rate ($k_{UAP} r_S X_H$), but its degradation rate does not decline significantly. Thus, the UAP are a major component in the effluent only when the cycle time is relatively short. On the contrary, the BAP are the most important component in the effluent when the cycle time is long. In both cases, a high influent S_S would increase the effluent soluble COD concentration in each cycle. For a long cycling time, the EPS become a large fraction of total VSS, in addition to X_H. For a short cycle period, the X_{STO} occupy a great fraction of total VSS. Moreover, X_I builds up continually. As a result, the increasing content of EPS, X_{STO}, and X_I would decrease the active biomass fraction of the total VSS. Hence, it is important to optimize a bioreactor operated at a high influent substrate concentration, in order to avoid a high effluent soluble COD (UAP and BAP) level as well as an accumulation of a high amount of non-active biomass (EPS, X_{STO}, and X_I), and also to reduce the associated high operating costs.

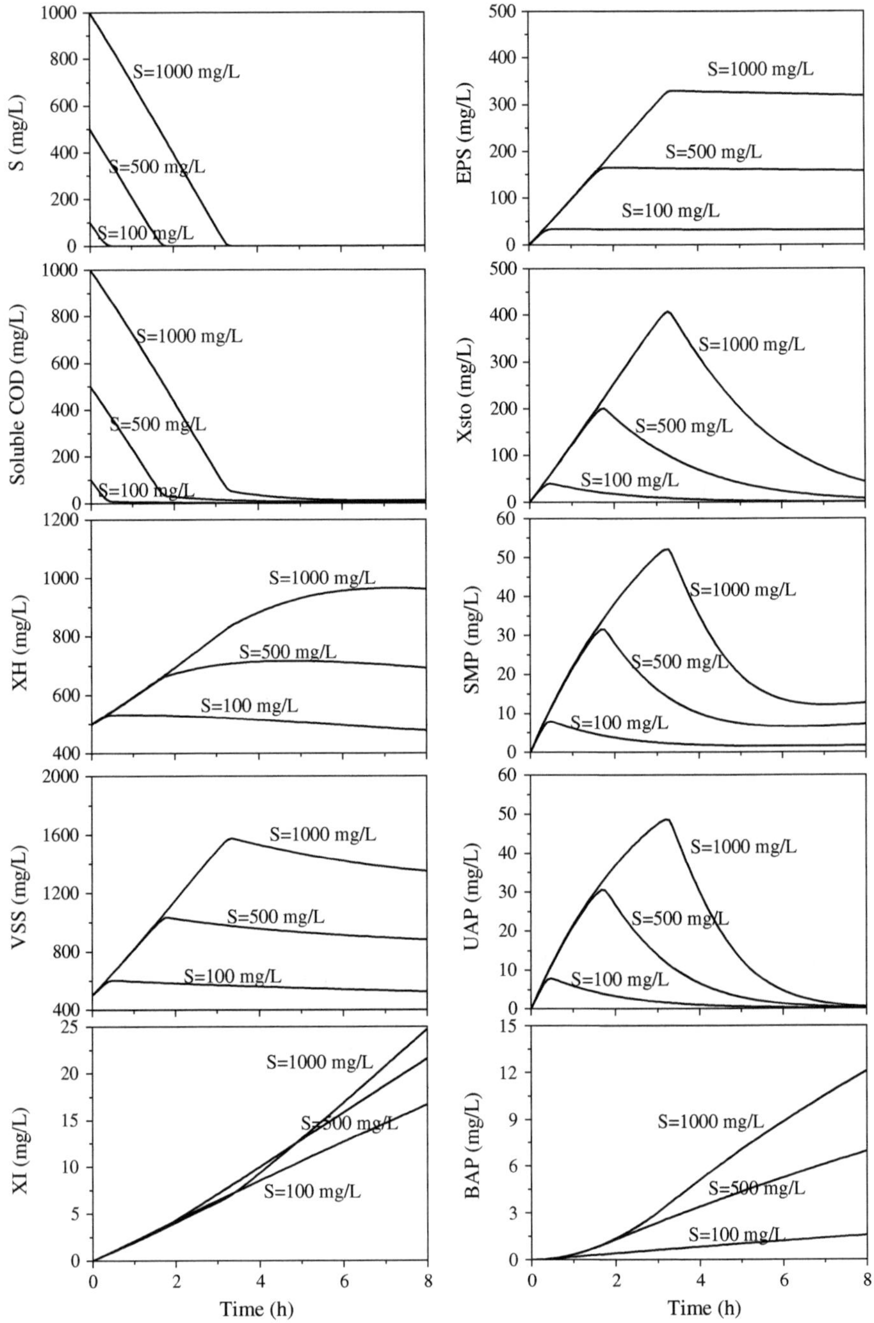

Fig. 9.14 Effect of substrate concentration on S_S consumption, soluble COD, active biomass, VSS, X_I, EPS, X_{STO}, SMP, UAP, and BAP concentration [From Ni and Yu (2010), reprinted with permission from Wiley-Blackwell]

9.2.4.5 Effect of Biomass Concentration on EPS, SMP, and X_{STO}

The formation of EPS, SMP, and X_{STO} has been shown to be closely linked to the quantity of active biomass present in a bioreactor. The EPS, SMP, and X_{STO} are produced at a rate proportional to the active biomass concentration, for instance, the EPS formation $=k_{EPS}r_{S}X_{H}$. Hence, an accumulation of active biomass in the bioreactor leads to an increased production rate of EPS, SMP, and X_{STO}. This is confirmed by the simulation results in this work. Figure 9.15 shows the model predictions for S_S, soluble COD, X_H, MLVSS, X_I, EPS, X_{STO}, SMP, UAP, and BAP profiles at three different active biomass concentrations. The EPS, UAP, BAP, and X_{STO} concentrations reach their peak values more rapidly at a higher biomass concentration. The production rates of EPS, SMP, and X_{STO} by the granular sludge increase with an increase in biomass concentration. The total X_I and BAP content increases with an increase in biomass concentration. The total content of EPS, UAP, and X_{STO}, however, does not show a direct correlation with the active biomass concentration. A lower active biomass concentration is not responsible for the lower concentration of EPS, UAP, and X_{STO}. In addition to the biomass concentration, as mentioned above, the substrate concentration is an important factor governing the production of these organic materials by microorganisms.

For an application point of view, a higher level of active biomass would enhance the effluent quality at a given HRT. The external substrate removal rate increases with the increasing active biomass content. A higher active biomass concentration would not result in a higher UAP content in the reactor (Fig. 9.15). Therefore, the effluent soluble COD concentration would decrease. Furthermore, a higher active biomass concentration would not result in a higher content of EPS and X_{STO}, but would increase the active biomass fraction of the total VSS.

9.2.5 Conclusions

The model established in this work describes the formation and utilization of microbial products, such as EPS, SMP, and X_{STO}, in an aerobic-granule-based bioreactor. The model simulates the experimental data from this complex system sufficiently well, as the simulated trends in OUR, EPS, SMP, X_{STO}, soluble COD, and VSS are similar to those measured in the experiments. Simulation results underline the importance of the initial substrate and biomass concentrations for the overall formation and consumption of EPS, SMP, and X_{STO} in the aerobic granules. A higher substrate concentration results in a greater concentration of EPS, SMP, and X_{STO}. An accumulation of biomass in a bioreactor leads to an increased production rate of EPS, SMP, and X_{STO}. However, there is no direct correlation between the biomass concentration with the total content of the EPS, SMP, and X_{STO}. The model can be used for process understanding and thus for optimization of the aerobic-granule-based reactors.

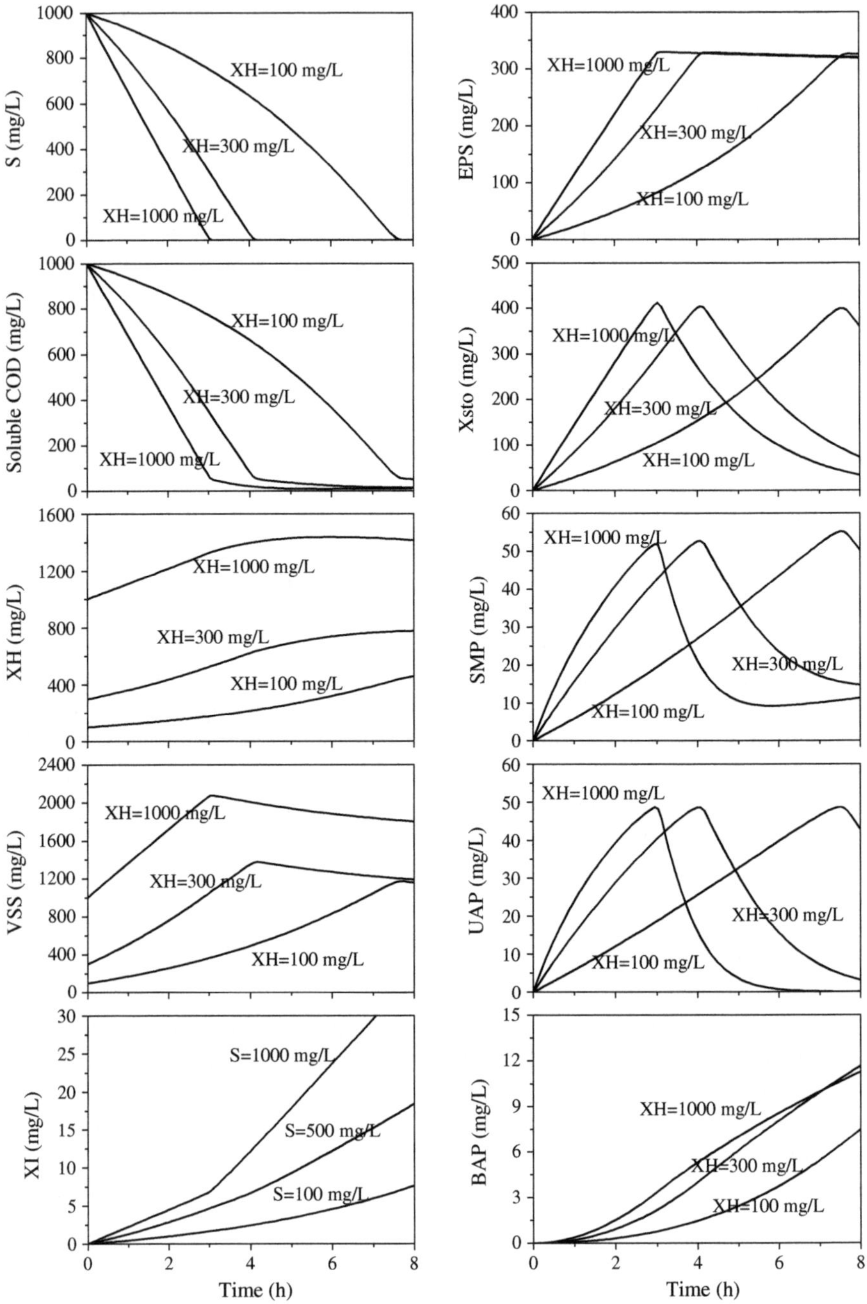

Fig. 9.15 Effect of biomass concentration on *S* consumption, soluble COD, active biomass, VSS, X_I, EPS, X_{STO}, SMP, UAP, and BAP concentration [From Ni and Yu (2010), reprinted with permission from Wiley-Blackwell]

References

APHA: Standard methods for the examination of water and wastewater, 19th edn. American Public Health Association, Washington, DC (1995)

Aquino, S.F., Stuckey, D.C.: Soluble microbial products formation in anaerobic chemostats in the presence of toxic compounds. Water Res. **38**, 255–266 (2004)

Ajbar, A., Alhumazi, K.: Microbial competition: study of global branching phenomena. AIChE J. **46**, 321–334 (2000)

Arrojo, B., Mosquera-Corral, A., Garrido, J.M., Mendez, R.: Aerobic granulation with industrial wastewater in sequencing batch reactors. Water Res. **38**, 3389–3399 (2004)

Avcioglu, E., Karahan, G., Orhon, D.: Estimation of stoichiometric and kinetic coefficients of ASM3 under aerobic and anoxic conditions via respirometry. Water Sci. Technol. **48**(8), 185–194 (2003)

Benjamin, S., Magbanua, Jr., Bowers, A.R.: Characterization of soluble microbial products (SMP) derived from glucose and phenol in dual substrate activated sludge bioreactors. Biotechnol. Bioeng. **93**, 862–870 (2006)

Beun, J.J., Heijnen, J.J., van Loosdrecht, M.C.M.: N-removal in a granular sludge sequencing batch airlift reactor. Biotechnol. Bioeng. **75**, 82–92 (2001)

Beun, J.J., Paletta, F., van Loosdrecht, M.C.M., Heijnen, J.J.: Stoichiometry and kinetics of poly-bhydroxybutyrate metabolism in aerobic, slow growing activated sludge cultures. Biotechnol. Bioeng. **67**, 379–389 (2000)

Beun, J.J., van Loosdrecht, M.C.M., Heijnen, J.J.: Aerobic granulation in a sequencing batch airlift reactor. Water Res. **36**, 702–712 (2002)

Beyenal, H., Leker, L., Tanyol, A.: Diffusion coefficients of phenol and oxygen in a biofilm of pseudomonas putida. AIChE J. **43**, 243–250 (1997)

de Kreuk, K., Heijnen, J.J., van Loosdrecht, M.C.M.: Simultaneous COD, nitrogen, and phosphate removal by aerobic granular sludge. Biotechnol. Bioeng. **90**, 761–769 (2005)

de Kreuk, M.K., Picioreanu, C., Hosseini, M., Xavier, J.B., van Loosdrecht, M.C.M.: Kinetic model of a granular sludge SBR—influences on nutrient removal. Biotechnol. Bioeng. **97**, 801–815 (2007)

de Kreuk, M.K., van Loosdrecht, M.C.M.: Formation of aerobic granules with domestic sewage. J. Environ. Engin. **132**, 694–697 (2006)

de Silva, D.G.V., Rittmann, B.E.: Nonsteady-state modeling of multispecies activated-sludge processes. Water Environ. Res. **72**, 554–565 (2000a)

de Silva, D.G.V., Rittmann, B.E.: Interpreting the response to loading changes in a mixed-culture completely stirred tank reactor. Water Environ. Res. **72**, 566–573 (2000b)

Gujer, W., Henze, M., Mino, T., van Loosdrecht, M.C.M.: Activated sludge model NO. 3. Water Sci. Technol. **39**, 183–193 (1999)

Henze, M., Grady, C.P.L. Jr., Gujer, W., Marais, G.V.R., Matsuo, T.: Activated sludge model No. 1. Scientific and Technical Report No. 1, IAWPRC, London (1987)

Horn, H., Neu, T.R., Wulkow, M.: Modelling the structure and function of extracellular polymeric substances in biofilms with new numerical techniques. Water Sci. Technol. **43**(6), 121–127 (2001)

Jendrossek, D., Selchow, O., Hoppert, M.: Poly(3-hydroxybutyrate) granules at the early stages of formation are localized close to the cytoplasmic membrane in *Caryophanon latum*. Appl. Environ. Microbiol. **73**, 586–593 (2007)

Jiang, H.L., Tay, J.H., Tay, S.T.L.: Aggregation of immobilized activated sludge cells into aerobically grown microbial granules for the aerobic biodegradation of phenol. Lett. Appl. Microbiol. **35**, 439–445 (2002)

Karahan, O., van Loosdrecht, M.C.M., Orhon, D.: Modeling the utilization of starch by activated sludge for simultaneous substrate storage and microbial growth. Biotechnol. Bioeng. **94**, 43–53 (2006)

Krishna, C., van Loosdrecht, M.C.M.: Substrate flux into storage and growth in relation to activated sludge modelling. Water Res. **33**, 3149–3161 (1999)

Laspidou, C.S., Rittmann, B.E.: A unified theory for extracellular polymeric substances, soluble microbial products, and active and inert biomass. Water Res. **36**, 2711–2720 (2002a)

Laspidou, C.S., Rittmann, B.E.: Non-steady state modeling of extracellular polymeric substances, soluble microbial products, and active and inert biomass. Water Res. **36**, 1983–1992 (2002b)

Lee, T.T., Wang, F.Y., Newell, R.B.: Distributed parameter approach to the dynamics of complex biological processes. AIChE J. **45**, 2245–2268 (1999)

Liu, Y.Q., Liu, Y., Tay, J.H.: The effects of extracellular polymeric substances on the formation and stability of biogranules. Appl. Microbiol. Biotechnol. **65**, 143–148 (2004)

Majone, M., Dircks, K., Beun, J.: Aerobic storage under dynamic conditions in activated sludge processes-the state of the art. Water Sci. Technol. **39**(1), 61–73 (1999)

McSwain, B.S., Irvine, R.L., Hausner, M., Wilderer, P.A.: Composition and distribution of extracellular polymeric substances in aerobic flocs and granular sludge. Appl. Environ. Microbiol. **71**, 1051–1057 (2005)

Ni, B.J., Fang, F., Rittmann, B.E., Yu, H.Q.: Modeling microbial products in activated sludge under feast-famine conditions. Environ. Sci. Techn. **43**, 2489–2497 (2009)

Ni, B.J., Yu, H.Q.: A new kinetic approach to microbial storage process. Appl. Microbiol. Biotechnol. **76**, 1431–1438 (2007)

Ni, B.J., Yu, H.Q.: Modeling and simulation of the formation and utilization of microbial products in aerobic granular sludge. AIChE J. **56**, 546–559 (2010)

Nichols, C.A.M., Garon, S., Bowman, J.P., Raguenes, G., Guezennec, J.: Production of exopolysaccharides by Antarctic marine bacterial isolates. J. Appl. Microbiol. **96**, 1057–1066 (2004)

Nicolella, C., van Loosdrecht, M.C.M., Heijnen, J.J.: Mass transfer and reaction in a biofilm airlift suspension reactor. Chem. Eng. Sci. **53**, 2743–2753 (1998)

Nopens, I., Koegst, T., Mahieu, K., Vanrolleghem, P.A.: PBM and activated sludge flocculation: from experimental data to calibrated model. AIChE J. **51**, 1548–1557 (2005)

Oehmen, A., Yuan, Z., Blackall, L.L., Keller, J.: Comparison of acetate and propionate uptake by polyphosphate accumulating organisms and glycogen accumulating organisms. Biotechnol. Bioeng. **91**, 162–168 (2005)

Peng, D., Bernet, N., Delgenes, J.P., Moletta, R.: Aerobic granular sludge-a case study. Water Res. **33**, 890–893 (1999)

Picioreanu, C., van Loosdrecht, M.C.M., Heijnen, J.J.: Mathematical modeling of biofilm structure with a hybrid differential-discrete cellular automaton approach. Biotechnol. Bioeng. **58**, 101–116 (1998)

Pratt, S., Yuan, Z., Keller, J.: Modelling aerobic carbon oxidation and storage by integrating respirometric, titrimetric, and off-gas CO_2 measurements. Biotechnol. Bioeng. **88**, 135–147 (2004)

Qin, L., Liu, Q.S., Yang, S.F., Tay, J.H., Liu, Y.: Stressful conditions-induced production of extracellular polysaccharides in aerobic granulation process. Civil. Eng. Res. **17**, 49–51 (2004)

Reichert, P.: Aquasim 2.0-User Manual, computer program for the identification and simulation of aquatic systems; EAWAG: Dübendorf, Switzerland, (ISBN 3 906484 16 5) (1998)

Schwarzenbeck, N.; Erley, R.; Mc Swain, B.S., Wilderer, P.A., Irvine, R.L.: Treatment of malting wastewater in a granular sludge sequencing batch reactor (SBR). Acta. Hydrochim. Hydrobiol. **32**, 16–24 (2004)

Sheng, G.P., Yu, H.Q., Li, X.Y.: Stability of sludge flocs under shear conditions: roles of extracellular polymeric substances (EPS). Biotechnol. Bioeng. **93**, 1095–1102 (2006)

Sin, G., Guisasola, A., de Pauw, D.J.W., Baeza, J.A., Carrera, J., Vanrolleghem, P.A.: A new approach for modelling simultaneous storage and growth processes for activated sludge systems under aerobic conditions. Biotechnol. Bioeng. **92**, 600–613 (2005)

Su, K.Z., Yu, H.Q.: Formation and characterization of aerobic granules in a sequencing batch reactor treating soybean-processing wastewater. Environ. Sci. Technol. **39**, 2818–2827 (2005)
van Loosdrecht, M.C.M., Pot, M., Heijnen, J.: Importance of bacterial storage polymers in bioprocesses. Water Res. **35**, 41–47 (1997)
Wanner, O., Reichert, P.: Mathematical modelling of mixed-culture bioflms. Biotechnol. Bioeng. **49**, 172–184 (1996)

Chapter 10
Microbial Products Formation in Autotrophic Granular Sludge

The formation mechanism, characterization, and mathematical modeling of the microbial products of autotrophs in the aerobic granular sludge were systemically investigated. The heterotrophic growth on the SMP in the nitrifying sludge grown on nonorganic carbon source was also evaluated. The aerobic growth of the nitrite-oxidizing bacteria (NOB) occurred at the expense of nitrite as an electron donor and resulted in the production of new biomass, UAP, EPS, and nitrate. These autotrophic microbial products could be utilized as the sole electron and carbon source for the heterotrophic growth in autotrophic systems.

10.1 Growth, Maintenance, and Product Formation of Autotrophs

10.1.1 Introduction

The activated sludge process is the most generally applied biological wastewater treatment method. Autotrophs are widely present in activated sludge, especially in BNR processes, in which nitrification and denitrification usually occur. Nitrification is a two-step reaction: ammonium is oxidized to nitrite by ammonium-oxidizing bacteria (AOB), and later nitrite is oxidized to nitrate by nitrite-oxidizing bacteria (NOB). These processes are respectively called nitritation and nitratation.

The growth rate of the autotrophs is considerably lower compared to that of the heterotrophs. Their growth rate is believed to be restricted by the limited energy generation from the oxidation of inorganic sources and also by the highly energy demanding anabolic reactions to synthesize cellular materials from inorganic carbon (Vadivelu 2006). Moreover, the nitratation rate becomes slower than the nitritation rate at higher temperatures (e.g., 30° C) (Hellinga et al. 1998). Finally, the nitratation process is the rate-limiting step when at a low dissolved oxygen

B.-J. Ni, *Formation, Characterization and Mathematical Modeling of the Aerobic Granular Sludge*, Springer Theses, DOI: 10.1007/978-3-642-31281-6_10,

concentration, since the oxygen affinity constant of the NOB is higher than that of the AOB (Picioreanu et al. 1997; Guisasola et al. 2005).

The energy of maintenance refers to the energy consumed for various cell survival activities other than biomass production (Vadivelu 2006). This includes re-synthesis of the damaged cellular materials, maintenance of the necessary concentration gradients across the cell membrane, cell motility, and similar processes. Despite the importance of the maintenance processes, the knowledge concerning in the maintenance energy requirement of the autotrophs is limited.

In activated sludge systems, most microorganisms produce EPS. They are sticky solid materials excreted by cells, and are involved in adhesion phenomena, formation of the matrix structure, controlling the microbial physiology, and the long-term stability of sludge (McSwain et al. 2005; Sheng et al. 2006; Al-Halbouni et al. 2008; Larsen et al. 2008). In addition to EPS, all bacteria convert a fraction of the organic substrate into SMP, which account for the bulk of the soluble organic carbon in reactor effluents (Benjamin et al. 2006; de Silva and Rittmann 2000a, b). SMP are defined as the pool of organic compounds that are released into the solution from heterotrophic substrate metabolism and biomass death (Laspidou and Rittmann 2002a). They have been classified into two groups based on the bacterial phase from which they were derived: UAP and BAP. Heterotrophic cells consume electron-donor substrate to build active biomass. At the same time, they produce bound EPS and UAP. Bound EPS are hydrolyzed to BAP, while active biomass undergoes endogenous death to form residual dead cells. Therefore, the SMP represent soluble EPS (Laspidou and Rittmann 2002a).

For the heterotrophs in activated sludge, extensive work has been done to explore their growth and associated product formation (Laspidou and Rittmann 2002a, b; McSwain et al. 2005; Sheng et al. 2006; Benjamin et al. 2006; Alpkvist et al. 2006). However, so far information on the growth, maintenance, and associated product formation of the autotrophs is still sparse, compared with the heterotrophic counterpart.

On the other hand, in the published models concerning nitrite oxidation (Koch et al. 2000, Jubany et al. 2005) the inhibitory influences of ammonia, and nitrous acid or inorganic carbon on nitrite oxidation are described (Wett and Rauch 2003), but organic carbon inhibition, biomass maintenance, and microbial products of the NOB have not taken into account. At a high initial substrate concentration, a model approach needs the inclusion of substrate inhibition. The incorporation of SMP brings model predictions into accord with empirical experience in terms of the soluble organic materials in reactor effluents. Therefore, in this work, a new model is formulated to describe the biomass growth, substrate inhibition, maintenance, death, and the formation EPS and SMP of the autotrophs, with NOB in activated sludge as an example. Also, experiments were conducted to calibrate and validate the modeling results. To the best of our knowledge, this work might be the first attempt to investigate the growth, maintenance, and product formation of the NOB in activated sludge. Hopefully, the results of this work might provide a better understanding of the activated sludge rich in autotrophs, and hence improve the design and operation of biological nitrogen removal systems.

10.1.2 Model Development

In this work, the unified theory for EPS and SMP developed by Laspidou and Rittmann (2002a, b) is integrated into our model. This new model describes the relationships among the four solid species: nitrite-oxidizing biomass, active heterotrophic bacteria, EPS, and residual inert biomass; four soluble species: nitrite, nitrate and two unique forms of SMP. The model employs the following symbols for concentration: nitrite (s_{NO_2}), nitrate (s_{NO_3}), nitrite-oxidizing biomass (X_N) active heterotrophic biomass (X_H), residual inert biomass (X_I), utilization-associated products (S_{UAP}), biomass-associated products (S_{BAP}), soluble microbial products ($SMP = S_{UAP} + S_{BAP}$), extracellular polymeric substances (X_{EPS}), and dissolved oxygen (S_O). It basically involves seven microbial processes: growth of X_N; maintenance of X_N; death of X_N; release of X_{EPS}; growth of X_H on S_{UAP}; growth of X_H on S_{BAP}; and death of X_H. The units for all species, except nitrite, nitrate, and DO, are oxygen demand or oxygen, which are directly proportional to electron equivalents (8 g O_2 per e^- equivalent).

Tables 10.1 and 10.2 summarize the stoichiometric and kinetic parameters for these reactions adopted from the IWA Activated Sludge Models (Henze et al. 2000). The structure of the established model is presented in a matrix format (Gujer and Larsen 1995), reflecting the basic stoichiometric relationships constituting the backbone of the model. In Table 10.1, model components are listed in the upper row and relevant stoichiometric coefficients are incorporated in appropriate cells of the matrix. The rightmost column of Table 10.2 gives the process rate expressions. In this way, the rate of change (generation or utilization) in a model component for a given biochemical process is obtained by multiplication of related process stoichiometry (Table 10.1) and kinetics (Table 10.2) (Gujer and Larsen 1995). Table 10.3 lists the definitions of the components in the developed model for a better understanding.

In this work, nitrite-N and nitrate–N are considered as nitrogen components in the model. Since the main objective of this work is to develop a new model to describe the biomass growth, substrate inhibition, maintenance, death, and the formation EPS and SMP of the autotrophs, the organic nitrogen in the biomass, EPS, and SMP are not taken into account, as it would significantly increase the complexity of numerical computation, but would not contribute to a better understanding of the processes. In this model, nitrite is assumed to be used by the NOB as their N source as well as energy source. However, since the biomass formula of the NOB is assumed to be $C_5H_7O_2N$ (Vadivelu 2006), the nitrite used as the nitrogen source for the NOB growth is 14 g N 160 g^{-1} COD, i.e., 0.0875 g N g^{-1} COD, which is much lower than that when the nitrite is used as the energy source in our case. Thus, the nitrogen use by the NOB growth is neglectable to simplify the modeling work. Furthermore, EPS and UAP have the same N content to that of the biomass X_N and X_H. Considering that EPS are changed to BAP, BAP have the same N content to EPS. Therefore, the nitrogen in BAP and

Table 10.1 Stoichiometric matrix for the established model

Component	S_O	S_{NO_2}	S_{NO_3}	S_{UAP}	S_{BAP}	X_{EPS}	X_N	X_H	X_I
Process	O_2	N	N	COD	COD	COD	COD	COD	COD
1	$-\frac{[1.14-k_{UAP}-k_{EPS}-Y_N(1-k_{UAP}-k_{EPS})]}{Y_N}$	$-\frac{1}{Y_N}$	$\frac{1}{Y_N}$	$\frac{k_{UAP}}{Y_N}$		$\frac{k_{EPS}}{Y_N}$	$1-k_{UAP}-k_{EPS}$		
2	-1.14	-1	1						
3							-1		1
4					1	-1			
5	$-\frac{1-Y_{H,SMP}}{Y_{H,SMP}}$			$-\frac{1}{Y_{H,SMP}}$				1	
6	$-\frac{1-Y_{H,SMP}}{Y_{H,SMP}}$				$-\frac{1}{Y_{H,SMP}}$			1	
7								-1	1

Table 10.2 Kinetic rates expressions for the reaction processes

	Process	Kinetics rates expressions
1.	Growth of X_N	$\mu_N \frac{S_{NO_2}}{K_{S,NO_2}+S_{NO_2}+\left(S_{NO_2}^2/K_{I,NO_2}\right)} \frac{S_O}{K_O+S_O} X_N$
2.	Maintenance of X_N	$m_N \frac{S_{NO_2}}{K_{S,NO_2}+S_{NO_2}+\left(S_{NO_2}^2/K_{I,NO_2}\right)} \frac{S_O}{K_O+S_O} X_N$
3.	Death of X_N	$b_N \frac{K_{S,NO_2}}{K_{S,NO_2}+S_{NO_2}} X_N$
4.	Release of X_{EPS}	$k_{hyd} X_{EPS}$
5.	Growth of X_H on S_{UAP}	$\mu_{H,UAP} \frac{S_{UAP}}{K_{UAP}+S_{UAP}} \frac{S_O}{K_O+S_O} X_H$
6.	Growth of X_H on S_{BAP}	$\mu_{H,BAP} \frac{S_{BAP}}{K_{BAP}+S_{BAP}} \frac{S_O}{K_O+S_O} X_H$
7.	Death of X_H	$b_H X_H$

Table 10.3 Definitions of the components in the developed model

Number	Component	Definition	Unit
Model dissolved components			
1	S_O	Dissolved oxygen	$g\ O_2\ m^{-3}$
2	S_{UAP}	Utilization-associated products, UAP	$g\ COD_{UAP}\ m^{-3}$
3	S_{BAP}	Biomass-associated products, BAP	$g\ COD_{BAP}\ m^{-3}$
4	S_{NO_2}	Nitrite nitrogen	$g\ N\ m^{-3}$
5	S_{NO_3}	Nitrate nitrogen	$g\ N\ m^{-3}$
Model particulate components			
6	X_H	Active heterotrophic organisms	$g\ COD_X\ m^{-3}$
7	X_N	Active nitrite-oxidizer bacteria, NOB	$g\ COD_X\ m^{-3}$
8	X_{EPS}	Extracellular polymeric substances	$g\ COD_{EPS}\ m^{-3}$
9	X_I	Inert, nonbiodegradable organics	$g\ COD_X\ m^{-3}$

UAP can be used as the nitrogen source by the heterotrophs for their growth on these SMP.

The aerobic growth of the NOB occurs at the expense of nitrite as an electron donor and results in the production of new biomass, UAP, EPS, and nitrate. In the microbial nitrite utilization process, part of donor electrons is used for the autotrophic biomass synthesis, while other donor electrons are diverted to the formation of UAP and EPS. The NOB growth is considered to be inhibited by substrate, and this is described with a Haldane-type equation (Eq. 10.1). Free nitrous acid is found to be inhibitory to the NOB (Jubany et al. 2005), and is in equilibrium with nitrite. Since both pH and temperature remain unchanged in a batch experiment, the inhibition coefficient can be expressed in terms of nitrite, K_{I,NO_2}.

$$\left(\frac{\mathrm{d}X_{\mathrm{N}}}{\mathrm{d}t}\right)_{\mathrm{Growth}} = (1 - k_{\mathrm{UAP}} - k_{\mathrm{EPS}})\mu_{\mathrm{N}} \frac{S_{\mathrm{NO_2}}}{K_{\mathrm{S,NO_2}} + S_{\mathrm{NO_2}} + \left(S_{\mathrm{NO_2}}^2 / K_{\mathrm{I,NO_2}}\right)} \frac{S_{\mathrm{O}}}{K_{\mathrm{O}} + S_{\mathrm{O}}} X_{\mathrm{N}} \tag{10.1}$$

Active microorganisms, including the NOB, require maintenance energy for their activity and chemical process. Maintenance is usually integrated into the substrate consumption rate by adding a maintenance coefficient (m) as follows (Horn and Hempel 1997):

$$\frac{\mathrm{d}S}{\mathrm{d}t} = -\left(\frac{\mu}{Y} + m\right)X \tag{10.2}$$

In the model matrix shown in Table 10.2, Eq. 10.2 is divided into a growth process and a substrate-dependent maintenance process with a Monod term. The kinetic rate r_{main} of the substrate-based maintenance is given as follows (van Loosdrecht and Henze 1999):

$$r_{\mathrm{main}} = m_{\mathrm{N}} X_{\mathrm{N}} \frac{S_{\mathrm{NO_2}}}{K_{\mathrm{S,NO_2}} + S_{\mathrm{NO_2}} + \left(S_{\mathrm{NO_2}}^2 / K_{\mathrm{I,NO_2}}\right)} \frac{S_{\mathrm{O}}}{K_{\mathrm{O}} + S_{\mathrm{O}}} \tag{10.3}$$

According to Laspidou and Rittmann (2002a, b), microbial growth results in the formation of UAP and EPS. UAP are released to the aqueous solution, while EPS are released as a solid to form the aggregate matrix. Hydrolysis of EPS produces soluble BAP. Substrate oxidation and respiration of the electrons reduce oxygen and generate the energy needed for the formation of active biomass, EPS, and UAP. In addition to nitrite oxidizers, a heterotrophic population should also be considered. Since both UAP and BAP are biodegradable, some of their electrons can also be used by the active heterotrophs. Utilization of SMP does not result in the formation of new SMP or EPS (Laspidou and Rittmann 2002a, b).

Equation 10.4 is the nonsteady-state mass balance for EPS in the NOB-rich culture. The first term on the right side is for EPS formation, while the second term is for EPS loss.

$$\frac{\mathrm{d}X_{\mathrm{EPS}}}{\mathrm{d}t} = \left(\frac{k_{\mathrm{EPS}}}{Y_{\mathrm{N}}} \mu_{\mathrm{N}} \frac{S_{\mathrm{NO_2}}}{K_{\mathrm{S,NO_2}} + S_{\mathrm{NO_2}} + \left(S_{\mathrm{NO_2}}^2 / K_{\mathrm{I,NO_2}}\right)} \frac{S_{\mathrm{O}}}{K_{\mathrm{O}} + S_{\mathrm{O}}} X_{\mathrm{N}}\right) - \left(k_{\mathrm{hyd}} X_{\mathrm{EPS}}\right) \tag{10.4}$$

SMP are divided into two categories (Laspidou and Rittmann 2002a, b): UAP produced as a direct result of the substrate utilization and BAP formed from the hydrolysis of EPS. Both UAP and BAP are biodegradable, and their degradation kinetics can be described with Monod-type expressions (Laspidou and Rittmann 2002a, b), for which BAP and UAP have distinct biodegradation kinetic parameters. Equations 10.5 and 10.6 are the nonsteady-state mass balances for UAP and BAP.

$$\frac{dS_{\mathrm{UAP}}}{dt} = \frac{k_{\mathrm{UAP}}}{Y_{\mathrm{N}}}\mu_{\mathrm{N}}\frac{S_{\mathrm{NO_2}}}{K_{\mathrm{S,NO_2}} + S_{\mathrm{NO_2}} + \left(S_{\mathrm{NO_2}}^2/K_{\mathrm{I,NO_2}}\right)}\frac{S_{\mathrm{O}}}{K_{\mathrm{O}} + S_{\mathrm{O}}}X_{\mathrm{N}} - \frac{\mu_{\mathrm{H,UAP}}}{Y_{\mathrm{H,SMP}}}\frac{S_{\mathrm{UAP}}}{K_{\mathrm{UAP}} + S_{\mathrm{UAP}}}\frac{S_{\mathrm{O}}}{K_{\mathrm{O}} + S_{\mathrm{O}}}X_{\mathrm{H}} \tag{10.5}$$

$$\frac{dS_{\mathrm{BAP}}}{dt} = k_{\mathrm{hyd}}X_{\mathrm{EPS}} - \frac{\mu_{\mathrm{H,BAP}}}{Y_{\mathrm{H,SMP}}}\frac{S_{\mathrm{BAP}}}{K_{\mathrm{BAP}} + S_{\mathrm{BAP}}}\frac{S_{\mathrm{O}}}{K_{\mathrm{O}} + S_{\mathrm{O}}}X_{\mathrm{H}} \tag{10.6}$$

The following equations are derived based on the model given in Tables 10.1 and 10.2. These equations are subsequently used for model simulation. Mass balance for oxygen and nitrite gives:

$$\mathrm{OUR} = \frac{1.14 - k_{\mathrm{UAP}} - k_{\mathrm{EPS}} - Y_{\mathrm{N}}(1 - k_{\mathrm{UAP}} - k_{\mathrm{EPS}})}{Y_{\mathrm{N}}}\frac{\mu_{\mathrm{N}}S_{\mathrm{NO_2}}}{K_{\mathrm{S,NO_2}} + S_{\mathrm{NO_2}} + \left(S_{\mathrm{NO_2}}^2/K_{\mathrm{I,NO_2}}\right)}\frac{S_{\mathrm{O}}X_{\mathrm{N}}}{K_{\mathrm{O}} + S_{\mathrm{O}}} + 1.14r_{\mathrm{main}} + \frac{1 - Y_{\mathrm{H,SMP}}}{Y_{\mathrm{H,SMP}}}\left(\frac{\mu_{\mathrm{UAP}}S_{\mathrm{UAP}}}{K_{\mathrm{UAP}} + S_{\mathrm{UAP}}} + \frac{\mu_{\mathrm{BAP}}S_{\mathrm{BAP}}}{K_{\mathrm{BAP}} + S_{\mathrm{BAP}}}\right)\frac{S_{\mathrm{O}}}{K_{\mathrm{O}} + S_{\mathrm{O}}}X_{\mathrm{H}} \tag{10.7}$$

where OUR (mg O_2 L^{-1} h^{-1}) is the overall oxygen uptake rate; the first and second terms on the right side are the oxygen consumption rates for the growth and maintenance processes of the NOB, respectively; and the third term represents the oxygen consumption rate due to the heterotrophic growth on SMP.

$$\mathrm{NUR} = \frac{1}{Y_{\mathrm{N}}}\mu_{\mathrm{N}}\frac{S_{\mathrm{NO_2}}}{K_{\mathrm{S,NO_2}} + S_{\mathrm{NO_2}} + \left(S_{\mathrm{NO_2}}^2/K_{\mathrm{I,NO_2}}\right)}\frac{S_{\mathrm{O}}}{K_{\mathrm{O}} + S_{\mathrm{O}}}X_{\mathrm{N}} + r_{\mathrm{main}} \tag{10.8}$$

where NUR (mg N L^{-1} h^{-1}) is the overall nitrite uptake rate; the terms on the right side respectively represent the nitrite consumption rate by the NOB for the growth (both energy generation and cell assimilation) and maintenance.

In addition, the death process of NOB is coupled to the substrate concentration with an inhibition expression (Table 10.2), in order to simulate the variations of the NOB death rate with the nitrite consumption. For the heterotrophs, the only carbon source for their growth is SMP, whose concentration is of a much low level in this NOB-enriched culture. Thus, the heterotrophic death rate would not change significantly. In this work, the expression of $b_{\mathrm{H}}X_{\mathrm{H}}$ is adapted to simplify modeling.

The simulation is performed with the AQUASIM software package (Reichert 1998), which offers a free definition of the biokinetic model, flow scheme, process control strategies, graphic support of the simulation, experimental data, and communication with spreadsheet programs.

10.1.3 Materials and Methods

10.1.3.1 Experimental Setup and Operation

A SBR was operated to selectively enrich the NOB. The SBR had a working volume of 2 L and was fed with a synthetic wastewater, in which nitrite was used as the sole energy source and bicarbonate as the sole carbon source. The reactor was operated with a cycle time of 6 h and a hydraulic retention time of 1 day at 22 ± 1 °C. Each cycle consisted of 270 min of aeration feeding, 20 min of aerobic reaction, 60 min of settling, and 10 min of decanting period. The pH of the mixed liquor was kept at around 7.3 through the dose of 1 M HCl and NaOH. The DO concentration of the mixed liquor was maintained within a range of 2.75–3.25 mg L^{-1} using an aeration ON/OFF controller. The synthetic wastewater comprised (L^{-1}): 4.93 g $NaNO_2$–N (1 g NO_2^-–N), 0.4 g $NaHCO_3$, 1 g KH_2PO_4, 1 g K_2HPO_4 and 2 mL of a stock solution containing trace elements. The trace element stock solution contained (L^{-1}): 1.25 g EDTA, 0.55 g $ZnSO_4 \cdot 7H_2O$, 0.40 g $CoCl_2 \cdot 6H_2O$, 1.275 g $MnCl_2 \cdot 4H_2O$, 0.40 g $CuSO_4 \cdot 5H_2O$, 0.05 g $Na_2MoO_4 \cdot 2H_2O$, 1.375 g $CaCl_2 \cdot 2H_2O$, 1.25 g $FeCl_3 \cdot 6H_2O$ and 44.4 g $MgSO_4 \cdot 7H_2O$.

10.1.3.2 Experiments

Experiments were conducted in 1–L reactors at 22 ± 1 °C and the medium pH was kept at around 7.3 through the addition of 1 M HCl and NaOH. The sludge was transferred from the SBR to one batch reactor after washed twice to clear background. The sludge was left in starvation for several hours, and then endogenous OUR was measured. Then, $NaNO_2$ at a predetermined concentration was added at pulse. The OUR values from this point to the depletion of substrate comprised endogenous and exogenous respiration. Hence, the exogenous OUR was calculated through subtracting the endogenous OUR value from the measured one. Initial sludge and substrate concentrations were variable, but the DO was kept at saturation level. Samples were analyzed for SMP, OUR, nitrite-N, nitrate–N, and volatile suspended solids.

10.1.3.3 Analytical Methods

The analysis of DO, OUR and sludge image was described in Chap. 3. Mixed liquid samples were centrifuged for 15 min at 12,000 rpm, and then were prefiltered through 0.45 μm acetate cellulose membranes to represent the SMP which was used for COD analyses. The EPS of the sludge samples were extracted using the CER technique (Dowex Marathon C, 20–50 mesh, sodium form, Fluka 91973)

Table 10.4 Stoichiometric evaluation of the experimental results

Parameter	Unit	Set 1	Set 2	Set 3	Set 4	Set 5
Nitrite feed (N)	mg N L^{-1}	99	204	300	410	507
Initial NOB sludge (VSS)	mg L^{-1}	247	251	252	248	251
Maximum oxygen consumption rate (OUR max)	mg O_2 L^{-1} h^{-1}	49.3	51.05	52.52	54.06	55.29
Maximum nitrite consumption rate (NUR max)	mg N L^{-1} h^{-1}	44.8	46.38	47.71	49.1	50.23

according to Chap. 7. Measurement of COD, nitrite-N, nitrate–N, MLSS, and MLVSS followed the Standard Methods (APHA 1995).

10.1.4 Results

10.1.4.1 Experimental Results

Experiments were carried out with nitrite as the sole substrate and energy source under different initial conditions. The nitrite-N analysis confirms that no nitrite is present in the reactor at the beginning and the end of the experiments. Table 10.4 summarizes the experimental results obtained from the five tests.

The OUR profiles are similar for all the five tests (Fig. 10.1a). Prior to the substrate addition, the OURs are low because of the activity of microbial maintenance. After the nitrite-N dose, the OURs increase rapidly and peak of 52.52 mg L^{-1} h^{-1}. Thereafter, the OURs decrease gradually due to the depletion of the exogenous nitrite-N. A longer time is needed for the culture to reach the endogenous phase if a higher amount of nitrite-N is dosed.

The measured nitrite and nitrate concentrations are plotted in Fig. 10.1b. All the batch tests exhibit a similar pattern of nitrification in terms of nitrite oxidization and nitrate formation. Nonetheless, the reactions proceed at different rates in relation to the nitrite loadings for the five tests. A higher initial nitrite concentration results in a longer nitrification period. In nitrification, the nitrite concentration decreases rapidly with time, suggesting its utilization by the AOB as an energy and nitrogen source. The concentration of nitrate as the end product increases with time, and becomes the dominant nitrogen species in the solution by the end of the tests.

The measured SMP (COD) results in the tests are shown in Fig. 10.1c. The SMP concentration increases slowly in the microbial growth period when nitrite is consumed at an increasing rate.

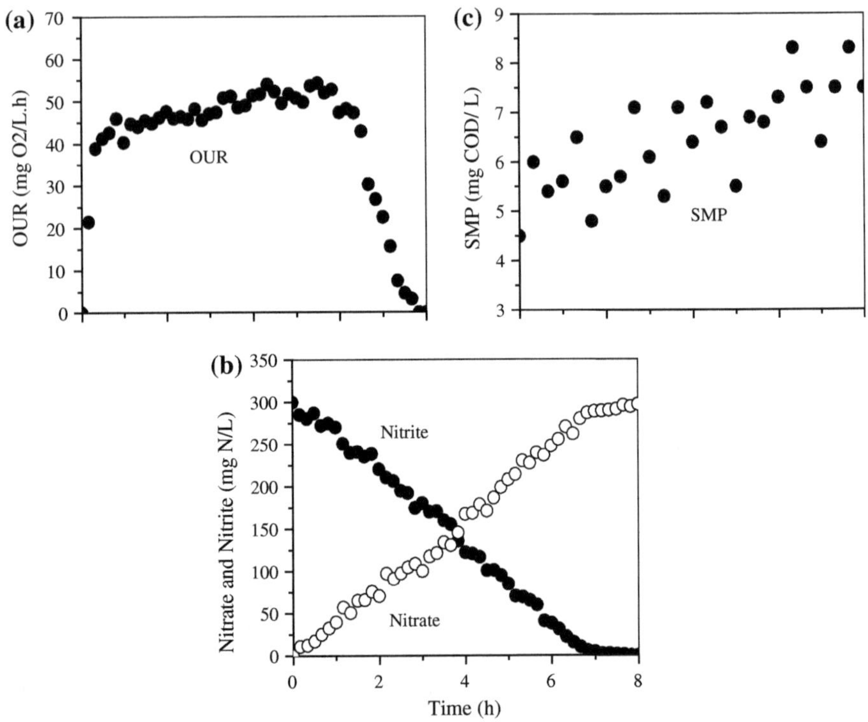

Fig. 10.1 Typical batch experimental results of Set 3: **a** total OUR response; **b** nitrite and nitrate profiles; and **c** SMP concentration (Reprinted from (Ni et al. 2008) with permission from Elsevier)

10.1.4.2 NOB Yield Determination

Petersen et al. (2003) found that only some parameter combinations are identifiable in the Monod-based activated sludge models. All of these combinations contain the biomass yield (Y_N) and the initial substrate concentration $S_{NO_2}(0)$. Therefore, it is not possible to obtain a set of identifiable parameters if Y_N and $S_{NO_2}(0)$ values are estimated together with the rest of the kinetic parameters. To solve this problem, each respirometric batch test should be conducted with a known $S_{NO_2}(0)$ and Y_N is determined individually with the five tests at different known initial nitrite concentrations.

The NOB yield is obtained by plotting the total oxygen consumption (OC) versus the initial nitrite concentration in each pulse (Fig. 10.2). The OC is calculated as the area under the exogenous OUR curve. The slope of the regression corresponds to the oxygen demand per unit of nitrogen and allows the calculation of $Y_N(OC = (1.14 - Y_N)S_{NO_2}(0))$. The Y_N obtained is 0.06 mg COD mg^{-1}N, which is in the same range for the NOB yield values reported in the literature (Koch et al. 2001; Jubany et al. 2005).

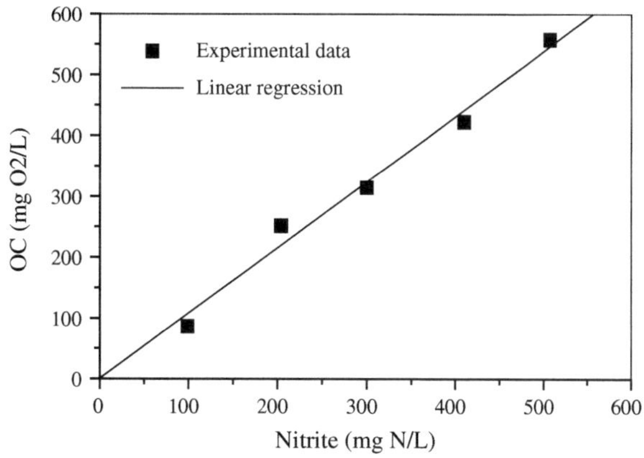

Fig. 10.2 Determination of Y_N: oxygen consumption (OC) as a function of the initial nitrite concentration (Reprinted from Ni et al. (2008) with permission from Elsevier)

10.1.4.3 Model Calibration and Validation

To initiate the calibration procedure, an initial guess of the involved parameters is necessary. Such initial values are obtained based on both measured results and data reported in the literature as shown in Table 10.5. Y_N is determined experimentally and independently. To simplify the calibration process, the model calibration is performed based on sensitivity analysis, and the strategy is to change as few constants as possible because of the limited variability of some parameters, because of the limited variability of some parameters (Xu and Hultman 1996). In the sensitivity analysis, the behavior of the model is evaluated as a consequence of a variation of the input parameters and is performed using a one-variable-at-a-time approach. The sensitivity coefficient represents the change in the output variable, resulting from a change in the input variable. The sensitivity analysis is conducted using the curve profiles of output with input variables. The procedure has been described in detail previously (Ni and Yu 2008).

In this study, six key parameters for microbial processes in this NOB-enriched culture which significantly affect model outputs, i.e., k_{UAP}, k_{EPS}, μ_N, m_N, $K_{S,}$ NO_2 and $K_{I,}$ NO_2, are adjusted based on the measured OUR, SMP, nitrite-N, and nitrate–N profiles. To make the calculated concentration profiles fit the measured data, the parameters of k_{UAP}, k_{EPS}, μ_N, m_N, $K_{S,}$ NO_2 and $K_{I,}$ NO_2 are changed based on the causality of the parameters to the model outputs. The initial active biomass concentrations of X_N and X_H in the model simulation are determined using the method of Vadivelu (2006). With the measurement of the initial EPS content, the initial value of X_I can be set as the difference among the total VSS and the sum of EPS, X_N ,and X_H.

Model calibration results are shown in Table 10.6. Model predictions fit the experimental data (Fig. 10.3), because the oxygen and nitrite consumption of the NOB are accurately described by the estimated parameters. Figure 10.3A also shows the model prediction for the NOB level profiles in the tests. The parameters obtained in the optimization are within the range of values reported in the literatures

Table 10.5 Initial parameters used in the established model (20 °C)

Parameter	Definition	Values	Unit	Source
Stoichiometry				
Y_N	Yield coefficient for NOB	0.06	g COD_X g^{-1} N_S	This study
k_{UAP}	Yield coefficient for UAP	0.009	g COD_{UAP} g^{-1} COD_S	Laspidou and Rittmann (2002b)
k_{EPS}	Yield coefficient for EPS	0.032	g COD_{EPS} g^{-1} COD_S	Laspidou and Rittmann (2002b)
$Y_{H,SMP}$	Yield for X_H on SMP	0.45	g COD_X g^{-1} COD_{SMP}	Laspidou and Rittmann (2002b)
Kinetics				
μ_N	Growth rate of NOB	0.031	h^{-1}	Moussa et al. (2005)
m_N	Maintenance rate of NOB	0.048	h^{-1}	Moussa et al. (2005)
b_N	Death rate of NOB	0.0083	h^{-1}	Moussa et al. (2005)
k_{hyd}	EPS hydrolysis rate	0.0071	h^{-1}	Laspidou and Rittmann (2002b)
$\mu_{H,UAP}$	Growth rate of X_H on UAP	0.053	h^{-1}	Laspidou and Rittmann (2002b)
$\mu_{H,BAP}$	Growth rate of X_H on BAP	0.0029	h^{-1}	Laspidou and Rittmann (2002b)
b_H	Death rate coefficient of X_H	0.0083	h^{-1}	Gujer et al. (1999)
K_{S,NO_2}	Affinity constant for nitrite	12.6	g N_S m^{-3}	Jubany et al. (2005)
K_{I,NO_2}	Nitrite inhibition constant	690	g N_S m^{-3}	Jubany et al. (2005)
K_{UAP}	Affinity constant for UAP	100	g COD_{UAP} m^{-3}	Laspidou and Rittmann (2002b)
K_{BAP}	Affinity constant for BAP	85	g COD_{BAP} m^{-3}	Laspidou and Rittmann (2002b)
K_O	Oxygen affinity constant	0.2	g COD_{O2} m^{-3}	Gujer et al. (1999)

Table 10.6 Parameters for optimum calibration of the experimental data

Parameter	Definition	Values	Unit
k_{UAP}	Yield coefficient for UAP	0.016	g COD_{UAP} g^{-1} COD_S
k_{PS}	Yield coefficient for EPS	0.069	g COD_{EPS} g^{-1} COD_S
μ_N	Maximum growth rate of NOB	0.031	h^{-1}
m_N	Maintenance coefficient of NOB	0.19	h^{-1}
K_S,NO_2	Biomass affinity constant for nitrite	21.33	g N_S m^{-3}
K_I,NO_2	Inhibition coefficient for nitrite	207.37	g N_S m^{-3}

(Henze et al. 1987; Koch et al. 2001; Jubany et al. 2005; Vadivelu, 2006). The half-saturation constant value is greater than the most of values reported (Henze et al. 1987; Gujer et al. 1999; Henze et al. 2000). This is attributed to the different

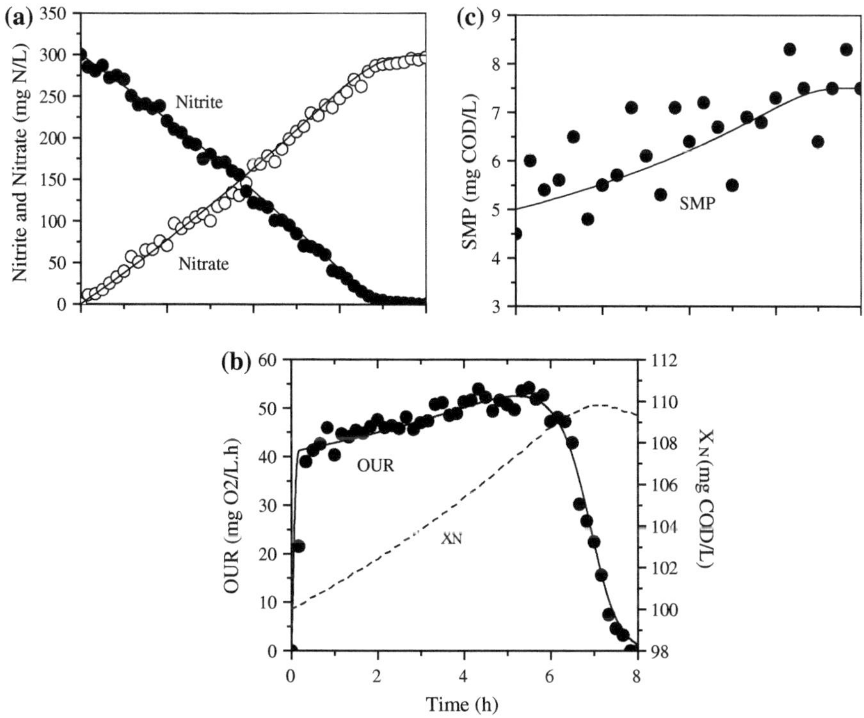

Fig. 10.3 Model calibration for the experimental data of Set 3: **a** OUR profile and predicted NOB concentration; **b** nitrite and nitrate responses; and **c** SMP concentration (Reprinted from Ni et al. (2008) with permission from Elsevier)

biomass acclimation. In the present work, the sludge grows at a high nitrogen loading, where nitrite build-up takes place. Different model structures maybe another reason for such a difference. Also, the variability between bibliographic parameters could be attributed to the different experiments conducted for their determination. It is known that microorganisms are usually less sensitive to a short-term nitrite pulse than a long-term nitrite exposure (Jubany et al. 2005).

The model validation is based on the comparison between the experimental results and model predictions with the same input model parameters. The OUR data of the three tests, which are not used for model calibration, are employed for model validation. Also, the initial experimental conditions are substantially different from those in Fig. 10.3. The range of the added external substrate concentrations (varying from 100 to 500 mg N L^{-1}) is extensive, and the concentrations are significantly different from each other (Fig. 10.4). This ensures testing the model against the data collected under much different conditions. The simulating results in Fig. 10.4 demonstrate that the model is able to predict the OUR profiles well. The good agreement between the measured and simulated results suggests the validity of the model developed in this work.

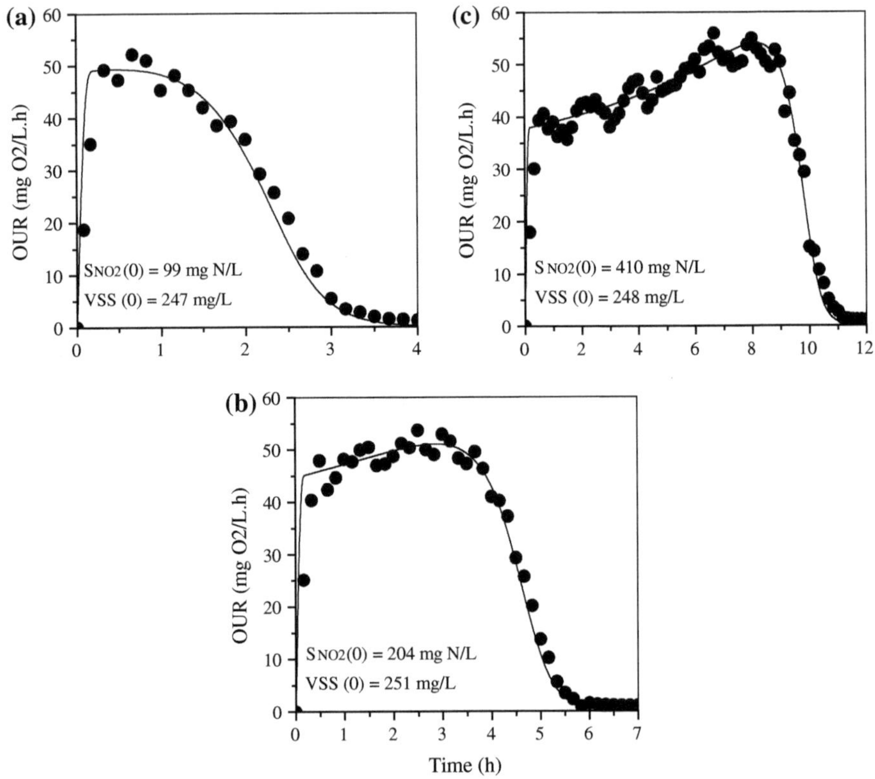

Fig. 10.4 Model validation of the OUR profiles (●-Measured,—Simulated) for the three sets of experimental results (Reprinted from Ni et al. (2008) with permission from Elsevier)

10.1.5 Discussion

10.1.5.1 Maintenance Energy Requirement of the NOB

In general, microbial energy generation is not always associated with the production of new biomass, and a great portion of the energy may be spent to meet the energy requirement of the cell maintenance processes. Microbial maintenance requests energy for various cell functions, in addition to the production of new biomass. The maintenance process may include re-synthesis of the damaged cellular materials, maintenance of the concentration gradients across cell membrane, cell motility, futile cycles, energy dissipation by proton leak, ATP hydrolysis, and other nongrowth-related functions (Vadivelu, 2006). Therefore, nitrite and oxygen are partially consumed for the maintenance of the NOB without an increase in cell. According to Eqs. 10.7 and 10.8, the oxygen consumption rate for maintenance $OURr_{\text{main}}$ is $1.14r_{\text{main}}$ and the nitrite consumption rate for maintenance NUR_{main} is r_{main}.

The m_N values presented in Table 10.6 clearly show that the NOB spend a considerable amount of energy for maintenance, and that the maintenance energy consumption rate is dependent on their specific growth rate. It is estimated for the NOB to spend 26.7 % of the energy generated on maintenance processes $(m_N/(\mu_N/Y_N + m_N) = 0.267)$ at their maximum growth rate. The OC and nitrite consumption (NC) for maintenance are 38.6 and 37.3 % of total content $(OC_{main}/OC = 0.386$ ***and*** $NC_{main}/NC = 0.373)$, respectively.

10.1.5.2 Apparent Growth Yield of the NOB

The apparent microbial yield is lower than their true yield due to maintenance energy consumption. For the enriched NOB in this study, the following equation shows the dependency of the apparent growth yield (Y_N,app) on the specific growth rate (μ_N):

$$\frac{Y_{N,app}}{Y_N} = \frac{\left(\frac{dx_N}{dt}\right)_{Growth} / Y_N}{\left(\frac{dx_N}{dt}\right)_{Growth} / Y_N + r_m} = \frac{(1 - k_{UAP} - k_{EPS})\mu_N/Y_N}{(1 - k_{UAP} - k_{EPS})\mu_N/Y_N + m_N} \tag{10.9}$$

With the determined m_N and μ_N values and the measured Y_N value, $Y_{N,app}$ is estimated as 0.044 mg COD biomass mg^{-1} N, which is in good agreement with the values widely reported (Vadivelu 2006). The apparent growth yield of the NOB is considered to be constant in a typical wastewater treatment plant from a practical point of view (Vadivelu 2006). This further implies that the maintenance processes (Row 2 in Tables 10.1 and 10.2) can be mathematically lumped into the growth process (Row 1 in Tables 10.1 and 10.2), with the replacement of the true growth yield (Y_N) by the apparent growth yield $Y_{N,app}$. A similar approach is commonly used for modeling activated sludge process, where the maintenance processes are not explicitly considered (Henze et al. 1987; Gujer et al. 1999; Henze et al. 2000).

However, it should be noted that the apparent growth yield of the NOB would vary greatly, attributing to the different maintenance energy requirements under different growth conditions. A variation of sludge retention time in a reactor and specific NOB growth rate would result in a different apparent NOB growth yield (Vadivelu 2006). For activated sludge, the NOB growth yield is reported to be in the range of 0.02–0.07 g VSS g^{-1} N (Vadivelu 2006). These wide ranges stand for the apparent yield, rather than the true growth yield, as they are calculated based on the assumption that all the energy generated from nitrite oxidation is used for growth. In practice, however, some of the energy should have been spent on maintenance processes. In this case, maintenance process should be incorporated into the model, as proposed in this work, to give a more accurate prediction for activated sludge.

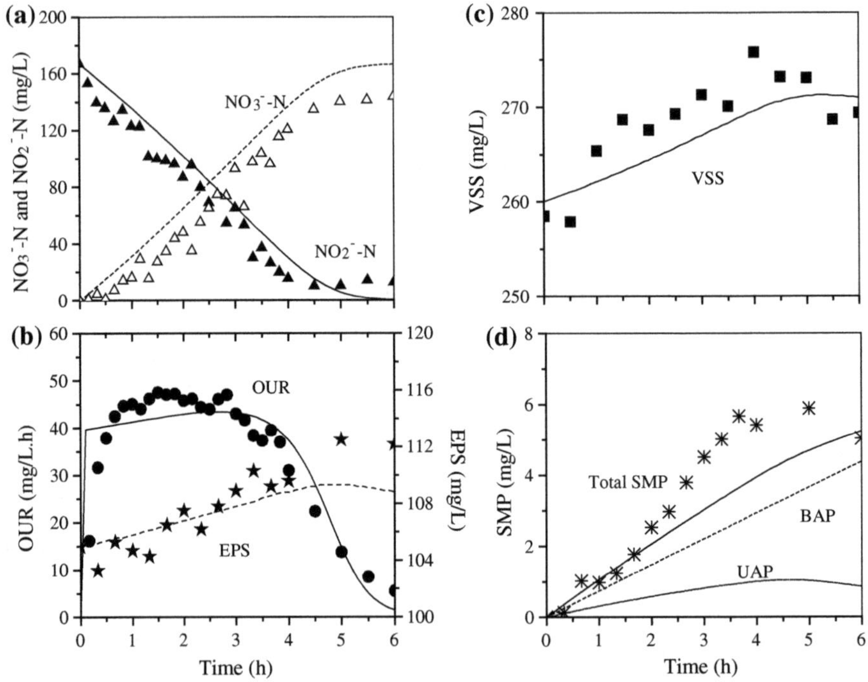

Fig. 10.5 Model simulation of OUR, nitrite, nitrate, formation, and hydrolysis of EPS, production, and degradation of SMP and VSS (Reprinted from Ni et al. (2008) with permission from Elsevier)

10.1.5.3 EPS and SMP Formation of the NOB

Figure 10.5 compares the simulation and experimental results for OUR, NO_2^-–N, NO_3^-–N, EPS, SMP, and VSS in another batch test with the calibrated input model parameters (Table 10.6). The experimental measurements of EPS and SMP in this batch test allow us to directly testify the model simulation results. As shown in Fig. 10.5, there is agreement between the model outputs and the experimental data, suggesting that the model established in this work properly captures the connection between the two types of microbial products and nitrite consumption in the NOB-enriched culture.

Attributed to the rapid consumption of nitrite by the active NOB for new synthesis and formation of UAP and EPS, the nitrite concentration decreases and approaches zero in the end (Fig. 10.5). On the contrary, the concentration of nitrate as the end product increases with time (Fig. 10.5a). Formation of UAP and EPS occurs only when nitrite is being utilized. The EPS concentration increases initially in the rapid growth period when nitrite is consumed at an increasing rate. After the depletion of the external nitrite, EPS are hydrolyzed, resulting in a little decline of EPS concentration (Fig. 10.5b). VSS initially increases and later decreases gradually after 5 h (Fig. 10.5d). Figure 10.5b shows that EPS occupy a significant fraction of total VSS. The UAP concentration initially increases when nitrite is consumed at an increasing rate, but later it decreases gradually to a low level, attributed to their partial biodegradation. However, the BAP build up slowly and continually because of the slow hydrolysis of EPS to BAP (Fig. 10.5c). The BAP also includes a negative term, i.e., the biodegradation of BAP. However, this term is a relatively minor loss, because BAP have slow biodegradation rate. The maximum specific degradation rate of BAP is 0.0029 h^{-1} and the half-maximum rate concentrations for BAP are 85 g COD_{BAP} m^{-3} (Table 10.5). The BAP are essentially not biodegradable. The BAP account for about 80–85 % of the total SMP content, while the UAP represent the residual SMP. Thus, the variation of the total SMP concentration is similar to the variation of the BAP concentration (Fig. 10.5c).

It should be noted that the hydrolysis of EPS was assumed to be the only source of BAP in the unified theory of Laspidou and Rittmann (2002a, b), on which the present model is based on. However, there are some disagreements in the literature regarding the unified theory (Aquino and Stuckey 2008). In addition to the soluble EPS, there are other BAP that might be released into the bulk solution as a result of lysis (i.e., endogenous organics and cell debris) and cell maintenance (i.e., turnover of intracellular components). Aquino and Stuckey (2008) considered that both soluble EPS and cell lysis products are the sources of BAP. In addition, soluble intracellular components may end up in the bulk solution due to the renewal and turnover of the internal structures or as a result of survival strategies adopted by microorganisms (Aquino and Stuckey 2004). Ramesh et al. (2006) compared the physicochemical characteristics of the SMP and soluble EPS from different sludge samples and concluded that that SMP are not identical to the soluble EPS. Because of the complexity of the microbial processes and the difficulty in modeling, information on the BAP formation of the autotrophs is still sparse. There are also controversial conclusions about the basic mechanisms in previous studies (Laspidou and Rittmann 2002a, b; Ramesh et al. 2006; Aquino and Stuckey 2008). Since no experimental procedure is available to accurately quantify the real source of BAP, the model of Laspidou and Rittmann (2002a, b) is adopted in our work. Furthermore, the agreement between the model outputs and the experimental measurements of OUR, NO_2^-–N, NO_3^-–N, EPS and SMP (see Fig. 10.5) suggests that the established model is able to capture the connection between the two types of microbial products and nitrite consumption in the NOB-enriched culture.

10.1.6 Conclusions

The growth, maintenance, death, and microbial product formation processes of the autotrophs, with the NOB as an example, in activated sludge are experimentally investigated with an NOB-enriched culture, and they are described well with a model established in our work. The agreement between the model outputs and the experimental measurements of OUR, NO_2^-–N, NO_3^-–N, EPS and SMP suggests that the established model is able to capture the connection between the two types of microbial products and nitrite consumption in the NOB-enriched culture. The results of this work give us a deeper look into the growth, maintenance, and product formation of the NOB in activated sludge. These results provide a better understanding of the activated sludge rich in autotrophs, and are beneficial to improving our design and operation of biological nitrogen-removal systems.

10.2 Nitrate Loss in the Nitrite Oxidation of Granular Sludge

10.2.1 Introduction

Nitrogen removal from wastewater has recently attracted an increasing interest (Vadivelu et al. 2006a b; Kampschreur et al. 2008; Zhou et al. 2008), and efforts have been made to improve the biological nitrogen removal efficiency and complete simultaneous removal of organic carbon and nitrogen (Yang et al. 2003; Ciudad et al. 2005; Walter et al. 2005). Nitrification is a two-step reaction: ammonium is oxidized to nitrite by AOB, and then nitrite is oxidized to nitrate by NOB. The two processes are respectively called nitritation and nitratation. Nitratation is the oxidation of nitrite as shown in the following reaction (Hao et al. 2002; Mosquera-Corral et al. 2005):

$$NO_2^- - N + 0.5O_2 \rightarrow NO_3^- - N \tag{10.10}$$

where oxidation of certain amount of nitrite-N will produce same amount of nitrate–N. Thus, no nitrate loss occurs in the conventional nitratation process when nitrite is used as the sole energy source and bicarbonate as the only carbon source (Vadivelu et al. 2006a b; Ni et al. 2008a).

However, it is a different case when aerobic granular sludge is used for wastewater treatment, in which simultaneous nitrification and denitrification (SND) may occur (Beun et al. 2001; de Kreuk et al. 2007). In this case, both an aerobic zone for nitrification and an anoxic substrate rich interior for denitrification coexist within a granule. The architecture of granular sludge is relevant to their microbial ecology, as they are macroscopic microbial consortia (Arrojo et al. 2004; Su and Yu 2005). Usually, aerobic granules display considerable heterogeneity, with respect to the microorganisms themselves and their physicochemical microenvironment (de Beer et al. 1994; Okabe et al. 1996). In nitrite-oxidizing granules, oxygen diffusion limitation creates both aerobic and anoxic microzones, allowing sequential utilization of electron acceptors such as oxygen and nitrate. On the other hand, an NOB-enriched culture has been demonstrated to be able to produce SMP in its nitrite consumption (Ni et al. 2008a). Thus, because of the coexistence of an outer aerobic shell and an inner anaerobic zone, it is likely to result in a nitrate loss for the heterotrophic denitrification with SMP as a carbon source, even though no other organic carbon sources are present in feeding solutions.

Understanding the pathways and mechanisms of the nitrate loss in nitrite oxidation of granular sludge and, consequently, the reactor performance is important from an engineering point of view. Therefore, in this study, the nitrogen transformation mechanisms in the second step of the nitrification process in aerobic granular sludge were investigated with both experimental and modeling approaches, after the autotrophic production and consumption of SMP were incorporated. Batch experiments were conducted to explore the features of nitrate loss in the granules. In addition, a mathematic model was formulated to describe the growth of NOB, the production and consumption of SMP, the anoxic heterotrophic growth with SMP and nitrate, the oxygen transfer and substrate diffusion in a nitrite-oxidizing granule. Hopefully, this work might provide a better understanding of the nitrite oxidation process occurring in aerobic granular sludge.

10.2.2 Materials and Methods

10.2.2.1 Experimental Setup and Operation

An SBR with an internal diameter of 7.0 cm and a height of 100 cm (working volume: 2.4 L) was used. It was fed with a synthetic wastewater, in which nitrite was used as the sole energy source and bicarbonate as the sole carbon source. The reactor was operated with a cycle time of 6 h. Each cycle consisted of 5 min of feeding, 340 min of aerobic reaction, 5 min of settling, and 10 min of decanting period. Effluent was drawn from the middle part of the reactor column, and the resulting HRT was 12 h. Air was introduced by an air-pump through an air diffusion from the reactor bottom. The DO concentration of the mixed liquor was maintained within a range of 2.75–3.25 mg L^{-1} using an aeration ON/OFF controller. The pH of the mixed liquor was kept at around 7.3 through the dose of 1 M HCl and NaOH. The synthetic wastewater comprised (L^{-1}): 4.93 g $NaNO_2$–N (1 g NO_2^-–N), 0.4 g $NaHCO_3$, 1 g KH_2PO_4, 1 g K_2HPO_4 and 2 mL of a stock solution containing trace elements. The trace element stock solution contained (L^{-1}): 1.25 g EDTA, 0.55 g $ZnSO_4 \cdot 7H_2O$, 0.40 g $CoCl_2 \cdot 6H_2O$, 1.275 g $MnCl_2 \cdot 4H_2O$, 0.40 g $CuSO_4 \cdot 5H_2O$, 0.05 g $Na_2MoO_4 \cdot 2H_2O$, 1.375 g $CaCl_2 \cdot 2H_2O$, 1.25 g $FeCl_3 \cdot 6H_2O$ and 44.4 g $MgSO_4 \cdot 7H_2O$.

10.2.2.2 Batch Experiments

Batch experiments were conducted in 1−L reactors at 22 ± 1 °C and the medium pH was kept at around 7.3 through the addition of 1 M HCl and NaOH. The granular sludge cultivated in the SBR was transferred to one batch reactor after washed twice with tap water. The sludge was left in starvation for several hours, and then endogenous OUR was measured. Then, $NaNO_2$ at a predetermined concentration was added at pulse. The OUR values from this point to the depletion of substrate comprised endogenous and exogenous respiration. Hence, the exogenous OUR was calculated through subtracting the endogenous OUR value from the measured one. Samples were analyzed for SMP, OUR, nitrite-N, nitrate–N, and volatile suspended solids.

10.2.2.3 Fluorescence in Situ Hybridization

The fluorescent in situ hybridization (FISH) method was employed for the analysis of microbial community structures. Sludge samples were fixed with 4 % fresh paraformaldehyde solution for 8 h at 4 °C. After freezing at −28 °C, slices with thicknesses of 40–50 μm were prepared with a cryostat (Leica, CM1850, Germany), and mounted onto a microscopic slide for observation. The 16S rRNA-targeted oligonucleotide probes used in this study are: EUB338 (labeled with FITC), a general probe for targeting all bacteria (Daims et al. 1999); NIT3 (labeled with ROX), a genus-specific probe for 16S rRNA regions of all hitherto-sequenced *Nitrobacter* strains. A confocal laser scanning microscope (CLSM; Carl Zeiss, LSM 510, Germany) coupled with an Ar ion laser (488 nm) and a HeNe laser (543 nm) was used to detect and record probe-stained cells.

10.2.2.4 Measurement of DO Profiles in Granules

Granules were sampled at the middle of an operation cycle for DO profile measurements. They were immediately transferred to a test chamber and a granule was carefully settled on the nylon net. A photolithography-fabricated microelectrode was employed to measure the DO distribution in granules (Liu et al. 2007. Two-point calibrations to the microelectrode were performed in the bulk solution before and after the measurements by saturating the solution with standard gas containing 0 % or 21 % oxygen, respectively. The changes of response were <5 %. A micromanipulator was used to adjust the fine positions of the microelectrode tip at a spatial resolution of better than 5 μm, and a microscope was used for precisely locating the granular surface. The movement of the microelectrode tip was perpendicular to the granular surface and the tip could be inserted into the granule readily. No buckling of the needle and no significant changes of the granule were observed in the measurements.

10.2.2.5 Self-Organizing Map

The SOM toolbox (version 2) for Matlab 7.0 was used in this study, as described in Chap. 7. The SOM model was applied to better understand the corresponding nitrate loss mechanisms in biological nitrite oxidation process by aerobic granular sludge.

10.2.2.6 Other Analysis

The MW distribution of SMP was measured using a GPC (Waters Co., USA), as described in Chap. 7. Mixed liquid samples were centrifuged for 15 min at 12,000 rpm, and then were prefiltered through 0.45 μm acetate cellulose membranes to represent the SMP which was used for COD analysis. Measurement of DO, COD, nitrite-N, nitrate–N, MLSS, and MLVSS followed the Standard Methods (APHA 1995).

10.2.3 Model Development

10.2.3.1 Model Descriptions

Modeling of biological nitrite oxidation process by granular sludge involves oxygen transfer, substrate diffusion within granules, and biological reactions. In this work, aerobic granules with a wide size of 0.21–1.32 mm are modeled with a consideration of substrate diffusion. The first step is its diffusion into the granules. Then, biological reactions involving each given model components occur within the granules. For biological reactions, the model describes the metabolism of the two microbial groups, i.e., NOB (X_{NOB}) and heterotrophic organisms (X_H). The active species present in one individual defines its bacterial group. In addition to growth, individuals of two species may have a fraction of inerts resulting from the endogenous respiration. A biological nitrite-oxidizing model is extended after including new kinetics of the formation and utilization of microbial products. For the heterotrophic growth, microbial products are the only source of an electron donor for their cell synthesis with oxygen or nitrate as an electron acceptor. In addition, the anoxic microbial endogenous respiration is also taken into account.

10.2.3.2 Biological Reactions

For the biological reactions involved in this mathematic model, the detailed model components, biological processes, kinetic rate equations, their stoichiometric interactions with components, and the stoichiometric and kinetic parameters are presented in matrix format in Tables 10.7, 10.8, 10.9, 10.10. The symbol system from ASMs (Henze et al. 2000) is followed and appropriately supplemented in the model developed in this work. The model employs the following symbols for concentration: nitrite (S_{NO_2}), nitrate (S_{NO_3}), NOB (X_N), active heterotrophic biomass (X_H), residual inert biomass (X_I), utilization-associated products (S_{UAP}), biomass-associated products (S_{BAP}), soluble microbial products ($S_{SMP} = S_{UAP} + S_{BAP}$), EPS ($X_{EPS}$), dissolved nitrogen ($S_{N2}$), and dissolved oxygen (S_O).

Related process kinetics and stoichiometrics to describe the interactions and transformations among model components are expressed compatibly with other activated sludge models reported previously (Henze et al. 2000). The process rate expressions for the heterotrophic organisms (X_H) and NOB ($X_{\rm NOB}$) are given in Table 10.7, reflecting the basic kinetic relationships constituting the model backbone. The relevant stoichiometric coefficients for these microorganisms are incorporated in appropriate matrix cells of Table 10.8. The changing rate (generation or utilization) of a model component for a given biochemical process can be obtained through multiplication of related process stoichiometrics and kinetics (Gujer and Larsen 1995). Suggested values for the kinetic and stoichiometric parameters are listed in Table 10.9a, which defines all the parameters used in our model, their symbols, and their units.

For NOB, part of the input external substrate (nitrite-N) is used for their biomass synthesis, which results in nitrate–N formation. The remaining nitrite-N electrons are diverted to the formation of utilization-associated products (UAP) and EPS. UAP are released to the aqueous solution, while EPS are released as a solid to form the aggregate matrix. Hydrolysis of EPS produces biomass-associated products (BAP), which are soluble. Substrate oxidation and respiration of the electrons to reduce O_2 and generate the energy needed to fuel for formation of active biomass, EPS, and UAP. Endogenous respiration of the active NOB biomass produces residual inert biomass (X_I), which is not biodegradable, with the consumption of electron acceptors, i.e., oxygen or nitrate.

It should be noted that since UAP and BAP are biodegradable, some of their electrons can also be used by the heterotrophic bacteria as an external substrate. Thus, for the heterotrophs, the released SMP of NOB are utilized by the active heterotrophic biomass in this model. The electrons in SMP are used for the heterotrophic biomass synthesis, while the other electrons are respired using oxygen or nitrate for energy generation. Similar to the NOB, the endogenous respiration of the active heterotrophic biomass produces X_I with the consumption of electron acceptors (oxygen or nitrate).

10.2.3.3 Oxygen Transfer

Before reaching the granule surface for diffusion and reaction, oxygen must transfer from gas phase to solid phase. The gas–liquid oxygen transfer rate is assumed to be proportional to the difference in oxygen concentrations between the gas–liquid interface, and the proportionality factor is volumetric oxygen transfer coefficient k_La (Nicolella et al. 1998). On the granule surface, oxygen transferred from the gas phase is equal to that diffused into granules, following a mass balance equation below:

$$\frac{\mathrm{d}S_{\rm O}(t)}{\mathrm{d}t} = k_{\rm L}a\left(S_{\rm O}^{*} - S_{\rm O}(t)\right) \tag{10.12}$$

Table 10.7 Stoichiometric matrix for the established model

Process	S_O O_2	S_{NO_2} N	S_{NO_3} N	S_{N2} N	S_{UAP} COD	S_{BAP} COD	X_{EPS} COD	X_N COD	X_H COD	X_I COD
1	$-\frac{[1.14-k_{UAP}-k_{EPS}-Y_N(1-k_{UAP}-k_{EPS})}{Y_N}$	$-\frac{1}{Y_N}$	$\frac{1}{Y_N}$		$\frac{k_{UAP}}{Y_N}$		$\frac{k_{EPS}}{Y_N}$	$1-k_{UAP}-k_{EPS}$		
2	$-(1-f_I)$							-1		f_I
3			$-\frac{(1-f_I)}{2.86}$	$\frac{1-f_I}{2.86}$				-1		f_I
4						1	-1			
5	$-\frac{1-Y_{H,UAP}}{Y_{H,UAP}}$				$-\frac{1}{Y_{H,UAP}}$				1	
6	$-\frac{1-Y_{H,BAP}}{Y_{H,BAP}}$					$-\frac{1}{Y_{H,BAP}}$			1	
7			$-\frac{1-Y_{H,UAP}}{2.86Y_{H,UAP}}$	$\frac{1-Y_{H,UAP}}{2.86Y_{H,UAP}}$	$-\frac{1}{Y_{H,UAP}}$				1	
8			$-\frac{1-Y_{H,BAP}}{2.86Y_{H,BAP}}$	$\frac{1-Y_{H,BAP}}{2.86Y_{H,BAP}}$		$-\frac{1}{Y_{H,BAP}}$			1	
9	$-(1-f_I)$								-1	f_I
10			$-\frac{(1-f_I)}{2.86}$	$\frac{1-f_I}{2.86}$					-1	f_I

Table 10.8 Kinetic rates expressions for the reaction processes

Process	Kinetics rates expressions
1. Aerobic growth of X_N	$\mu_N \frac{S_{NO_2}}{K_{NO_2}+S_{NO_2}+\left(S_{NO_2}^2/K_{I,NO_2}\right)} \frac{S_O}{K_O+S_O} X_N$
2. Aerobic respiration of X_N	$b_N \frac{S_O}{K_O+S_O} X_N$
3. Anoxic respiration of X_N	$b_N \eta_{NOx} \frac{K_O}{K_O+S_O} \frac{S_{NO_2}+S_{NO_3}}{K_{NO_x}+S_{NO_2}+S_{NO_3}} X_N$
4. Release of X_{EPS}	$k_{hyd} X_{EPS}$
5. Aerobic growth of X_H on S_{UAP}	$\mu_{H,UAP} \frac{S_{UAP}}{K_{UAP}+S_{UAP}} \frac{S_O}{K_O+S_O} X_H$
6. Aerobic growth of X_H on S_{BAP}	$\mu_{H,BAP} \frac{S_{BAP}}{K_{BAP}+S_{BAP}} \frac{S_O}{K_O+S_O} X_H$
7. Anoxic growth of X_H on S_{UAP}	$\mu_{H,UAP} \eta_{NO_x} \frac{S_{UAP}}{K_{UAP}+S_{UAP}} \frac{K_O}{K_O+S_O} \frac{S_{NO_2}+S_{NO_3}}{K_{NO_x}+S_{NO_2}+S_{NO_3}} X_H$
8. Anoxic growth of X_H on S_{BAP}	$\mu_{H,BAP} \eta_{NO_x} \frac{S_{BAP}}{K_{BAP}+S_{BAP}} \frac{K_O}{K_O+S_O} \frac{S_{NO_2}+S_{NO_3}}{K_{NO_x}+S_{NO_2}+S_{NO_3}} X_H$
9. Aerobic respiration of X_H	$b_H \frac{S_O}{K_O+S_O} X_H$
10. Anoxic respiration of X_H	$b_H \eta_{NO_x} \frac{K_O}{K_O+S_O} \frac{S_{NO_2}+S_{NO_3}}{K_{NO_x}+S_{NO_2}+S_{NO_3}} X_H nn$

where $NO_2{}^*$ is the maximum oxygen solubility in liquid phase; S_O is the oxygen concentration on granule surface, equal to that in the bulk liquid, because the liquid–solid oxygen transfer resistance is ignored (Pras 1989; Su and Yu 2005).

10.2.3.4 Substrate Diffusion

For the soluble components involved in the biological reactions, the first step is their diffusion into the granules where the reactions take place. Discretization in time of the partial differential equation describing the reaction–diffusion kinetics in a spherical particle (i.e., granule) is described as the following equation:

$$D_i \cdot \left(\frac{d^2 S_i(r)}{dr^2} + \frac{2}{r} \cdot \frac{dS_i(r)}{dr} \right) + r_i = 0 \tag{10.13}$$

with two boundary conditions:

$$\frac{dS_i}{dr} = 0, \quad \text{at}\, r = 0$$

$$S_i = S_{i,\text{sur}}, \quad \text{at}\, r = R$$

where S_i is the substrate concentration of component i in the particle, r_i is the volumetric substrate conversion rate in the particle, and D_i is the diffusion coefficient of substrate i, r is the distance from the granule center.

The number of nitrite-oxidizing granules and their diameter has an effect on the simulation results, because they influence the overall liquid/granule interfacial area (de Kreuk et al. 2007). The number of granules (*PNo*) in the reactor was calculated according to Equation 10.14 (Gonzalez-Gil et al. 2001:

Table 10.9 **a** Initial parameters used in the established model **b**. Calibrated parameters used in the established model

Parameter	Definition	Values	Unit	Source
Stoichiometry				
Y_N	Yield coefficient for NOB	0.08	g COD g^{-1} N	(1)
k_{UAP}	Yield coefficient for UAP	0.05	g COD g^{-1} COD	(2)
k_{EPS}	Yield coefficient for EPS	0.18	g COD g^{-1} COD	(2)
$Y_{H,UAP}$	Yield coefficient for X_H on UAP	0.45	g COD g^{-1} COD	(2)
$Y_{H,BAP}$	Yield coefficient for X_H on BAP	0.45	g COD g^{-1} COD	(2)
f	Fraction of X_I in biomass respiration	0.2	g COD g^{-1} COD	(3)
Kinetics				
μ_N	Maximum growth rate of NOB	0.019	h^{-1}	(1)
b_N	Respiration rate coefficient of NOB	0.0059	h^{-1}	(1)
k_{hyd}	EPS hydrolysis rate coefficient	0.0071	h^{-1}	(2)
$\mu_{H,UAP}$	Maximum growth rate of X_H on UAP	0.053	h^{-1}	(2)
$\mu_{H,BAP}$	Maximum growth rate of X_H on BAP	0.0029	h^{-1}	(2)
b_H	Respiration rate coefficient of X_H	0.0083	h^{-1}	(3)
η_{NOx}	Anoxic reduction factor	0.6	–	(3)
K_{NO_2}	Biomass affinity constant for nitrite	12.6	g N m^{-3}	(1)
K_{I,NO_2}	Inhibition coefficient for nitrite	690	g N m^{-3}	(1)
K_{UAP}	Biomass affinity constant for UAP	100	g COD_{UAP} m^{-3}	(2)
K_{BAP}	Biomass affinity constant for BAP	85	g COD_{BAP} m^{-3}	(2)
K_O	Dissolve oxygen affinity constant	0.2	g O_2 m^{-3}	(3)
K_{NOx}	Biomass affinity constant for S_{NOx}	0.50	g N m^{-3}	(3)
Granules size				
R_{min}	Minimum radius of the granules	0.21	mm	(4)

(continued)

Table 10.9 (continued)

Parameter	Definition	Values	Unit	Source
R_{max}	Maximum radius of the granules	1.32	mm	(4)
R_{mean}	Mean radius of the nitratation granules	0.65	mm	(4)
Mass diffusion				
D^{NO_2}	Effective diffusivity of NO_2	0.000576	$dm^2\ h^{-1}$	(5)
D^{NO_3}	Effective diffusivity of NO_3	0.000576	$dm^2\ h^{-1}$	(5)
D^{UAP}	Effective diffusivity of UAP	0.000576	$dm^2\ h^{-1}$	(5)
D^{BAP}	Effective diffusivity of BAP	0.000576	$dm^2\ h^{-1}$	(5)
D^{O2}	Effective diffusivity of S_O	0.0023	$dm^2\ h^{-1}$	(6)
Density of solid phase				
ρ_{XH}	Density of X_H	35,000	$g\ COD\ m^{-3}$	(7)
ρ_{XI}	Density of X_I	35,000	$g\ COD\ m^{-3}$	(7)
ρ_{XN}	Density of X_N	35,000	$g\ COD\ m^{-3}$	(7)
ρ_{XEPS}	Density of X_{EPS}	4,000	$g\ COD\ m^{-3}$	(6)
Parameter	Definition	Values	Unit	Source
Y_N	Yield coefficient for NOB	0.055	$g\ COD\ g^{-1}\ N$	(4)
K_{UAP}	Yield coefficient for UAP	0.11	$g\ COD\ g^{-1}\ COD$	(4)
μN	Maximum growth rate of NOB	0.048	h^{-1}	(4)
ηNOx	Anoxic reduction factor	0.72	–	(4)
K_{NO_2}	Biomass affinity constant for nitrite	15.4	$g\ N\ m^{-3}$	(4)
K_{I,NO_2}	Inhibition coefficient for nitrite	810	$g\ N\ m^{-3}$	(4)

(1) Jubany et al. (2005); (2) Laspidou and Rittmann (2002); (3) Henze (1999); (4) This study; (5) Beun et al. (2001); (6) Horn et al. (2001); (7) de Kreuk et al. (2007)

$$PN_0 = \frac{S_{TX,R} \times V_{TR}}{S_{TX,G} \times V_{TG}} \tag{10.14}$$

where $S_{\mathrm{TX,G}}$ and $S_{\mathrm{TX,R}}$ are the total biomass concentration in the granules and reactor, respectively; V_{TG} is the total volume of the granules; and V_{TR} refers to the total effective reactor volume.

The nitrite-oxidizing granules in this study had a radius distribution between 0.21 mm and 1.32 mm with an average of 0.65 mm. It is true that granules with a wide range in size were formed in the reactor and that their size distribution also changed in time. However, the granule size distribution was not taken into account in this model, as it would significantly increase the complexity of numerical computation and it is not expected to contribute to a better understanding of the system (de Kreuk et al. 2007). Therefore, in the simulation the radius was chosen to be 0.65 mm, which was the most representative for the granules. Parameters for the nitrite-oxidizing granules are shown in Table 10.9a.

10.2.4 Results and Discussion

10.2.4.1 Nitrate Loss in the Nitrite Oxidation Process in Aerobic Granules

Experiments were carried out with nitrite as the sole substrate and energy source under different initial conditions. The nitrite-N analysis confirms that no nitrite was present in the reactor at the beginning and the end of experiments. The measured nitrite and nitrate concentrations are plotted in Fig. 10.6. The two batch tests exhibited a similar pattern of nitrification in terms of nitrite oxidization and nitrate formation. Nonetheless, the reactions preceded at different rates in relation to the nitrite loadings for the two tests. A higher initial nitrite concentration resulted in a longer nitrification period. In nitrification, the nitrite concentration decreased rapidly with time, suggesting its utilization by the NOB as an energy and nitrogen source. The concentration of nitrate, as the end product, increased with time, and became the dominant nitrogen species in the solution by the end of tests.

The most interesting findings were that the produced nitrate–N concentrations by the granular sludge at the end of tests were substantially lower than the initial added nitrite-N concentrations in both batch tests. As shown in Fig. 10.6a, there was approximately 25 mg N L^{-1} of nitrate–N loss at 4 h after the dose of 95 mg N L^{-1} nitrite-N. The similar phenomenon was also observed in another batch test as shown in Fig. 10.6b. The nitrate–N loss in this experiment was approximately 40 mg N L^{-1}. In both batch tests with different initial nitrite-N concentrations, the content of nitrate–N loss increased with the consumption of nitrite (Fig. 10.6). Furthermore, the content of nitrate–N loss in the reactor increased with an increase in the initial nitrite-N concentration. This shows that nitrate loss was enhanced by the microbial utilization of nitrite with aerobic granular sludge.

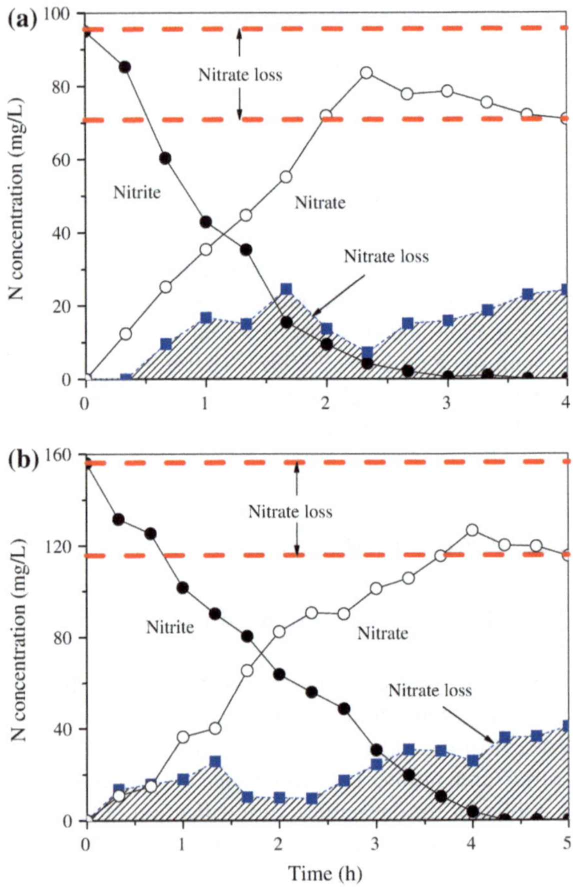

Fig. 10.6 Nitrate loss in the nitrite oxidation process by the aerobic granular sludge in two typical batch experiments: **a** 95 mg N L^{-1} of initial nitrite concentration; and **b** 160 mg N L^{-1} of initial nitrite concentration (From Ni et al. (2011), reprinted with permission from Wiley-Blackwell)

10.2.4.2 Roles of NOB, SMP and DO in the Nitrate Loss in Nitrite-Oxidizing Granules

It is hypothesized that oxygen diffusion limitation causes an anoxic microzone in aerobic granules (de Kreuk et al. 2007; Ni et al. 2008b), which allows sequential heterotrophic utilization of electron acceptors (nitrate) with the SMP released by NOB as a carbon source. This might result in the nitrate loss attributed to denitrification. To identify this nitrate loss mechanism, the diversity of active bacteria in the nitrite-oxidizing granule was firstly characterized by FISH.

Figure 10.7 illustrates the FISH images of the nitrite-oxidizing granules. The diversity of active bacteria in the granules was characterized. The CLSM image of all bacteria (hybridized with FITC-labeled EUBMIX) in the granules is shown in Fig. 10.7b (green) and the NOB image (NIT3, labeled with ROX) is shown in Fig. 10.7c (red). Figures 10.7b and c are spliced to show the active microbial fractions (Fig. 10.7d). NOB populations were most closely related to *Nitrobacter*

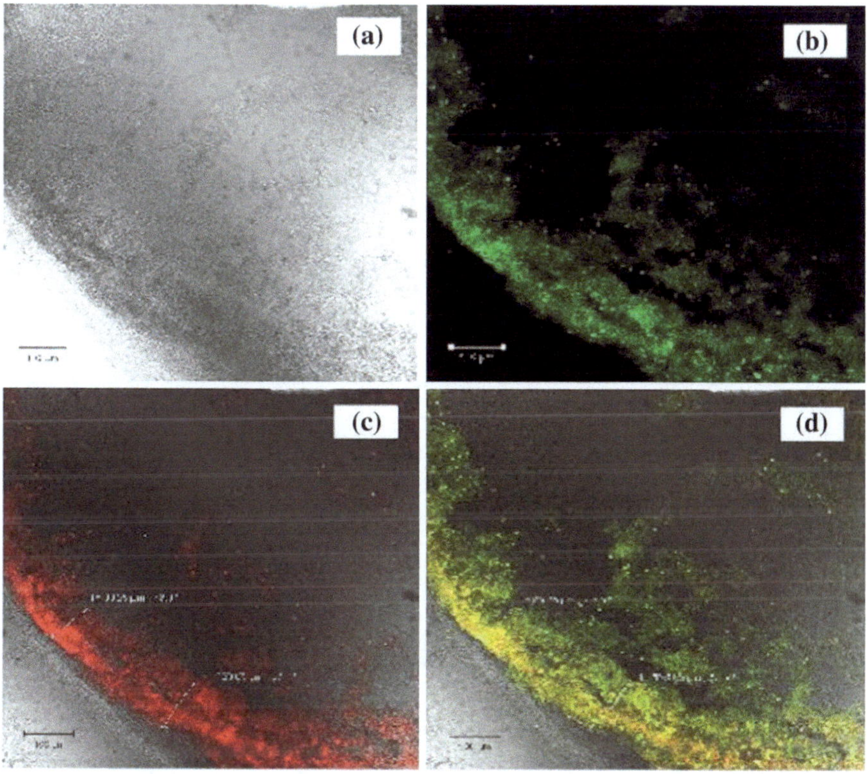

Fig. 10.7 Images of a cross-section through the nitrite-oxidizing granule: **a** optical microscopy image of the granular slice; **b** CLSM image of all bacteria (hybridized with FITC-labeled EUBMIX, *green*); **c** CLSM image of NOB (NIT3, *red*); and **d** the spliced image of **b** and **c** (*yellow* for NOB) (From Ni et al. (2011), reprinted with permission from Wiley-Blackwell)

(red), which were found to be the dominative microbial species. *Nitrobacter* is regarded to be the most important nitrite oxidizers in biological wastewater treatment reactors (Nogueira and Melo 2006; Kindaichi et al. 2006). However, there also exist heterotrophic organisms other than NOB in this granular sludge. The NOB accounted for about 85 % of the total bacteria, while heterotrophs accounted for approximately 15 % of the total bacteria in the granules. In addition, NOB were crowded in an outer layer of granules, while heterotrophic cells were mainly located inside the granules. Microbial products, not contained in the feed solution, were likely to be the energy and carbon sources for the heterotrophic growth with nitrate as an electron acceptor inside the granular sludge.

The SMP productions of the nitrite-oxidizing granules were then characterized using the GPC measurement. Two sets of time courses of SMP formation are shown in Fig. 10.8. The SMP increased gradually from 4 mg L^{-1} to 16 mg L^{-1} during the microbial growth period when nitrite was consumed at an increasing

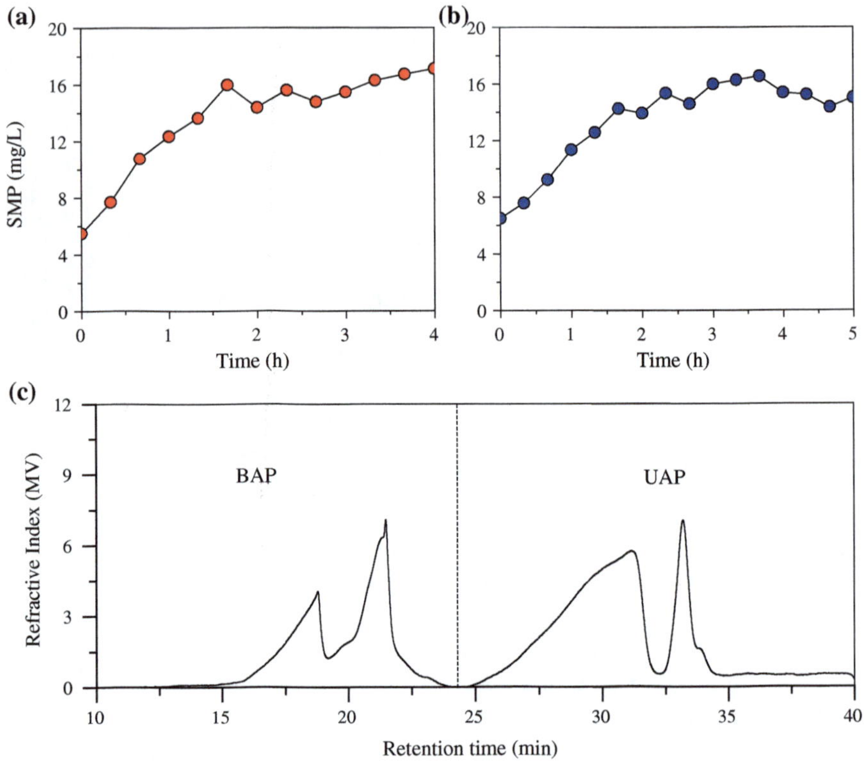

Fig. 10.8 SMP production in the nitrite oxidation process by granular sludge in two batch experiments: **a** 95 mg N L^{-1} of initial nitrite concentration for SMP formation; **b** 160 mg N L^{-1} of initial nitrite concentration for SMP production; and **c** GPC measurement of SMP sample (From Ni et al. (2011), reprinted with permission from Wiley-Blackwell)

rate (Figs. 10.8a and b). The external substrate consumption (see Fig. 10.6) resulted in an increase in the amount of total SMP. The SMP chromatogram shown in Fig. 10.8c shows that different molecular size fractions were identified in the samples. The SMP had four main peaks with a retention time of approximately 19, 22, 32 and 33.5 min at MW of 800,000, 200,000, <1,000 and <500 Da, respectively. This results clearly indicate that the SMP mainly contained small carbonaceous compounds derived from the original substrate in the microbial growth, as UAP (Laspidou and Rittmann 2002a, b), and cellular macromolecules generated in the endogenous phase, as BAP (Jarusutthirak and Amy 2006). Both UAP and BAP are biodegradable, but have distinct heterotrophic biodegradation kinetics, attributed to their significantly different molecular weights and chemical natures.

In order to get insight of the oxygen limitation in the nitrite-oxidizing granule, the DO distribution from bulk solution to a granule center is illustrated in Fig. 10.9. Point 0 on the x axis indicates the granule surface. The DO concentration decreased monotonically from 3.5 to 1.4 mg O_2 L^{-1} in the 500 μm thick

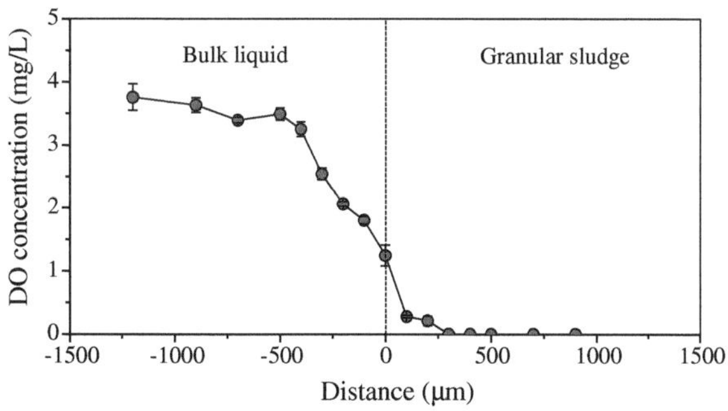

Fig. 10.9 DO microprofiles of the nitrite-oxidizing granules. Point 0 on the x axis indicates the granule surface. Negative depth values (to the *left* of 0) are in the bulk, and positive depth values (*right* of zero) are in the granule (From Ni et al. (2011), reprinted with permission from Wiley-Blackwell)

diffusion boundary layer above the granule surface. The DO level fell continuously in the upper layer of the granule (250 μm thick), and approached zero in the inner part. The oxygen-depleted zone was found in the granule. Biological nitrite oxidation is an oxygen consumption process. The NOB consumed oxygen, resulting in a DO gradient. Therefore, the main active NOB in the granule was located in the upper 250 μm layer, because oxygen was depleted in the inner part. This observation is in accordance with the FISH results. Since DO was not detected at the granule center, denitrification might occur in the granule. These results illuminate the nitrate loss mechanisms in biological nitrite oxidation process by aerobic granules. The proposed pathways of the nitrate loss mechanisms in biological nitrite oxidation of granular sludge were described in Fig. 10.9. Oxygen diffusion limitation resulted in anoxic microenvironments in the nitrite-oxidizing granules. Thus, heterotrophs could use the released SMP by NOB as a carbon source and nitrate as an electron acceptor for their anoxic growth, which resulted in nitrate losses in the nitrite oxidizing process of nitrite-oxidizing granules.

10.2.4.3 Modeling the Nitrate Loss in Nitrite-Oxidizing Granules

Calibration of a model using experimental data involves a process of adjusting model coefficient values, so that the results produced by the model with these coefficients closely match the measured data. With regard to the stoichiometric and kinetic parameters, values for the most of these coefficients are taken from the previous studies (Henze et al. 2000; Laspidou and Rittmann, 2002b; Moussa et al. 2005; Pambrun et al. 2006) in order to reduce the number of parameters to be calibrated. These values are derived from extensive experiments and model applications and are considered to be applicable. However, in our model, a number of new biological processes were introduced as shown in Tables 10.7 and 10.8. Thus, key parameters governing the biological processes should be estimated by fitting the simulation results to the experimental data. Six key parameters for microbial processes in the NOB granules, i.e., Y_N, k_{UAP}, μ_N, η_{NOx}, K_{NO_2} and K_{I,NO_2}, were adjusted based on the measured OUR, SMP, nitrite-N, and nitrate–N profiles. To make the calculated concentration profiles fit the measured data, these parameters were changed based on the causality of parameters to the model outputs. The standard deviation for parameter calibration defined globally for all data points is set to 50 % to ensure the validity of the parameter values obtained.

The calibrated parameter values are summarized in Table 10.9b, whereas the simulation results are shown in Fig. 10.10a. The model predictions match the experimental data well. The Y_N obtained is 0.055 mg COD mg^{-1}N, which is lower than the NOB yield values reported in the literature (Henze et al. 2000; Jubany et al. 2005; Pambrun et al. 2006), because the electrons from the external substrate (nitrite) were also for formation of SMP, in addition to the new biomass synthesis. The μ_N values obtained in the optimization are within the range of values reported in the literatures (Jubany et al. 2005; Pambrun et al. 2006). The values of the half-saturation constant and the substrate inhibition constant for NOB are greater than the values reported (Carrera et al. 2004; Jubany et al. 2005; Baquerizo et al. 2005; Pambrun et al. 2006). This is attributed to the different biomass acclimation. In the present work, the granular sludge was applied for nitrite oxidation. The acclimatization of the NOB population in different sludge systems might lead to an increase in the value of the inhibition threshold. Different model structures could be another reason for such a difference. Also, the variability between bibliographic parameters could be attributed to the different experiments conducted for their determination.

The model is then validated for the set of parameters shown in Table 10.9b, in order to evaluate whether the model is able to describe the experimental results from independent batch experiments. The evaluation also helps to identify the relationship among the different components of the model. The experimental data for the model evaluation were different from those for parameter calibration. Figure 10.10b compares the simulation and experimental results for OUR, nitrite-N, nitrate–N, and SMP. Model and experimental data agree well for the four parameters in Fig. 10.10b. The mean of the differences between the simulated and measured results is low, confirming the validity of the model.

Evaluation of the experimental data with our model indicates that initially the OUR increased rapidly when the electron flow directly went to respiration after the substrate was dosed and then decreased gradually due to the depletion of the exogenous substrate. The sharp bending point in the OUR corresponds with the complete removal of nitrite. The model is also able to predict the nitrate loss (Fig. 10.10). The nitrite-N concentration decreased rapidly with the elapsing time and approaches nil in the end, suggesting its utilization by the NOB as an energy and nitrogen source for new synthesis and formation of microbial products. The concentration of nitrate as the end product increased and became the dominant nitrogen compound in the solution with an amount of nitrogen loss. Model evaluation results demonstrate that the model is able to appropriately simulate the nitrate loss mechanisms in biological nitrite oxidation process in aerobic granular sludge.

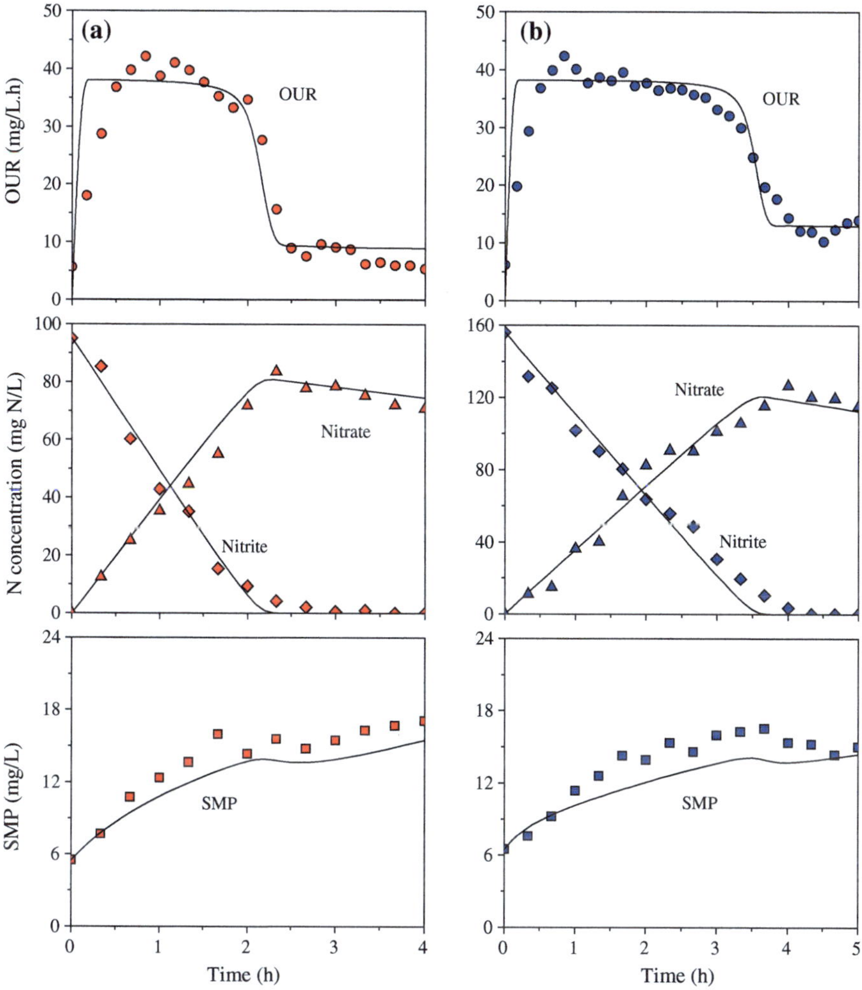

Fig. 10.10 Model calibration and validation for the experimental data: **a** 95 mg N L^{-1} of initial nitrite concentration for model calibration; and **b** 160 mg N L^{-1} of initial nitrite concentration for model validation (*dot* for measurement, and *line* for simulation) (From Ni et al. (2011), reprinted with permission from Wiley-Blackwell)

10.2.4.4 Evaluation of the Nitrate Loss in Granules Using SOM

The SOM model was applied to identify the relationship between the nitrate loss in nitrite-oxidizing granules and related key parameters. The component planes for each variable of the SOM model are shown in Fig. 10.11. The unified distance matrix (U-matrix) representation of the SOM visualizes the distances among the map neurons. The distances among the neighboring map neurons are calculated, and subsequently visualized by applying colorized shade scaling among them. For

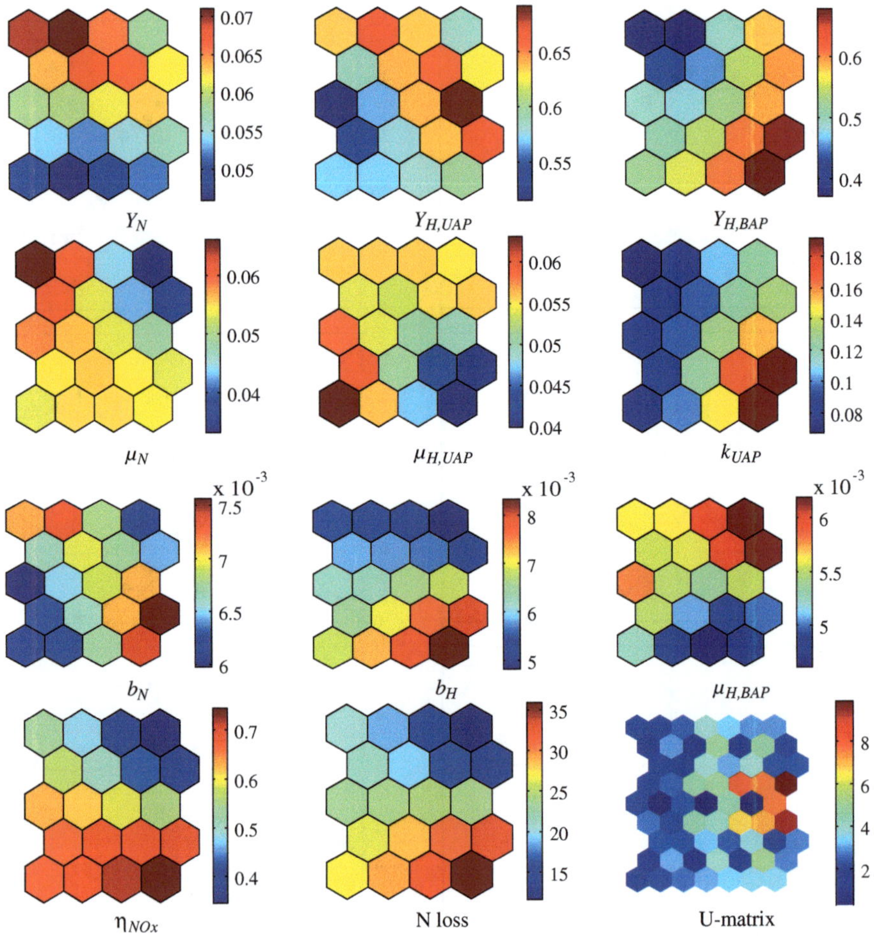

Fig. 10.11 Abstract visualization of the relationships among the nitrate loss (mg N L^{-1}) and Y_N (g COD_X g^{-1} N), $Y_{H,UAP}$ (g COD_X g^{-1} COD_{UAP}), $Y_{H,BAP}$ (g COD_X g^{-1} COD_{BAP}), μ_N (h^{-1}), $\mu_{H,UAP}$ (h^{-1}), k_{UAP} (g COD_{UAP} g^{-1} COD_S), b_N (h^{-1}), b_H (h^{-1}), $\mu_{H,BAP}$ (h^{-1}), and η_{NOx} (—) using an SOM model

example, the carmine shades are associated with the high relative component value of the corresponding weight vector. This helps to identify and subsequently illustratively show the clusters in the input data.

The high nitrate loss in the nitrite-oxidizing granules is linked to the high anoxic reduction factor η_{NOx}, the high heterotrophic endogenous respiration rate b_H and the low NOB yield Y_N values. Nitrate loss in the nitrite-oxidizing granules does not relate to the maximum growth rate of NOB and yield coefficient for X_H on SMP significantly. The low NOB yield value is apparently related to the high nitrate loss as more substrate electrons are diverted to UAP, resulting in more carbon source for the nitrate denitrification. In contrast, the anoxic reduction factor

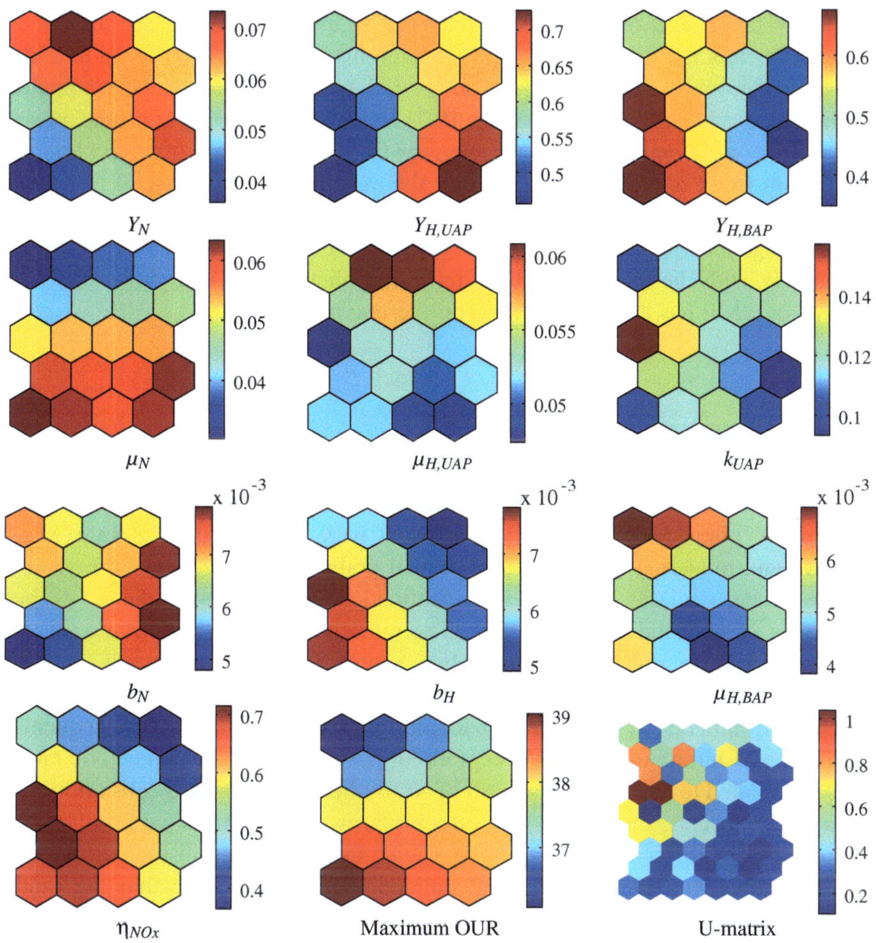

Fig. 10.12 Abstract visualization of the relationships between the maximum OUR (mg O_2 L^{-1} h^{-1}) and Y_N (g COD_X g^{-1} N), $Y_{H,\ UAP}$ (g COD_X g^{-1} COD_{UAP}), $Y_{H,\ BAP}$ (g COD_X g^{-1} COD_{BAP}), μ_N (h^{-1}), $\mu_{H,\ UAP}$ (h^{-1}), k_{UAP} (g COD_{UAP} g^{-1} COD_S), b_N (h^{-1}), b_H (h^{-1}), $\mu_{H,\ BAP}$ (h^{-1}), and η_{NOx} (—) using an SOM model

is positively correlated with the nitrate loss. The elevated η_{NOx} can enhance the denitrifying rate, and accordingly nitrate is used for denitrification at a higher rate. No correlation between the maximum growth rate of X_H on SMP and the nitrate loss could be found (Fig. 10.11). In addition, the visualization of the relationship between nitrate loss and b_N demonstrates that the respiration rate coefficient of NOB has no substantial impact on the nitrate loss in the granules. Therefore, it can be found that the nitrate loss is largely influenced by the anoxic reduction factor and the heterotrophic endogenous respiration rate. The parameters of the maximum growth rate of NOB and yield coefficient for X_H on SMP seem to be of less importance.

The visualization of relationships among the maximum OUR and the related key parameters is shown in Fig. 10.12. A high maximum OUR is linked to a high maximum growth rate of NOB, because NOB were found to be the dominative microbial species in this granule (Fig. 10.7) and NOB could only consume oxygen as an electron acceptor for their aerobic growth. Unlike the case of nitrate loss, the maximum OUR is greatly influenced by the maximum growth rate of NOB, and has no significant correlation with the anoxic reduction factor and the heterotrophic endogenous respiration rate. The yield coefficient for X_H on SMP correlates negatively with the maximum OUR, indicating that an elevated yield coefficient for X_H on SMP has a negative effect on the maximum OUR.

10.2.5 Conclusions

In this work, the nitrate loss mechanisms in the second step of the nitrification process by granular sludge, nitrite oxidation, were elucidated with both experimental, and modeling approaches. Microbial population diversity and abundance in the nitrite-oxidizing granules was identified to be 85 % NOB and 15 % heterotrophs of the total active bacteria. The released SMP in the nitrite oxidation process could be utilized as the carbon source for the heterotrophic growth. Microelectrode tests showed that the oxygen diffusion limitation caused anoxic microenvironments in the nitrite-oxidizing granules and allowed a sequential utilization of nitrate as electron acceptors for denitrification. A mathematic model was formulated to describe the growth of NOB, the production and consumption of microbial products, the anoxic heterotrophic growth with microbial products and nitrate (attributing to nitrate loss), the oxygen transfer and the substrate diffusion in the granules. Model calibration and validation results indicate that the model developed in this work was capable of simulating the nitrate loss mechanisms in biological nitrite oxidation process by aerobic granular sludge. Nitrate loss was largely influenced by the anoxic reduction factor and the heterotrophic endogenous respiration rate.

References

Al-Halbouni, D., Traber, J., Lyko, S., Wintgens, T., Melin, T., Tacke, D., Janot, A., Dott, W., Hollender, J.: Correlation of EPS content in activated sludge at different sludge retention times with membrane fouling phenomena. Water Res. **42**, 1475–1488 (2008)

Arrojo, B., Mosquera-Corral, A., Garrido, J.M., Mendez, R.: Aerobic granulation with industrial wastewater in sequencing batch reactors. Water Res. **38**, 3389–3399 (2004)

Alpkvist, E., Picioreanu, C., van Loosdrecht, M.C.M., Heyden, A.: Three-dimensional biofilm model with individual cells and continuum EPS matrix. Biotechnol. Bioeng. **94**, 961–979 (2006)

APHA, Standard methods for the examination of water and wastewater. 19th edn. American Public Health Association, Washington (1995)

Aquino, S.F., Stuckey, D.C.: Soluble microbial product formation in anaerobic chemostats in the presence of toxic compounds. Water Res. **38**, 255–266 (2004)

Aquino, S.F., Stuckey, D.C.: Integrated model of the production of soluble microbial products (SMP) and extracellular polymeric substances (EPS) in anaerobic chemostats during transient conditions. Biochem. Eng. J. **38**, 138–146 (2008)

Baquerizo, G., Maestre, J.P., Sakuma, T., Deshusses, M.A., Gamisans, X., Gabriel, D., Lafuente, J.: A detailed model of a biofilter for ammonia removal: Model parameters analysis and model validation. Chem. Eng. J. **113**, 205–214 (2005)

Beun, J.J., Heijnen, J.J., van Loosdrecht, M.C.M.: N-removal in a granular sludge sequencing batch airlift reactor. Biotechnol. Bioeng. **75**, 82–92 (2001)

Benjamin, S., Magbanua, Jr, Bowers, A.R.: Characterization of soluble microbial products (SMP) derived from glucose and phenol in dual substrate activated sludge bioreactors. Biotechnol. Bioeng. **93**, 862–870 (2006)

Carrera, J., Jubany, I., Carvallo, L., Chamy, R., Lafuente, J.: Kinetic models for nitrification inhibition by ammonium and nitrite in a suspended and an immobilised biomass systems. Process Biochem. **39**, 1159–1165 (2004)

Ciudad, G., Rubilar, O., Munoz, P., Ruiz, G., Chamy, R., Vergara, C., Jeison, D.: Partial nitrification of high ammonia concentration wastewater as a part of a shortcut biological nitrogen removal process. Process Biochem. **40**, 1715–1719 (2005)

Daims, H., Bruhl, A., Amann, R., Schleifer, K.H., Wagner, M.: The domain-specific probe EUB338 is insufficient for the detection of all bacteria: Development and evaluation of a more comprehensive probe set. Syst. Appl. Microbiol. **22**, 434–444 (1999)

de Beer, D., Stoodley, P., Lewandowski, Z.: Liquid flow in heterogeneous biofilms. Biotechnol. Bioeng. **44**, 636–641 (1994)

de Kreuk, M.K., Picioreanu, C., Hosseini, M., Xavier, J.B., van Loosdrecht, M.C.M.: Kinetic model of a granular sludge SBR—Influences on nutrient removal. Biotechnol. Bioeng. **97**, 801–815 (2007)

de Silva, D.G.V., Rittmann, B.E.: Nonsteady-state modeling of multispecies activated-sludge processes. Water Environ. Res. **72**, 554–565 (2000a)

de Silva, D.G.V., Rittmann, B.E.: Interpreting the response to loading changes in a mixed-culture completely stirred tank reactor. Water Environ. Res. **72**, 566–573 (2000b)

Gonzalez-Gil, G., Seghezzo, L., Lettinga, G., Kleerebezem, R.: Kinetics and mass-transfer phenomena in anaerobic granular sludge. Biotechnol. Bioeng. **73**, 125–134 (2001)

Guisasola, A., Jubany, I., Baeza, J.A., Carrera, J., Lafuente, J.: Respirometric estimation of the oxygen affinity constants for biological ammonium and nitrite oxidation. J. Chem. Technol. Biotechnol. **80**, 388–396 (2005)

Gujer, W., Henze, M., Mino, T., van Loosdrecht, M.C.M.: Activated sludge model No. 3. Water Sci. Technol. **39**, 183–193 (1999)

Gujer, W., Larsen, T.A.: The implementation of biokinetics and conservation principles in ASIM. Water Sci. Technol. **31**, 257–266 (1995)

Hao, X.D., Heijnen, J.J., van Loosdrecht, M.C.M.: Model-based evaluation of kinetic, biofilm and process parameters in a one reactor Ammonium removal (CANON) process. Biotechnol. Bioeng. **77**, 266–277 (2002)

Hellinga, C., Schellen, A.A.J.C., Mulder, J.W., van Loosdrecht, M.C.M., Heijnen, J.J.: The SHARON process: an innovative method for nitrogen removal from ammonium-rich waste waster. Water Sci. Technol. **37**, 135–142 (1998)

Henze, M., Grady, C.P.L.Jr., Gujer, W., Marais, G.V.R., Matsuo, T.: Activated sludge model No. 1. Scientific and Technical Report No. 1, IAWPRC, London (1987)

Henze, M., Gujer, W., Mino, T., van Loosdrecht, M.C.M.: Activated Sludge Models ASM1, ASM2, ASM2d, and ASM3. IWA Scientific and Technical Report No. 9. IWA Publishing, London (2000)

Horn, H., Hempel, D.: Growth and decay in an auto-/heterotrophic biofilm. Water Res. **31**, 2243–2252 (1997)

Horn, H., Neu, T.R., Wulkow, M.: Modelling the structure and function of extracellular polymeric substances in biofilms with new numerical techniques. Water Sci. Technol. **43**, 121–127 (2001)

Jarusutthirak, C., Amy, G.: Role of soluble microbial products (SMP) in membrane fouling and flux decline. Environ. Sci. Technol. **40**, 969–974 (2006)

Jubany, I., Baeza, A.J., Carrera, J., Lafuente, J.: Respirometric calibration and validation of a biological nitrite oxidation model including biomass growth and substrate inhibition. Water Res. **39**, 4574–4584 (2005)

Kampschreur, M.J., Tan, N.C.G., Kleerebezem, R., Picioreanu, C., Jetten, M.S.M., van Loosdrecht, M.C.M.: Effect of dynamic process conditions on nitrogen oxides emission from a nitrifying culture. Environ. Sci. Technol. **42**, 429–435 (2008)

Kindaichi, T., Kawano, Y., Ito, T., Satoh, H., Okabe, S.: Population dynamics and in situ kinetics of nitrifying bacteria in autotrophic nitrifying biofilms as determined by real-time quantitative PCR. Biotechnol. Bioeng. **94**, 1111–1121 (2006)

Koch, G., Egli, K., van der Meer, J.R., Siegrist, H.: Mathematical modelling of autotrophic denitrification in a nitrifying biofilm of a rotating biological contactor. Water Sci. Technol. **41**, 191–198 (2000)

Koch, G., Kuhni, M., Siegrist, H.: Calibration and validation of an ASM3-based steady-state model for activated sludge systems—Part I: Prediction of nitrogen removal and sludge production. Water Res. **35**, 2235–2245 (2001)

Larsen, P., Nielsen, J.L., Svendsen, T.C., Nielsen, P.H.: Adhesion characteristics of nitrifying bacteria in activated sludge. Water Res. (2008). doi:10.1016/j.watres.2008.02.015

Laspidou, C.S., Rittmann, B.E.: A unified theory for extracellular polymeric substances, soluble microbial products, and active and inert biomass. Water Res. **36**, 2711–2720 (2002a)

Laspidou, C.S., Rittmann, B.E.: Non-steady state modeling of extracellular polymeric substances, soluble microbial products, and active and inert biomass. Water Res. **36**, 1983–1992 (2002b)

Liu, S.Y., Liu, G., Tian, Y.C., Chen, Y.P., Yu, H.Q., Fang, F.: An innovative microelectrode fabricated using photolithography for measuring dissolved oxygen distributions in aerobic granules. Environ. Sci. Technol. **41**, 5447–5452 (2007)

Mosquera-Corral, A., Gonzalez, F., Campos, J.L., Mendez, R.: Partial nitrification in a SHARON reactor in the presence of salts and organic carbon compounds. Process Biochem. **40**, 3109–3118 (2005)

McSwain, B.S., Irvine, R.L., Hausner, M., Wilderer, P.A.: Composition and distribution of extracellular polymeric substances in aerobic flocs and granular sludge. Appl. Environ. Microbiol. **71**, 1051–1057 (2005)

Moussa, M.S., Hooijmans, C.M., Lubberding, H.J., Gijzen, H.J., van Loosdrecht, M.C.M.: Modelling nitrification, heterotrophic growth and predation in activated sludge. Water Res. **39**, 5080–5098 (2005)

Ni, B.J., Yu, H.Q.: An approach for modeling two-step denitrification in activated sludge systems. Chem. Eng. Sci. **63**, 1449–1459 (2008)

Ni, B.J., Fang, F., Xie, W.M., Yu, H.Q.: Growth, maintenance and product formation of autotrophs in activated sludge: Taking the nitrite-oxidizing bacteria as an example. Water Res. **42**, 4261–4270 (2008a)

Ni, B.J., Yu, H.Q., Sun, Y.J.: Modeling simultaneous autotrophic and heterotrophic growth in aerobic granules. Water Res. **42**, 1583–1594 (2008b)

Ni, B.J., Xie, W.M., Chen, Y.P., Fang, F., Liu, S.Y., Ren, T.T., Sheng, G.P., Yu, H.Q., Liu, G., Tian, Y.C.: Heterotrophs grown on the soluble microbial products (SMP) released by autotrophs are responsible for the nitrogen loss in nitrifying granular sludge. Biotechnol. Bioeng. **108**, 2844–2852 (2011)

Nicolella, C., van Loosdrecht, M.C.M., Heijnen, J.J.: Mass transfer and reaction in a biofilm airlift suspension reactor. Chem. Eng. Sci. **53**, 2743–2753 (1998)

Nogueira, R., Melo, L.F.: Competition between *Nitrospira* spp. and *Nitrobacter* spp. in nitrite-oxidizing bioreactors. Biotechnol. Bioeng. **95**, 169–175 (2006)

Okabe, S., Hiratia, K., Ozawa, Y., Watanabe, Y.: Spatial microbial distributions of nitrifiers and heterotrophs in mixed-population biofilms. Biotechnol. Bioeng. **50**, 24–35 (1996)

Pambrun, V., Paul, E., Sprandio, M.: Modeling the partial nitrification in sequencing batch reactor for biomass adapted to high ammonia concentrations. Biotechnol. Bioeng. **95**, 120–131 (2006)

Petersen, B., Gernaey, K., Devisscher, M., Dochain, D., Vanrolleghem, P.: A simplified method to assess structurally identifiable parameters in Monod-based activated sludge models. Water Res. **37**, 2893–2904 (2003)

Picioreanu, C., van Loosdrecht, M.C.M., Heijnen, J.J.: Modelling the effect of oxygen concentration on nitrite accumulation in a biofilm airlift suspension reactor. Water Sci. Technol. **36**, 147–156 (1997)

Pras, N.: Further kinetic characterization of alginate-entrapped cells of *Mucuna pruriens* L. Biotechnol. Bioeng. **33**, 1461–1468 (1989)

Ramesh, A., Lee, D.J., Hong, S.G.: Soluble microbial products (SMP) and soluble extracellular polymeric substances (EPS) from wastewater sludge. Appl. Microbiol. Biotechnol. **73**, 219–225 (2006)

Reichert, P.: Aquasim 2.0-User Manual, Computer Program for the Identification and Simulation of Aquatic Systems; EAWAG: Dübendorf, (ISBN 3 906484 16 5) (1998)

Sheng, G.P., Yu, H.Q., Li, X.Y.: Stability of sludge flocs under shear conditions: roles of extracellular polymeric substances (EPS). Biotechnol. Bioeng. **93**, 1095–1102 (2006)

Su, K.Z., Yu, H.Q.: Gas holdup and oxygen transfer in an aerobic granule-based sequencing batch reactor. Biochem. Eng. J. **25**, 209–215 (2005)

Vadivelu, V.M.: Kinetics and stoichiometric characterization of enriched *Nitrosomonas* and *Nitrobacter* cultures by decoupling the growth and energy generation processes. PhD Thesis. The University of Queensland (2006)

Vadivelu, V.M., Yuan, Z., Fux, C., Keller, J.: The inhibitory effects of free nitrous acid on the energy generation and growth processes of an enriched *Nitrobacter* culture. Environ. Sci. Technol. **40**, 4442–4448 (2006a)

Vadivelu, M.V., Yuan, Z., Fux, C., Keller, J.: Stoichiometric and kinetic characterization of *Nitrobacter* in mixed culture by decoupling the growth and energy generation processes. Biotechnol. Bioeng. **94**, 1176–1188 (2006b)

van Loosdrecht, M.C.M., Henze, M.: Maintenance, endogeneous respiration, lysis, decay and predation. Water Sci. Technol. **39**, 107–117 (1999)

Walter, B., Haase, C., Rabiger, N.: Combined nitrification/denitrification in a membrane reactor. Water Res. **39**, 2781–2788 (2005)

Wett, B., Rauch, W.: The role of inorganic carbon limitation in biological removal of extremely ammonia concentrated wastewater. Water Res. **37**, 1100–1110 (2003)

Xu, S.L., Hultman, B.: Experiences in wastewater characterization and model calibration for the activated sludge process. Water Sci. Technol. **33**, 89–98 (1996)

Yang, S.F., Tay, J.H., Liu, Y.: A novel granular sludge sequencing batch reactor for removal of organic and nitrogen from wastewater. J. Biotechnol. **106**, 77–86 (2003)

Zhou, Y., Pijuan, M., Zeng, R.J., Yuan, Z.: Free nitrous acid inhibition on nitrous oxide reduction by a denitrifying-enhanced biological phosphorus removal sludge. Environ. Sci. Technol. **42**, 8260–8265 (2008)

Chapter 11
Granulation in Pilot-Scale Reactor with Municipal Wastewater

Aerobic granulation of activated sludge was achieved in a pilot-scale sequencing batch reactor (SBR) for the treatment of low-strength municipal wastewater (<200 mg L^{-1} of COD, chemical oxygen demand). The volume exchange ratio and settling time of an SBR were found to be two key factors in the granulation of activated sludge grown on the low-strength municipal wastewater. After operation of 300 days, the mixed liquor suspended solids (MLSS) concentration in the SBR reached 9.5 g L^{-1} and consisted of approximate 85 % granular sludge. The average total COD removal efficiency kept at 90 % and NH_4^+–N was almost completely depleted (~95 %) after the formation of aerobic granules. The granules (with a diameter over 0.212 mm) had a diameter ranging from 0.2 to 0.8 mm and had good settling ability with a settling velocity of 18–40 m h^{-1}. Three bacterial morphologies of rod, coccus, and filament coexisted in the granules. Mathematical modeling was performed to get insight into this pilot-scale granule-based reactor. The modified IWA activated sludge model No 3 (ASM3) was able to adequately describe the pilot-scale SBR dynamics during its cyclic operation.

11.1 Introduction

Aerobic granules could be applied for the simultaneous removal of organic matter, nitrogen, and phosphorus (de Kreuk et al. 2005), and decomposition of toxic wastewaters (Jiang et al. 2002). Aerobic-granule-based SBR have been proven to be applicable for the treatment of wastewaters from various industries, such as malting (Schwarzenbeck et al. 2004), dairy (Arrojo et al. 2004), and soybean-processing (Su and Yu 2005). Many factors are involved in the granulation of activated sludge. Contributing factors include carbon source, hydrodynamic shear force, feast–famine regime, feeding strategy, dissolved oxygen, reactor configuration, pH, temperature, volume exchange ratio, and settling time (de Bruin et al. 2004; de

B.-J. Ni, *Formation, Characterization and Mathematical Modeling of the Aerobic Granular Sludge*, Springer Theses, DOI: 10.1007/978-3-642-31281-6_11,

Kreuk and van Loosdrecht 2004; Liu and Tay 2004; Su and Yu 2005; de Kreuk 2006). Because of the excellent settling capacity of aerobic granules, the land area needed for a municipal wastewater treatment plant could be reduced by 80 %, and accordingly the construction investment would be cut down, when a granular sludge system, instead of a conventional activated sludge system, was applied (de Bruin et al. 2004).

It shall be noticed that most of studies concerning aerobic granule have been focused on well-controlled lab-scale reactors with high- or middle-strength synthetic wastewaters (Beun et al. 1999; Peng et al. 1999; Moy et al. 2002; de Kreuk and van Loosdrecht 2004; Arrojo et al. 2004; Schwarzenbeck et al. 2004; Su and Yu 2005). Operating parameters and experience harvested in these studies may be applicable to the utilization of aerobic granule process for the treatment of municipal wastewater, which is usually characterized as a low-strength wastewater. How to cultivate active and compact aerobic granules with such a low-strength municipal wastewater is crucial for its full application. de Kreuk and van Loosdrecht (2006) had tried to cultivate aerobic granules by using presettled sewage as influent. After 20 days of operation at a high COD loading, heterogeneous aerobic granular structures were observed, with a sludge volume index (SVI) after 10 min settling of 38 mL g^{-1} and an average diameter of 1.1 mm. This led to a pilot-scale plant setup at a WWTP in Ede, the Netherlands. Two bubble columns with a diameter of 0.6 m and height of 6 m were used. After about 1 year cultivation, dry weight concentration was 10 g L^{-1}, of which 80 % consisted of granular sludge with a diameter larger than 0.212 mm. The influent concentration to the SBR in their work was in a range of 270–400 mg COD L^{-1}. These results provided valuable information for potential application of aerobic granules for the treatment of domestic sewage and municipal wastewater.

In China, with the recent rapid economic growth and urbanization, a large number of municipal wastewater treatment plants have to be constructed. The increasing population of China results in serious shortage of land area. Thus, aerobic granule process might be the promising technique for municipal wastewater treatment. However, the COD concentration of municipal wastewater in China is typically lower than 200 mg L^{-1}. Such a low substrate level would severely influence the granule formation, and could be bottleneck of wide application of this innovative process in China and other countries (Moy et al. 2002). Granulation of activated sludge on such a low-strength municipal wastewater has not been reported yet.

Therefore, the main objective of this work was to demonstrate the feasibility of cultivating aerobic granules with a low-strength wastewater and applying for this process for the effective treatment of municipal wastewater at pilot scale, to explore the crucial factors for the granulation of activated sludge with low-concentration substrate, and to model the substrate removal dynamics in a granule-based SBR.

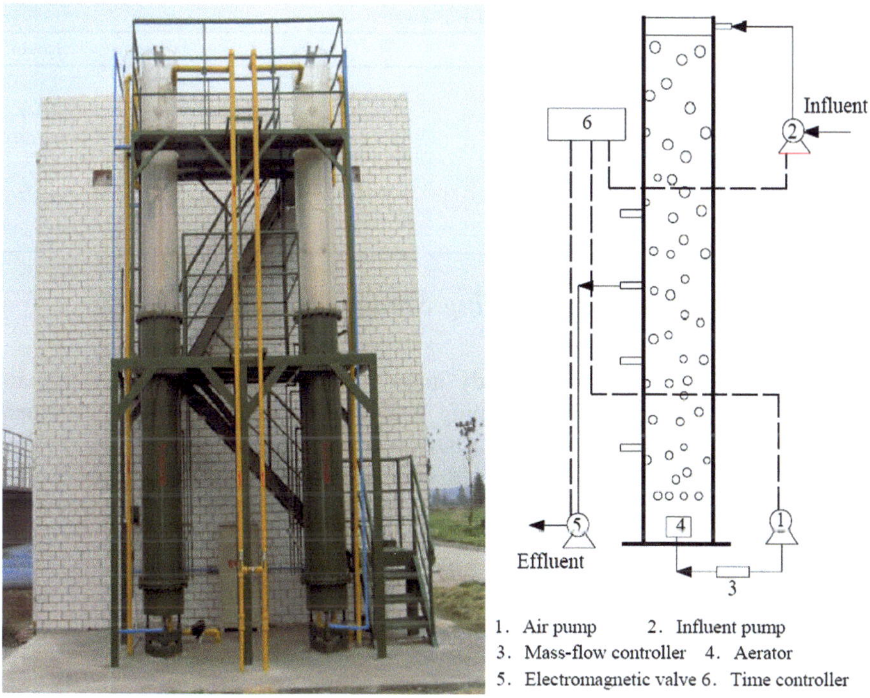

Fig. 11.1 The pilot-scale (1 m^3) SBR reactor for cultivating aerobic granules with low-strength municipal wastewater in Zhuzhuanjing WWTP. The left photograph shows the upper compartment (Plexglas), as well as the lower compartment (steel). The right schematic picture shows the SBR reactor configuration and flow chart (Reprinted from Ni et al. (2009) with permission from Elsevier)

11.2 Materials and Methods

11.2.1 Pilot-Scale SBR

A pilot-scale SBR was constructed in the Zhuzhuanjing WWTP, which was located in eastern Hefei, China. The columned SBR had a working volume of 1 m^3 with an internal diameter of 0.5 m and a height of 6 m (Fig. 11.1). The reactor was aerated by using a fine bubble aerator. The DO concentration was kept around 2 mg L^{-1} with a proportional integral derivative controller connected to an air compressor and an aeration ON/OFF controller. The reactor was operated in a fill-draw mode. Wastewater was introduced from the top of the reactor. The hydraulic retention time was 6–8 h and the effluent ratio varied from 50 to 70 % during the entire operation period.

Table 11.1 Composition of the municipal wastewater

Component	Level
Total COD (mg L^{-1})	95–200
Soluble COD (mg L^{-1})	35–120
NH_4^+–N (mg L^{-1})	10–40
Total N (mg L^{-1})	12–50
pH	6.5–7.5
VSS (mg L^{-1})	90–200

11.2.2 WasteWater and Seeding Sludge

The influent wastewater was directly introduced to the SBR from the primary sedimentation tank in the Zhuzhuanjing WWTP. The composition of the influent was given in Table 11.1. The COD level in this wastewater was below 200 mg L^{-1}. The SBR was inoculated with activated sludge taken from an aeration tank from the Zhuzhuanjing WWTP. The seeding sludge had a sludge age of 10 days, an MLSS concentration of 6.6 g L^{-1} and a sludge volume index (SVI) of 75.5 mL g^{-1}. Its specific gravity and the settling velocity were 1.004 and 6.8 m h^{-1}, respectively. Activated sludge of 700 L was seeded to the SBR, resulting in an initial MLSS concentration of 4.0 g L^{-1} in the reactor.

11.2.3 Operating Strategy

After inoculation, the SBR was initially operated in successive cycles of 4 h each. One cycle consisted of 5 min of influent addition, 185–200 min of aeration, 15–30 min of settling, and 20 min of effluent withdrawal. Effluent was withdrawn from the port at 3 m from the reactor bottom. After a 1-month operation, the reactor was operated in a cycle of 3 h, which consisted of 5 min of influent addition, 125–140 min of aeration, 15–30 min of settling, and 20 min of effluent withdrawal. Effluent was withdrawn from the port at 2.5 m from the reactor bottom. After granulation occurred (on day 80), the reactor was operated in a cycle of 3 h and the effluent was withdrawn from the port at 2.0 m of height from the reactor bottom.

11.2.4 Analysis

Granule formation was monitored by measuring SVI and determining the granule size distribution, which was done by sieving a sludge sample through a 0.212 mm sieve and measuring the total dry weight of biomass on the sieves. Biomass with a diameter larger than 0.212 mm was considered as granular sludge and its amount was expressed as the percentage of total dry weight in the sample (de Kreuk and van Loosdrecht 2006). The SVI_{10} and SVI_{30} were determined by reaching the volume of biomass in a 100 mL sample, after a settling period of 10 and 30 min, respectively, and then calculated with the biomass dry weight in this sample.

The granule morphology was analyzed by using an image analysis system (Image-Pro Express 4.0, Media Cybernetics) with an Olympus CX41 microscope and a digital camera (Olympus C5050 Zoom). The microstructure and predominant bacterial morphologies of the granules were observed using a scanning electron microscopy (SEM, Quantn 200, FEI Ltd.).

The EPS of granules were extracted using the CER method described by Zhang et al. (1999). The carbohydrate concentration in EPS was determined as glucose equivalent using enthrone–sulfuric acid method (Dubois et al. 1956), whereas the protein concentration was measured as bovine albumin equivalent using the Lowry method (Lowry et al. 1951).

The granule structure, particularly the distribution of microbial cells and EPS within granules, was examined with a CLSM (LSM 5 Pascal, Zeiss Inc.). For florescent staining of both cells and EPS, two probes were applied collectively: SYTO9 (25 μm, Molecular Probe, Eugene, OR) to target all microbes, and ConA-TRITC lectin (250 mg L^{-1}, Sigma) to target the polysaccharides with D-glucose or D-mannose (Chu and Li 2005). When excited by a laser at proper wavelengths, the SYTO9 and ConA-TRITC probes emitted green light and red light, respectively. The granules were placed and stained in slide wells and were incubated in a moisture chamber (a 50 mL conical centrifuge tube) in dark at room temperature for 20 min. After incubation, the samples were carefully rinsed with filtered phosphate buffered saline three times to remove any unbound probes. The stained samples were visualized on two channels with corresponding excitations and emissions for SYTO9 (488 nm, BP 515-530) and ConA-TRITC (543 nm, LP560). A stack scanning was performed to render the distribution of microbial cells and EPS.

The hydrophobic nature of the granular sludge was determined by measuring contact angle following the axisymmetric drop shape method proposed by Duncan-Hewitt et al. (1989) by using a contact angle analyzer (Powereach JC2000A, Zhongchen Co.). The settling velocity of the granules was measured by recording the time taken for an individual granule to fall from a certain height in a measuring cylinder. A column with a diameter of 4.0 cm and a height of 40 cm was used to determine the settling velocity of granules in water. Measurement of MLSS, MLVSS, effluent SS, NH_4^+–N, and COD was performed according to the Standard Methods (APHA 1995).

11.3 Mathematic Modeling

11.3.1 Model Description

For modeling the pilot-scale aerobic granular SBR, the ASM3 (Gujer et al. 1999) was extended with a consideration of aerobic/anoxic simultaneous storage and growth. The extended model had 11 model components: heterotrophic microorganisms (X_H), autotrophic microorganisms (X_A), storage products of heterotrophic

Table 11.2 Kinetic and stoichiometric parameters used in the model

Parameter	Definition	Values	Unit
Y_{STO}	Yield coefficient for X_H storage	0.81	g COD g^{-1} COD
$Y_{H,S}$	Yield coefficient for X_H growth on S_S	0.58	g COD g^{-1} COD
$Y_{H,STO}$	Yield coefficient for X_H growth on X_{STO}	0.68	g COD g^{-1} COD
Y_A	Yield coefficient for X_A growth	0.24	g COD g^{-1} COD
η_{NOx}	Anoxic reduction factor	0.55	–
f_I	Fraction of X_I in respiration	0.2	g COD g^{-1} COD
i_{NBM}	Nitrogen content of biomass	0.07	g N g^{-1} COD
i_{NXI}	Nitrogen content of X_I	0.02	g N g^{-1} COD
k_{STO}	Maximum storage rate of X_H	0.25	h^{-1}
k_H	Maximum hydrolysis rate	0.125	h^{-1}
$\mu_{H,STO}$	Maximum growth rate of X_H on X_{STO}	0.11	h^{-1}
$\mu_{H,S}$	Maximum growth rate of X_H on S_S	0.58	h^{-1}
K_S	Biomass affinity constant for S_S	11.38	g COD m^{-3}
K_{STO}	Biomass affinity constant for X_{STO}	1.0	g COD g^{-1} COD
K_X	Hydrolysis affinity constant	1.0	g COD g^{-1} COD
K_{O_2}	Dissolve oxygen affinity constant	0.20	g O_2 m^{-3}
b_{STO}	Respiration rate coefficient of X_{STO}	0.016	h^{-1}
b_H	Respiration rate coefficient of X_H	0.016	h^{-1}
μ_A	Maximum growth rate of X_A	0.05	h^{-1}
b_A	Respiration rate coefficient of X_A	0.0063	h^{-1}
K_AO_2	Oxygen affinity constant for X_A	0.50	g O2 m^{-3}
K_{NH_4}	Biomass NH_4 affinity constant	0.01	g N m^{-3}
K_{A,NH_4}	NH_4 affinity constant for X_A	1.0	g N m^{-3}
K_{NO_3}	Biomass NO_3 affinity constant	0.50	g N m^{-3}

microorganisms (X_{STO}), readily biodegradable substrate (S_S), slowly biodegradable substrate (X_S), residual inert organics (X_I), soluble inert organics (S_I), ammonia–N (S_{NH_4}), nitrate–N (S_{NO_3}), nitrogen (S_{N_2}), and dissolved oxygen (S_O). It had 14 microbial processes: hydrolysis; aerobic storage; anoxic storage; aerobic growth on S_S; anoxic growth on S_S; aerobic growth on X_{STO}; anoxic growth on X_{STO}; aerobic endogenous respiration; anoxic endogenous respiration; aerobic respiration of X_{STO}; anoxic respiration of X_{STO}; aerobic growth; aerobic endogenous respiration; and anoxic endogenous respiration. The detailed information about the model development could be found in our previous paper (Ni et al. 2008). Table 11.2 defined all the parameters used in the developed model, their symbols, and their units.

X_S was hydrolyzed with a rate characterized by surface reaction kinetics as in ASM3 (Eq. 11.1). Hydrolysis process was presumably a rate-limiting step for partial readily biodegradable substrate utilization processes.

$$\frac{dX_S}{dt} = -k_H \frac{X_S/X_H}{K_X + X_S/X_H} X_H \tag{11.1}$$

Heterotrophs primarily utilized the readily biodegradable substrate (S_S) available via storage and growth. OURs attributed to the primary heterotrophic substrate consumption (OUR_{growS}) and substrate storage (OUR_{STO}) were respectively defined by the following process kinetic equations:

$$OUR_{growS} = \mu_{H,S} X_H \frac{1 - Y_{H,S}}{Y_{H,S}} \frac{S_S}{K_S + S_S} \frac{S_O}{K_{O_2} + S_O} \frac{S_{NH_4}}{K_{NH_4} + S_{NH_4}} \tag{11.2}$$

$$OUR_{STO} = k_{STO} X_H \frac{1 - Y_{STO}}{Y_{STO}} \frac{S_S}{K_S + S_S} \frac{S_O}{K_{O_2} + S_O} \tag{11.3}$$

Oxygen utilization attributed to the autotrophic NH_4^+–N consumption was defined by the equation below:

$$OUR_{growX_A} = \mu_A X_A \frac{4.57 - Y_A}{Y_A} \frac{S_{NH_4}}{K_{A,NH_4} + S_{NH_4}} \frac{S_O}{K_{A,O_2} + S_O} \tag{11.4}$$

After the consumption of the primary S_S, the secondary heterotrophic growth occurred on the stored X_{STO} at the famine phase. Oxygen utilization attributed to the secondary aerobic heterotrophic substrate (X_{STO}) consumption ($OUR_{growSTO}$) was defined by Eq. 11.4.

$$OUR_{growSTO} = \mu_{H,STO} X_H \frac{1 - Y_{H,STO}}{Y_{H,STO}} \frac{K_H}{K_H + S_H} \frac{X_{STO}/X_H}{K_{STO} + X_{STO}/X_H} \frac{S_{NH_4}}{K_{NH_4} + S_{NH_4}} \frac{S_O}{K_{O_2} + S_O} \tag{11.5}$$

In the AQUASIM program (Reichert 1998), the aerobic-granule SBR was modeled with a biofilm compartment connected to a completely mixed compartment with an advective link, and a high recirculation flow rate ($Q_{recirculation}$) was incorporated from the biofilm compartment to the completely mixed one to ensure the same concentrations in the liquid of the biofilm compartment and in the completely mixed one (Beun et al. 2001). To simulate the effluent draw, another completely mixed compartment was introduced and connected to the first one with an advective link (Beun et al. 2001). Aeration was simulated by introducing a gas compartment (i.e., completely mixed compartment) connected with the biofilm one with a diffusive link.

The granules grown in the SBR had a size distribution of 0.2–0.8 mm with an average diameter of 0.32 mm, and their size distribution keeps changing in time. However, the size distribution was not taken into account in this work, as it would significantly increased the complexity of numerical computation and it was not expected to contribute to a better understanding of the system. Therefore, in the simulation the granule diameter was chosen to be 0.32 mm, the average diameter of the granules in this SBR.

11.3.2 Wastewater Characterization

For modeling purpose, the initial fractionation of organic matter in the municipal wastewater was performed with the procedures developed based on the standard Dutch STOWA guidelines (Roeleveld and van Loosdrecht 2002). However, in this work some minor modifications had been made according to Sahlstedt et al. (2003). The biodegradable COD (BCOD) in the influent was the sum of the readily biodegradable soluble COD (S_S) and the slowly biodegradable particulate COD (X_S). In the guidelines of Roeleveld and van Loosdrecht (2002) the fraction BCOD was determined from a biological oxygen demand (BOD) analysis where the BOD was measured as a function of time. The choice for the BOD-analysis was made because it was still a widely used in municipal wastewater treatment plants. The biodegradable COD (BCOD, $S_S + X_S$) concentration was calculated from the formula proposed by Grady et al. (1999). The inert fraction S_I had to be determined independently and subtracted from the soluble COD to give the fraction S_S. S_I was evaluated from the inert COD in the effluent. The concentrations of X_S and X_I were estimated based on the BOD measurements considered as described by Makinia et al. (2005). With the presented methodology for the characterization of the organic fractions S_I, S_S, and X_S, the fraction X_I was found to be the difference between the total COD and the other fractions. The full organic fractionation was carried out on each consecutive day of the long-term simulations.

11.4 Results and Discussion

11.4.1 Performance of the Pilot-Scale SBR

The pilot-scale SBR reactor had been continuously operated for more than 300 days, and the operating results are shown in Fig. 11.2. In the initial days, the effluent ratio was stepwise increased from 50 to 70 %, and the loading rate was correspondingly increased from 0.6 to 1.0 kg COD m^{-3} day^{-1}. A settling period of 30 min was applied in the initial 10 days to prevent severe sludge washout and was then decreased gradually to 15 min in the subsequent operation. As a result of the improvement of settling ability, the MLSS kept increasing despite of the excess sludge discharge, even after the settling time was reduced to 15 min in February, 2007.

The sludge properties, including MLVSS/MLSS and MLSS concentration, are shown in Figs. 11.2a and b. After March of 2007, the MLSS was stabilized at 8.0–10.0 g L^{-1}, which was much higher than that of a conventional activated sludge treatment plant. The ratio of MLVSS/MLSS increased continuously from 50 % of the seeding sludge to 70–80 % of the granules. The reactor performance was continuously improved in terms of total COD and NH_4^+–N removal efficiencies during the operation (Figs. 11.2c, d). This improvement became obvious after

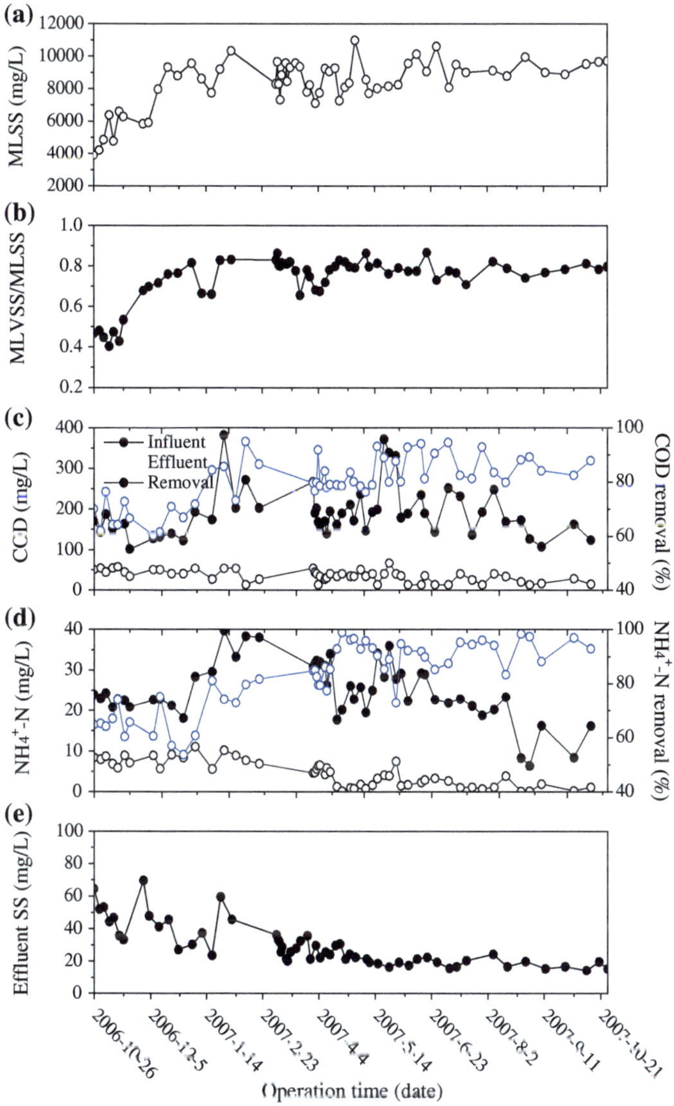

Fig. 11.2 Operating parameters and performance of the pilot-scale SBR: **a** MLSS; **b** MLVSS/MLSS; **c** influent and effluent total COD levels as well as total COD removal efficiency; **d** influent and effluent NH_4^+–N concentrations as well as NH_4^+–N removal efficiency; and **e** effluent SS concentrations (Reprinted from Ni et al. (2009) with permission from Elsevier)

around 3 months of operation, and the total COD removal efficiency was kept at 85–95 %, while the NH_4^+–N removal efficiency was stabled at 90–99 % afterwards, even with the disturbance of the influent nutrient concentration.

As shown in Fig. 11.2e, the effluent SS concentrations fluctuated within a range of 10–80 mg L^{-1}. At the beginning of granulation process, to avoid severe washing out of the seed activated sludge, settling time was sufficiently long to allow most of sludge to settle. After 1 month of operation, effluent was withdrawn from the port at 2.5 m (instead of 3 m) of height from the reactor bottom and small bioparticles were washed out. This resulted in a selection pressure for bioparticle size in sedimentation process. Correspondingly, the effluent SS reached 70 mg L^{-1}, but decreased gradually (Fig. 11.2e). After the granulation occurred on day 80, the effluent was withdrawn from the port at 2.0 m of height from the reactor bottom, resulting in a slight increase in SS concentration again. Thereafter, the effluent SS concentration decreased gradually and became stable at about 15 mg L^{-1} (Fig. 11.2e).

Since the main objective of this work was to demonstrate the feasibility of cultivating aerobic granules with a low-strength wastewater and to explore the crucial factors for the granulation of activated sludge with low-concentration substrate in a pilot-scale reactor, anoxic phase was not included in SBR cycles to favor the conversion (N-removal) efficiency. As a result, a small fraction of nitrate produced from the removal of NH_4^+–N was used for denitrification (Fig. 11.6d). However, after successful granulation of activated sludge the reactor performance could be readily optimized for total N-removal, when an anoxic phase was introduced and the simultaneous nitrification–denitrification by granular sludge was promoted.

11.4.2 Granulation of Activated Sludge

Images of the sludge in the granulation process are shown in Fig. 11.3. The seeding sludge had a fluffy, irregular, and loose-structure morphology (Fig. 11.3a). The sludge color gradually changed from brown to yellow with the process of experiment (Fig. 11.3). After 80 days of operation, small granules with a diameter of about 0.3 mm were observed (Fig. 11.3b). Thereafter, the number and average diameter of the granules increased slowly (Fig. 11.3c). At day 120, the granules had a mean diameter of about 0.4 mm (Fig. 11.3c). Afterwards, the granules kept growing in a much lower speed up to the end of the operation with a diameter ranging from 0.2 to 0.8 mm (Fig. 11.3d).

Figure 11.4 illustrates the profiles of granule fractions (the amount of granules as percentage of the total dry weight), and SVI_{10} and SVI_{30} of the sludge in the granulation process. The formation of granules was a gradual process, as evidenced by the increase in granule fractions (Fig. 11.4a). After 300 days of operation, the MLSS concentration in the SBR consisted of approximate 85 % granular sludge. Figure 11.4b shows that the SVI_{10} decreased gradually from 100 to 35 ml g^{-1}, whereas the SVI_{30} decreased gradually from 65 to 35 ml g^{-1}. The mean size of bioparticles gradually increased from 0.03 to 0.45 mm in the granulation process (Fig. 11.4c).

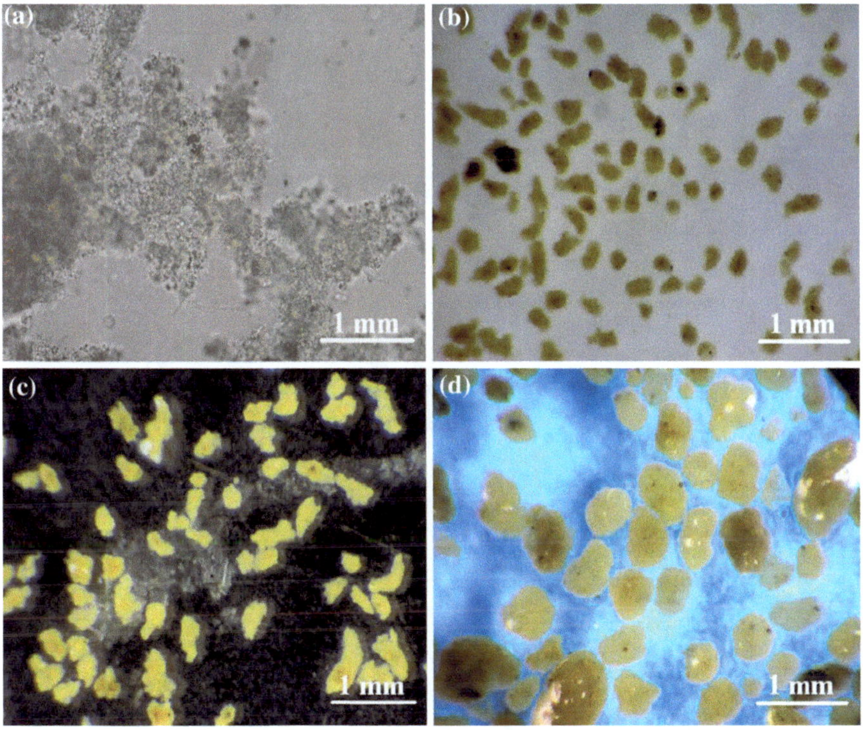

Fig. 11.3 Images of sludge in the granulation process: **a** seeding sludge; **b** sludge on day 80; **c** sludge on day 120; and **d** sludge on day 200 (Reprinted from Ni et al. (2009) with permission from Elsevier)

11.4.3 Characteristics of the Granular Sludge

The size of the granules, in a range of 0.2–0.8 mm, was relatively smaller compared with the granules feeding with high-strength wastewaters (Table 11.3). However, these granules played an important role in maintaining the dynamic sludge balance and stability of this pilot scale reactor. Figure 11.5a shows the porous inner structure of the granules. Under the SEM, three types of bacterial, i.e., rod, coccus, and filament, with irregular morphologies were found to in the granules (Fig. 11.5a). They showed excellent settleability with a settling velocity in a range of 18–40 m h^{-1}, allowing a greater biomass accumulation in the reactor and more effective sludge-effluent separation in the system.

In this study, each gram of aerobic granules (in VSS) contained 62.2 ± 5.8 mg carbohydrates and 26.5 ± 4.5 mg proteins, with a ratio of proteins/carbohydrates about 0.4. The granulation of microbes was greatly related to their ability to produce EPS. EPS can mediate both cohesion and adhesion of cells, and plays a crucial role in maintaining structural integrity in a community of immobilized cells (Liu and Tay 2001). The granules consist of much EPS with a higher fraction of

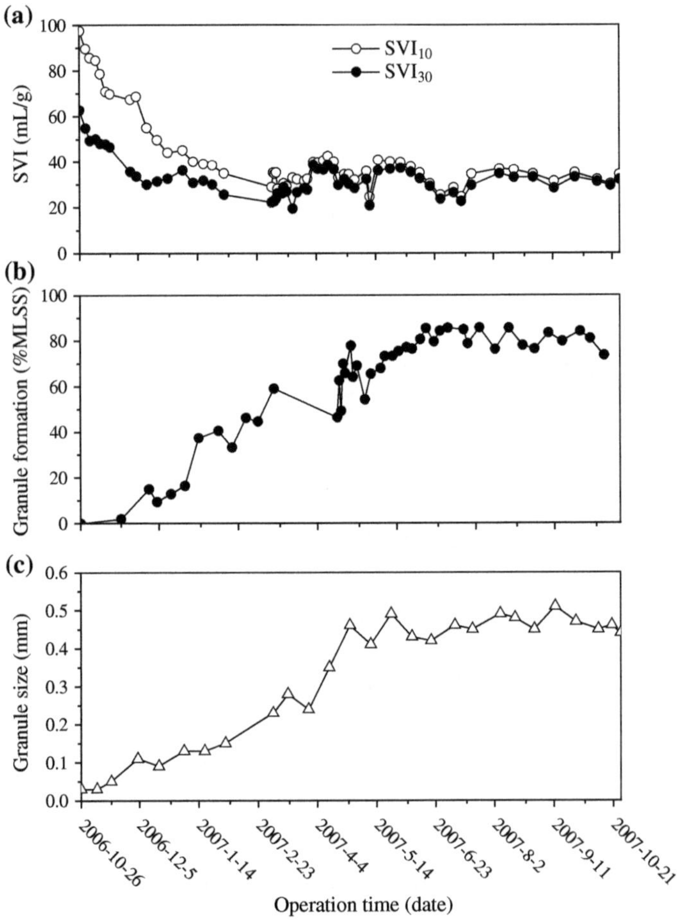

Fig. 11.4 Granulation process in the pilot-scale SBR: **a** variations of the SVI_{10} and SVI_{30}; **b** increase in the amount of aerobic granules (diameter >0.212 mm) as percentage of the total dry weight; and **c** mean granular size profiles (Reprinted from Ni et al. (2009) with permission from Elsevier)

carbohydrates than proteins. It was supposed that the gel-forming carbohydrates played an important role in construction of a stable structure for sludge granulation (McSwain et al. 2004; Vargas-Garcia et al. 2003; Yang et al. 2005). Thus, an increase in both EPS abundance and carbohydrate content should be essential to the formation of the granules.

The CLSM examination after fluorescent staining indicated different patterns of cell and EPS distributions in the aerobic granules. Figure 11.5b shows that cells and EPS were well mixed. The EPS were distributed throughout the granules, while microbes were mainly situated in the outer granule layer. The spatial distributions of EPS and cells within the granules provided an insight into their structures and

Table 11.3 Comparison among various aerobic granules

Influent COD (mg L^{-1})	SVI (mL g^{-1})	Settling velocity (m h^{-1})	Diameter (mm)
600	80-100	–	0.3–0.5
830	–	>24	4.6
300	–	>10	2.5
500	40	–	0.52
500–3000	60	20	0.25–4.0
2000	30.8	36.6	1.24
1250	23	18–31	0.5–1.2
270–400	60	–	0.212–0.6
500	50–75	–	0.4–1.9
1000–7000	40–130	24.2–36.4	1.2–1.9
<200	35	18–40	0.2–0.8

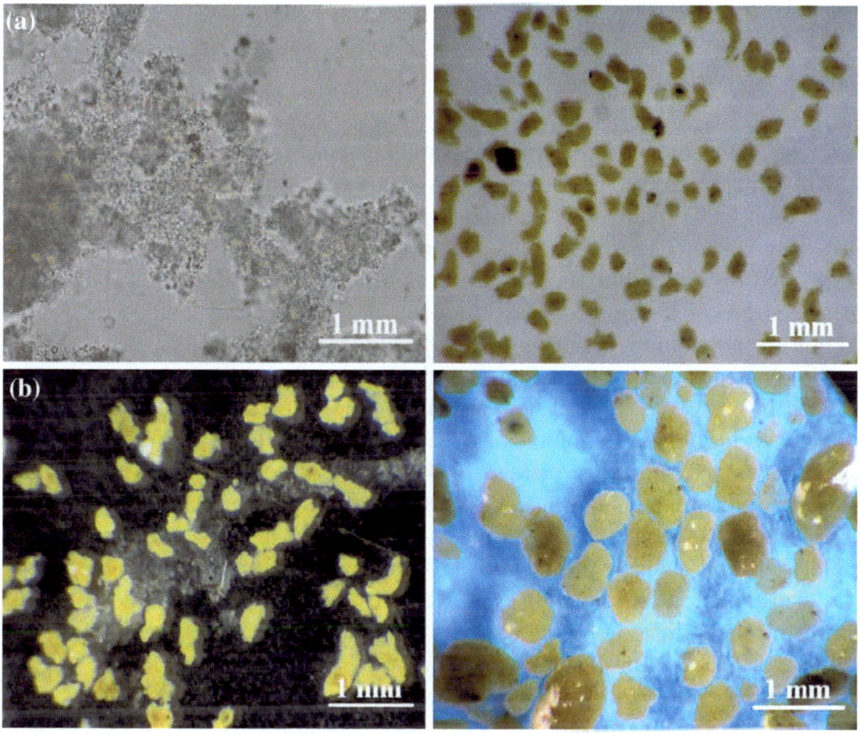

Fig. 11.5 **a** SEM images of the granular sludge in the pilot-scale SBR; and **b** CLSM images of the 50-μm cryosections through the center of the granules. Cells are stained with SYTO9 (*green*) and polysaccharides are stained with ConA (*red*) (Reprinted from Ni et al. (2009) with permission from Elsevier)

related granulation mechanisms. The formation of aerobic granules was mainly attributable to the growth of microbes together with the conglutination of EPS under the hydrodynamic conditions. When the small and dense microbial aggregates continued to grow, granules were formed. However, the growth of the granules in size increased the mass transfer limitation of substrates into the granules. Active cells would grow and most of them might stay close with the outer granule layer in the range of about 50 μm (Fig. 11.5b). Due to substrate limitation, cells inside the granules might decay and leak out cellular substances, which constituted the EPS matrix in the granules (Fig. 11.5b).

After the formation of the granules, the sludge contact angle increased from $31 \pm 3^{\circ}$ of the seeding sludge to $42 \pm 5^{\circ}$. From a thermodynamic point of view, an increase in the hydrophobicity of cell surface causes a decrease in the excess Gibbs energy of the surface (Bos et al. 1999), which is in favor of the formation of the granules. Therefore, a higher hydrophobicity of the cell surfaces in granules would result in a more strengthened cell-to-cell interaction, and is beneficial for the formation of more dense and stable structure (Thaveesri et al. 1995).

11.4.4 Modeling

After an inclusion of aerobic/anoxic simultaneous storage and growth into the ASM3 (Ni et al. 2008), the developed model was able to adequately describe the OUR, total COD removal and ammonia dynamics during a cycle of the granular SBR (Fig. 11.6). The oxygen uptake showed a good agreement between the predicted and observed data (Fig. 11.6a). The difference between the simulated and measured NH_4^+–N value for was less than 15 %, suggesting a good model prediction for ammonia removal (Fig. 11.6b).

Simultaneous carbon oxidation and nitrification took place when oxygen was utilized in one cycle (Fig. 11.6). The only several minutes lasting initial peak in OUR was due to the oxidation of S_S and NH_4^+–N. Then, nitrification and the oxidation of the substrate released by hydrolysis accounted for the subsequent oxygen utilization. This transition for oxygen consumption curve could be described well with the developed model. The total COD concentration decreased rapidly and continuously for the initial 20 min (Fig. 11.6), as the external substrate was consumed. In nitrification, the NH_4^+–N concentration decreased rapidly, suggesting its utilization by the nitrifiers as an energy and nitrogen source, which resulted in NO_3^-–N production (Fig. 11.6).

Competition between them for DO and space in the aerobic granules occurred when DO was below a sufficient level. For the autotrophs this was disadvantageous, because they had to grow under oxygen-diffusion-limitation conditions, which slowed the nitrification process. In a discontinuously fed reactor like this granule-based SBR, organic substrate was able to completely penetrate into granules because of the temporarily high substrate concentration in liquor, but oxygen was present only on the outer granule layers. However, the autotrophs would not successfully outcompete the heterotrophs on the outer granule layers, because their maximum growth rate was lower than that of the heterotrophs. The further hydrolysis of suspended solids adsorbed on the granules would also cause

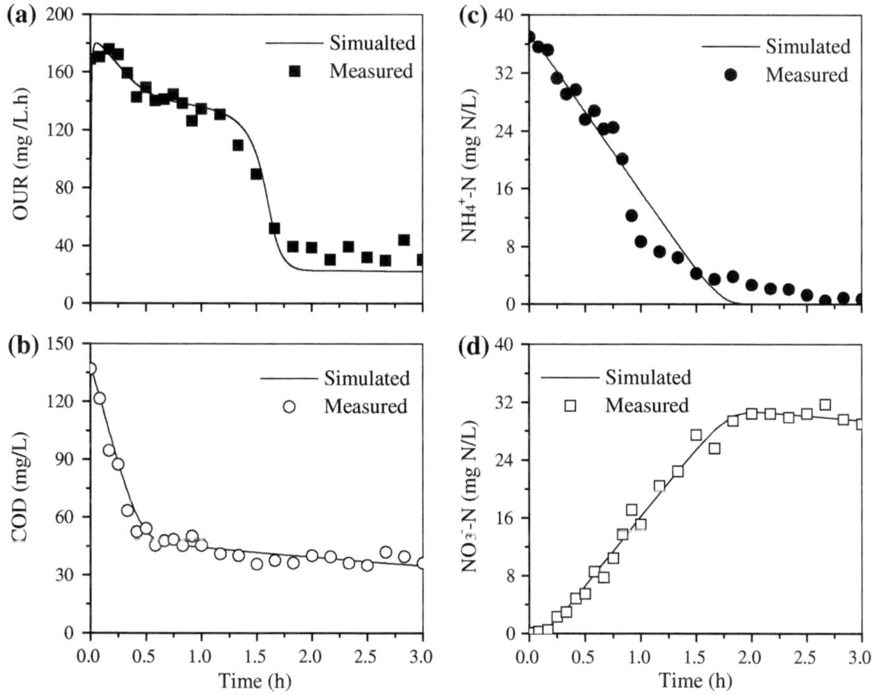

Fig. 11.6 Modeling of OUR, total COD removal, ammonia and nitrate profiles in a cyclic operation of the pilot-scale SBR (Reprinted from Ni et al. (2009) with permission from Elsevier)

heterotrophic growth in this outer layer and competition for the autotrophs. Consequently, it was expected that the distribution order of microorganisms in granules was heterotrophs–autotrophs–heterotrphs from the outside to the inside with these feast–famine regimes.

11.4.5 Key Factors in Granulation with a Low-Strength Municipal Wastewater

Aerobic granulation is a microbial phenomenon induced by selection pressure through changing microbial surface properties and metabolic behavior (Liu and Tay 2004). The important role of organic loading rate (OLR) in the formation of aerobic granules have been recognized (Liu and Tay 2004; de Kreuk 2006). A relatively high OLR facilitates the granulation process (Moy et al. 2002; de Kreuk 2006).

For an SBR system, the OLR is dependent on both influent substrate concentration and hydraulic loading rate (HLR). Thus, if a low-concentration wastewater is used as the substrate and a usual HLR is employed, the resulting OLR will not be sufficiently high to keep sludge increase for granulation. On the other hand, an increased OLR, attributed to an elevated HLR, will compensate for the low substrate concentration, and this in turn favors the granulation process. In order to keep a high OLR and accordingly to achieve sludge granulation in this pilot-scale SBR for the treatment of municipal wastewater, its HLR has to be kept at a high level. The HLR of an SBR is governed by its volume exchange ratio (VER) and cycle period. In this work, the cycle period was set at 3 h because of the low total COD level in the wastewater. In this case, keeping a high level of VER was the only way to elevate its HLR. The VER of the pilot-scale SBR was gradually increased from 50 to 75 %, which was much higher than the values (e.g., 50 %) reported for aerobic granulation on high- or middle-strength synthetic wastewaters (Beun et al. 1999; Peng et al. 1999; Moy et al. 2002; de Kreuk and van Loosdrecht 2004; Arrojo et al. 2004; Schwarzenbeck et al. 2004; Su and Yu 2005). In addition, it was also higher than the VER values (e.g., 40 %) used for municipal wastewater treatment plants where no granulation occurs (Irvine and Ketchum 1988; Grady et al. 1999). Therefore, the HLR of the pilot-scale SBR was deliberately elevated to maintain a sufficiently high level of OLR for granulation (e.g., 1.0 kg COD m^{-3} day^{-1} in this work).

The settling time is another key factor in the sludge granulation with a low-concentration wastewater. In an SBR, wastewater is treated in successive cycles, each of which lasts several hours. At the end of each cycle, the biomass is settled before the effluent is withdrawn. The settling time acts as a major hydraulic selection pressure on sludge. A short settling time preferentially selects for the growth of rapidly settling bioparticles and the bioparticles with a poor settleability are washed out. A lower settling time would result in a higher settling velocity for the retained bioparticles, and vice versa. The readily retainable bioparticles in the SBR ensures a more rapid and efficient granulation. In previous studies, a short settling time has been commonly applied to enhance the aerobic granulation in SBRs (Beun et al. 1999; Moy et al. 2002; Arrojo et al. 2004; Schwarzenbeck et al. 2004; Su and Yu 2005). At a long settling time, poorly settling sludge flocs can not be withdrawn effectively and they may in turn outcompete granule-forming bio-particles on nutrients for growth. As a result, aerobic granulation can not be achieved in an SBR operated at a longer settling time. This suggests that the settling time is a crucial factor in the granulation of activated sludge.

As demonstrated by the experimental results above, the VER and settling time could be manipulated as effective selection pressure for a successful granulation of activated sludge in an SBR. For the low-strength municipal wastewater, granulation could be achieved and accelerated when the SBR was operated at a high VER and a low settling time.

11.4.6 Significance of this Work

To the best of our knowledge, this work might be the first attempt to cultivate aerobic granules on wastewater with an average substrate concentration less than 170 mg COD L^{-1} and to apply them for the effective pilot-scale treatment of municipal wastewater. This study presents a systematic approach, a useful operating strategy and important process parameters for the cultivation of aerobic granules and the treatment of a low-strength wastewater. The VER and settling time were recognized as the crucial parameters in the formation of aerobic granules on this type of wastewater in SBRs.

Aerobic granular sludge process is a promising technology from both engineering and economic points of view. Performance of municipal wastewater treatment plants can be improved by using aerobic granular sludge in ways that allows high conversion rates and efficient biomass separation to minimize the reactor volume. As a result, the land area and investment needed for a wastewater treatment plant can be significantly reduced. The information provided here will be useful for the design, development, and application of aerobic granular sludge process for the treatment of municipal wastewater and domestic sewage.

11.5 Conclusions

Granulation of activated sludge was achieved in a pilot-scale SBR system for the treatment of a low-strength municipal wastewater. After operation of 300 days, the activated sludge in the reactor was predominated with granular sludge (approximate 85 %). The diameter size of the granules was a range of 0.2–0.8 mm and they had good settling ability with a settling velocity of 18–40 m h^{-1}. The granules grown on the wastewater were heterogeneous and their carbon and ammonium removal capacity was high. The volume exchange ratio and settling time were found to be two key factors in the granulation of activated sludge grown on such a low-concentration wastewater in an SBR. A modified ASM3 with a consideration of aerobic/anoxic simultaneous storage and growth was able to model the pilot-scale SBR dynamics.

References

APHA: Standard methods for the examination of water and wastewater, 19th edn. American Public Health Association, Washington, DC (1995)

Arrojo, B., Mosquera-Corral, A., Garrido, J.M., Mendez, R.: Aerobic granulation with industrial wastewater in sequencing batch reactors. Water Res. **38**, 3389–3399 (2004)

Beun, J.J., Hendriks, A., van Loosdrecht, M.C.M., Morgenroth, M., Wilderer, P.A., Heijnen, J.J.: Aerobic granulation in a sequencing batch reactor. Water Res. **33**, 2283–2290 (1999)

Beun, J.J., Heijnen, J.J., van Loosdrecht, M.C.M.: N-removal in a granular sludge sequencing batch airlift reactor. Biotechnol. Bioeng. **75**, 82–92 (2001)

Bos, R., van der Mei, H.C., Busscher, H.J.: Physico-chemistry of initial microbial adhesive interactions—its mechanisms and methods for study. FEMS Microbiol. Rev. **23**, 179–230 (1999)

Chu, H.P., Li, X.Y.: Membrane fouling in a membrane bioreactor (MBR): sludge formation and fouling characteristics. Biotechnol. Bioeng. **90**, 323–331 (2005) de Bruin, L.M.M., de Kreuk, M.K., van Der Roest, H.F.R., van Loosdrecht, M.C.M., Uijterlinde, C.: Aerobic granular sludge technology, alternative for activated sludge technology. Water Sci. Technol. **49**, 1–9 (2004)

de Kreuk, M.K.: Aerobic granular sludge scaling up a new technology. PhD thesis, Delft University of Technology, Delft (2006)

de Kreuk, M.K., van Loosdrecht, M.C.M.: Selection of slow growing organisms as a means for improving aerobic granular sludge stability. Water Sci. Technol. **49**, 9–19 (2004)

de Kreuk, M.K., van Loosdrecht, M.C.M.: Formation of aerobic granules with domestic sewage. J. Environ. Eng. **132**, 694–697 (2006)

de Kreuk, M.K., Heijnen, J.J., van Loosdrecht, M.C.M.: Simultaneous COD, nitrogen, and phosphate removal by aerobic granular sludge. Biotechnol. Bioeng. **90**, 761–769 (2005)

Dubois, M., Gilles, K.A., Hamilton, J.K., Rebers, P.A., Smith, F.: Colorimetric method for determination sugars and related substance. Anal. Chem. **28**, 350–356 (1956)

Duncan-Hewitt, W.C., Policova, Z., Cheng, P., Vargha-Butler, E.I., Neumann, A.W.: Semi-automatic measurement of contact angles on cell layers by a modified axis symmetric drop shape analysis. Colloids Surf. **42**, 391–402 (1989)

Grady Jr., C.P.L., Daigger, G.T., Lim, H.C.: Biological wastewater treatment. Revised and expanded, 2nd edn. Marcel Dekker, New York (1999)

Gujer, W., Henze, M., Mino, T., van Loosdrecht, M.C.M.: Activated sludge model No. 3. Water Sci. Technol. **39**, 183–193 (1999)

Irvine, R.L., Ketchum Jr., L.H.: Sequencing batch reactors for biological wastewater treatment. Critical Rev. Environ. Control **18**, 255–294 (1988)

Jiang, H.L., Tay, J.H., Tay, S.T.L.: Aggregation of immobilized activated sludge cells into aerobically grown microbial granules for the aerobic biodegradation of phenol. Lett. Appl. Microbiol. **35**, 439–445 (2002)

Liu, Y., Tay, J.H.: Detachment forces and their influence on the structure and metabolic behaviour of biofilms. World J. Microb. Biotechnol. **17**, 111–117 (2001)

Liu, Y., Tay, J.H.: State of the art of biogranulation technology for wastewater treatment. Biotechnol. Adv. **22**, 533–563 (2004)

Lowry, O.H., Farr, A.L., Randall, R.J.: Protein measurement with the folin phenol reagent. J. Biol. Chem. **193**, 265–275 (1951)

Makinia, J., Rosenwinkel, K.H., Spering, V.: Long-term simulation of the activated sludge process at the Hanover-Gummerwald pilot WWTP. Water Res. **39**, 1489–1502 (2005)

McSwain, B.S., Irvine, R.L., Wilderer, P.A.: The effect of intermittent feeding on aerobic granule structure. Water Sci. Technol. **49**, 19–25 (2004)

Moy, B.Y.P., Tay, J.H., Toh, S.K., Liu, Y., Tay, S.T.L.: High organic loading influences the physical characteristics of aerobic sludge granules. Lett. Appl. Microbiol. **34**, 407–412 (2002)

Ni, B.J., Xie, W.M., Yu, H.Q., Wang, Y.Z., Wang, G., Dai, X.L.: Granulation of activated sludge in a pilot-scale sequencing batch reactor for the treatment of low-strength municipal wastewater. Water Res. **43**, 751–761 (2009)

Ni, B.J., Yu, H.Q., Sun, Y.J.: Modeling simultaneous autotrophic and heterotrophic growth in aerobic granules. Water Res. **42**, 1583–1594 (2008)

Peng, D., Bernet, N., Delgenes, J.P., Moletta, R.: Aerobic granular sludge-a case study. Water Res. **33**, 890–893 (1999)

Reichert, P.: AQUASIM 2.0-user manual, computer program for the identification and simulation of aquatic systems. Swiss Federal Institute for Environmental Science and Technology (EAWAG), Dübendorf (1998)

Roeleveld, P.J., Van Loosdrecht, M.C.M.: Experience with guidelines for wastewater characterisation in The Netherlands. Water Sci. Technol. **45**, 77–87 (2002)

Sahlstedt, K.E., Aurola, A.M., Fred, T.: Practical modelling of a large activated sludge DN-process with ASM3. In: Proceedings of the 9th IWA Specialized Conference on Design, Operation and Economics of Large Wastewater Treatment Plants, pp. 141–148. Praha, Czech Republic, 1–4 Sept 2003

Schwarzenbeck, N., Erley, R., Mc Swain, B.S., Wilderer, P.A., Irvine, R.L.: Treatment of malting wastewater in a granular sludge sequencing batch reactor (SBR). Acta Hydrochim. Hydrobiol. **32**, 16–24 (2004)

Su, K.Z., Yu, H.Q.: Formation and characterization of aerobic granules in a sequencing batch reactor treating soybean-processing wastewater. Environ. Sci. Technol. **39**, 2818–2827 (2005)

Thaveesri, J., Daffonchio, D., Liessens, B., Vandemeren, P., Verstraete, W.G.: Granulation and sludge bed stability in upflow anaerobic sludge bed reactors in relation to surface thermodynamics. Appl. Environ. Microbiol. **61**, 3681–3686 (1995)

Vargas-Garcia, M.C., Lopez, M.J., Elorrieta, M.A., Suarez, F., Moreno, J.: Properties of polysaccharide produced by *Azotobacter vinelandii* cultured on 4-hydroxylbenzolic acid. J. Appl. Microbiol. **94**, 388–395 (2003)

Yang, S.F., Tay, J.H., Liu, Y.: Effect of substrate nitrogen/chemical oxygen demand ratio on the formation of aerobic granules. J. Environ. Eng. ASCE **131**, 86–92 (2005)

Zhang, X.Q., Bishop, P.L., Kinkle, B.K.: Comparison of extraction methods for quantifying extracellular polymers in biofilms. Water Sci. Technol. **39**, 211–218 (1999)

Chapter 12
Start-up of the Anammox Process by Seeding Aerobic Granular Sludge

The accelerating start-up of anaerobic ammonium oxidation (anammox) process in an upflow anaerobic sludge blanket (UASB) reactor was achieved by seeding aerobic granules. The structure and physicochemical properties of the anammox granules and the reactor performance were characterized. Microbial community analysis was performed to reveal the composition and dynamics of the microbial consortium based on denaturing gradient gel electrophoresis (DGGE) analysis and 16S rRNA genes amplified from the granular sludge. The results demonstrate that aerobic granular sludge could be effectively used to inoculate anammox reactors.

12.1 Microbial and Physicochemical Characteristics of Anammox Granules

12.1.1 Introduction

The anammox (ANaerobic AMMonia OXidation) process is a recently discovered biological nitrogen removal process, in which the ammonia is oxidized to nitrogen gas while nitrite is used as the electron acceptor (Jetten et al. 1997; van Dongen et al. 2001; Zhang et al. 2007). In contrast to heterotrophic denitrification (van de Graaf et al. 1996; Jetten et al. 2005), the anammox process does not require the dose of external electron donors (e.g., methanol) due to their chemolithoauto-trophic lifestyle. Furthermore, if this process is combined with a partial nitrification step, only half of the ammonium needs to be nitrified to nitrite, which can subsequently be converted into nitrogen gas with the remaining ammonium through the anammox process. This reduces the oxygen demand in the system and leads to further reduction in operational costs (van der Star et al. 2007).

The anammox bacteria have a slow growth rate (Strous et al. 1999a, b) with their doubling time at best estimated at 7–11 d (Strous et al. 1999a, b; van der Star

B.-J. Ni, *Formation, Characterization and Mathematical Modeling of the Aerobic Granular Sludge*, Springer Theses, DOI: 10.1007/978-3-642-31281-6_12,

et al. 2008). The yield of the anammox bacteria have been determined to be 0.066 C-mol biomass mol^{-1} ammonium consumed, and the maximum ammonium consumption rate is ~ 45 nmol mg^{-1} protein min^{-1} (Strous et al. 1999a, b). Given the slow growth rate and low yield, the enrichment of anammox bacteria from a mixed inoculum requires the optimization of conditions favorable for the anammox bacteria, and generally takes 200–300 days (Dapena-Mora et al. 2004a; van der Star et al. 2007; Jetten et al. 2005). Thus, conditions that would reduce the start-up time of anammox reactors would positively effect the implementation of the process. Several sources of inocula, such as activated sludge (Dapena-Mora et al. 2004a), nitrifying activated sludge (van der Star et al. 2007), and anaerobic sludge (Jetten et al. 2005), have been used for the start-up of anammox reactors with start-up time as long as 1000 d (van der Star et al. 2007). Given the slow growth rate, very efficient biomass retention seems to be essential to retain the anammox bacteria within the reactor systems during enrichment (Strous et al. 2002).

Aerobic granules have been reported to have high microbial diversity (Xavier et al. 2007) and compact structure with very good settling properties. Thus, granular sludge is an efficient means of biomass retention. These properties including interspecies competition and mass transfer result in the stratification of microbial species with anoxic pockets in the interior of the granules that may be suitable to harbor anammox bacteria. Therefore, the main objective of this study was to investigate the feasibility of start-up of the anammox process by using a UASB reactor and seeding the reactor with aerobic granular sludge. After the successful start-up and the formation of anammox granules, the structure and physicochemical properties of the anammox granules, and the reactor performance were characterized. Microbial community analysis revealed that the dominant anammox species was related to a species of anammox bacteria present in anammox biofilms. It is expected that results of this study will be used for a shorter start-up and more stable operation of anammox reactors in full-scale treatment of nitrogen-rich wastewaters.

12.1.2 Materials and Methods

12.1.2.1 Reactor, Seeding Aerobic Granules, and Wastewater

A plexiglas UASB reactor (2.1 L, 10 cm in inner diameter) and a gas–solid separator portion (2.6 L) were used in this study. The seeding aerobic granular sludge was taken from a lab-scale sequencing batch reactor treating soybean-processing wastewater (Su and Yu 2005). The volatile suspended solids (VSS) concentration of the seeding aerobic granules was 11.12 ± 0.15 g L^{-1}. Sludge (1.0 L) was seeded to the UASB reactor, resulting in an initial VSS level of 7.85 g L^{-1} in the reactor.

A synthetic wastewater was used in this experiment. It consisted of $(NH_4)_2SO_4$, 25–200 mg N L^{-1}; $NaNO_2$, 25–220 mg N L^{-1}; $CaCl_2{\cdot}2H_2O$, 0.18 g L^{-1}; $MgSO_4{\cdot}7H_2O$, 0.12 g L^{-1}; KH_2PO_4, 0.027 g L^{-1}; $KHCO_3$, 0.5 g L^{-1}; and 1 mL L^{-1} trace element solution I and II, in which trace element solution I: EDTA, 5.0 g L^{-1}; $FeSO_4$, 5.0 g L^{-1}; and trace element solution II: EDTA, 5.0 g L^{-1}; $ZnSO_4{\cdot}7H_2O$, 0.43 g L^{-1}; $CoCl_2{\cdot}6H_2O$, 0.24 g L^{-1}; $MnCl_2{\cdot}4H_2O$, 0.99 g L^{-1}; $CuSO_4{\cdot}5H_2O$, 0.25 g L^{-1}; $Na_2MoO_4{\cdot}2H_2O$, 0.22 g L^{-1}; $NiCl_2{\cdot}6H_2O$, 0.19 g L^{-1}; $Na_2SeO_4{\cdot}10H_2O$, 0.21 g L^{-1}; H_3BO_4, 0.014 g L^{-1}. The influent pH value was adjusted to 7.0 by adding 3 M NaOH or 3 M HCl.

12.1.2.2 Operation

After inoculation, the reactor was sparged with nitrogen gas for 30 min at a flow rate of 0.1 L min^{-1}. In the start-up, the UASB reactor was initially fed with the synthetic wastewater with a relatively low nitrogen loading rate (NLR) of 0.042 g L^{-1} per day at an HRT of 30 h for 1 month. Later, the HRT was shortened to 24 h for another 1 month of operation. Thereafter, the HRT was kept constant at 24 h, but the NLR was elevated through increasing the influent ammonium and nitrite concentrations, as shown in Fig. 12.1. The ammonium and nitrite concentrations in the influent were increased up to 0.21 g L^{-1} of NH_4^+–N and 0.24 g L^{-1} of NO_2^-–N, respectively. The applied Nitrogen loading rate was increased from 0.042 to 0.5 g L^{-1} per day. The reactor temperature was maintained at 30°C using a ribbon heater and a temperature controller throughout the experiments.

12.1.2.3 DNA Extraction

For DNA isolation, samples (2 mL) were centrifuged, the supernatant discarded and resuspended in 0.75 ml (120 mM sodium phosphate buffer) and glass beads (~0.3 g, 0.25 mm diameter) were added. Afterwards, cells were disrupted by bead-beating for 30 s. For DNA isolation, the procedures follow the protocol described by Kowalchuk et al. (2004). DNA was then suspended in 50 μL ultra-pure water, and kept at 4°C for 24 h until further analysis. DNA quality was assessed by 1% agarose gel electrophoresis, and concentration measured with the NanoDrop ND-1000 (Isogen Life science).

12.1.2.4 16S rRNA Sequences and Phylogenetic Analysis

A primer combination of Pla46F (*Escherichia coli* positions 46–63) forward and universal reverse (630R) (*E. coli* positions 1529–1545) primers was used for the preferential amplification of 16S rRNA genes of the members of the Planctomycetes (Schmid et al. 2003). PCR fragments were cloned with the pGEM®-T Easy cloning kit (Promega) and XL1 Bleu *E. coli* competent cells. Plasmid DNA was

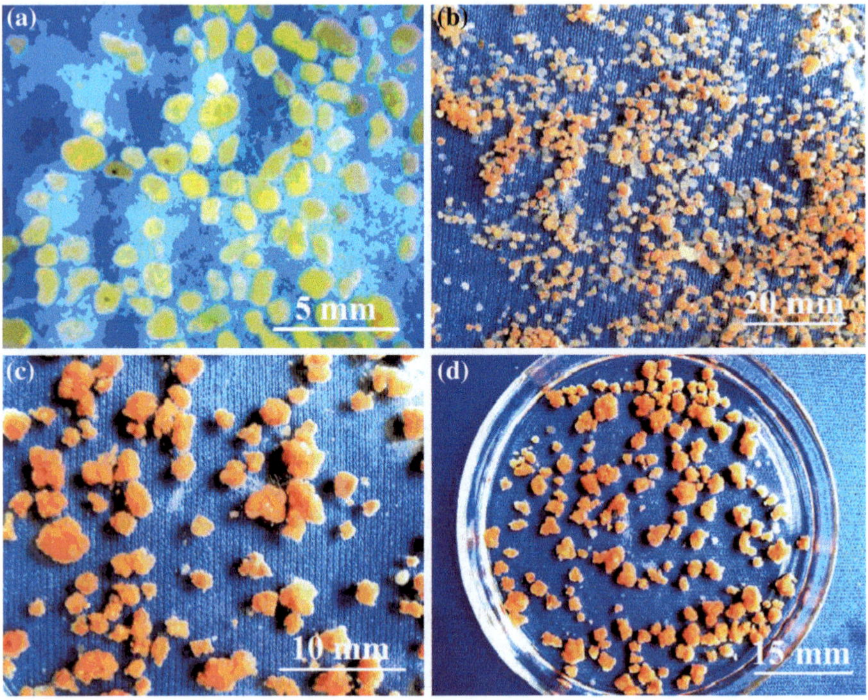

Fig. 12.1 Images of sludge samples in the granulation process: **a** seeding aerobic granules; **b** granules after 140 days of operation; **c** granules on 250th day; and **d** granules on day 300th (Reproduced from Ni et al. (2010) with permission from American Society for Microbiology)

isolated and purified with the Gene JETTM Plasmid Miniprep kit (Fermentas life sciences). The complete sequences of the 16S rRNA gene fragments were determined by using M13 forward and reverse primers targeting vector sequences adjacent to the multiple cloning site. Phylogenetic analyses were performed with maximum likelihood, neighbor joining, and maximum parsimony methods with 50 % sequence conservation filters for Planctomycetes. Sequencing and retrieval of the cloned 16S rRNA genes and phylogenetic analyses were performed with MEGA 4 as described in (Kartal et al. 2007).

12.1.2.5 Real-Time PCR Assay

A primer combination of Amx694F (*Escherichia coli* positions 694–713: GGGGAGAGTGG AACTTCGG) forward and reverse Amx960R (*E. coli* positions 960-979: GCTCGCACAAGCGGTGGAGC) primers was used for quantifying anammox bacteria.

Real time gradient PCR was performed with an iCycler iQ5 thermocycler & real-time detection system (Bio-Rad). For the standard curves, PCR fragments were cloned with the pGEM®-T Easy cloning kit (Promega) and XL1 Bleu *E.coli* competent cells. The plasmids were isolated and purified with the Gene JETTM Plasmid Miniprep kit (Fermentas life sciences). Each PCR mixture (25 μL) was composed of 12.5 μL of 1 × SYBR Green PCR master mix (Finnzymes, Finland), 1 μL of each forward and reverse primers (20 *p*mol μL^{-1}), either 1μL of template DNA or 10^0–10^9 copies per well of the standard vector plasmid of the clone Candidatus *Kuenenia stuttgartiensis* (AF375995) from one single cell reactor, i.e., the type of the newly described membrane bioreactor in which anammox bacteria grow as a single cellular suspension rather than biofilm aggregates (8). PCR amplification and detection were performed in MicroAmp Optical 96-well reaction plates. The PCR temperature program was initiated with 3 min at 96°C, followed by 40 cycles of 1 min at 96°C, 1 min at different annealing temperature and 1 min at 72°C. For 16S rRNA primers the annealing temperature was increasing from 55 to 65°C using the iQ™ Custom SyBR® Green Supermix (Biorad). A melting curve analysis for SYBR Green assay was done after amplification to distinguish the targeted PCR product from the non-targeted PCR product (Tsushima et al. 2007).

12.1.2.6 Fish

Biomass (1 mL) was harvested from the enrichment culture and hybridized with fluorescent probes Amx368, the probe was purchased as Cy3 labeled derivatives from Thermo Electron Corporation (Ulm, Germany). The hybridization conditions and approaches were performed as described in (Kartal et al. 2007).

12.1.2.7 Other Analysis

The morphology of the anammox granules was analyzed by using an image analysis system (Image-Pro Express 4.0, Media Cybernetics) with an Olympus CX41 microscope and a digital camera (Olympus C5050 Zoom). Their micro-structure and predominant bacterial morphologies were observed using a scanning electron microscopy (SEM, Quantn 200, FEI Ltd.). The structure, particularly the distribution of microbial cells and EPS within granules, was examined using a CLSM (LSM 5 Pascal, Zeiss Inc.). The hydrophobic nature of the granules was determined by measuring the contact angle following the axisymmetric drop shape method described by Sheng et al. (2005). The extraction and determination of EPS in granules were conducted according to Sheng and Yu (2006). Their settling velocity was measured by recording the time taken for an individual granule to fall from a certain height in a measuring cylinder. Measurement of total suspended solids (TSS), VSS, NH_4^+–N, NO_2^-–N, and NO_3^-–N was performed according to the Standard Methods (APHA 1995).

12.1.2.8 Nucleotide Sequence Accession Numbers

The 16S rDNA was deposited with Genbank as GQ175382.

12.1.3 Results and Discussion

12.1.3.1 Formation of the Anammox Granules

A 2.1-L UASB reactor was seeded with 7.9 g VSS L^{-1} aerobic granular sludge. The aerobic granules were regular in shape and yellow. Their average diameter was around 1.2 mm (Fig. 12.1a). One week after seeding the UASB reactor, the granules began to shrink and then partially disintegrated to diameters of 0.3–0.5 mm probably due to the anaerobic conditions. In the subsequent 7–40 days after seeding a part (about 50 %) of the suspended biomass was washed out and new granulation occurred. About 100 days after seeding, the outer surface of the particles turned red in color, suggesting that the anammox microorganisms became dominant in the reactor. In this phase, the out cells began to grow and particles became bigger. Anammox granules with henna color and diameters of 0.7–0.8 mm were visible at the reactor bottom after 160 days of operation (Fig. 12.1b). These granules with diameters of 0.7–0.8 mm developed rapidly, and large red granules with diameters over 2.5 mm were formed after 250 days of operation (Fig. 12.1c). After 250–280 days rapid growth slowed, followed by decrease in the granule formation. As a result, stable red granules formed a stratified layer in the reactor (Fig. 12.1d).

12.1.3.2 Performance of the Anammox Reactor

The anammox reactor was continuously operated for more than 300 days, and the results are shown in Fig. 12.2. From the 20th day, the NH_4^+–N removal efficiency gradually increased (Fig. 12.2a) reaching a maximum of 90 % after 200 days of operation. The total nitrogen removal efficiency was 94 % after 250 days of operation.

As depicted in Figs. 12.1 and 12.2, the anammox reactor start-up was achieved within 8 months as shown by the morphology changes (color and size) of granules in the reactor with high nitrogen removal efficiencies. This is quite similar to the fastest start-up reported in other studies (van de Graaf et al. 1996; Strous et al. 1999a, b; van Dongen et al. 2001; Dapena-Mora et al. 2004a, b; Third et al. 2005; Kartal et al. 2008). The start-up times by seeding aerobic granules in this work (150 days) were similar to those of seeding with activated sludge and anaerobic digestion sludge (150 days, Chamchoi and Nitisoravut 2007) with a higher nitrogen removal rates of 0.47 kg N m^{-3} d^{-1} compared to 0.09 kg N m^{-3} d^{-1}. The start-up times in this work (150 days) were more shortened than those of

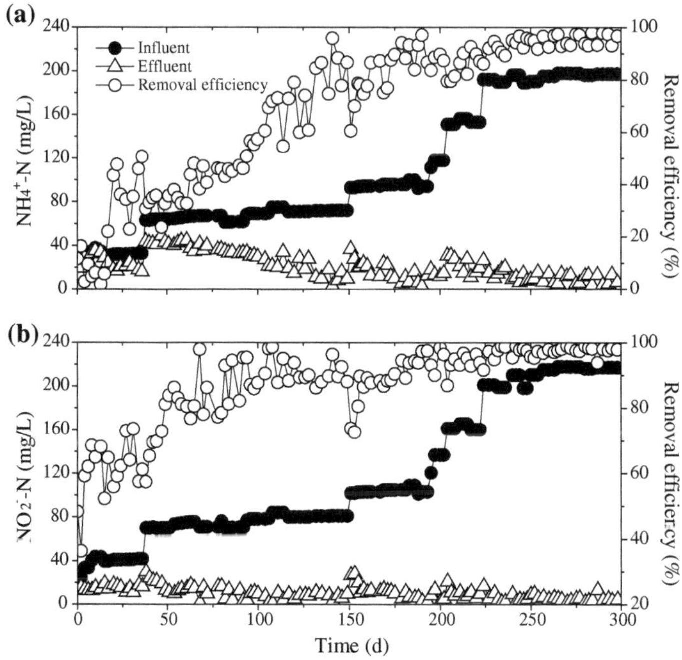

Fig. 12.2 Reactor performance: **a** concentration profiles of NH_4^+–N; and **b** concentration profiles of NO_2^-–N (Reproduced from Ni et al. (2010) with permission from American Society for Microbiology)

seeding with nitrification sludge (250 days, Pynaert et al. 2004) or denitrification sludge (392 days, Tsushima et al. 2007).

Based on the observation and studies on the microstructure of the granules, the granule formation after seeding with aerobic granules can be considered to proceed as disintegrated granular core mechanism: (i) the seeded aerobic granules disintegrated into small particles due to the anaerobic conditions; (ii) ongoing growth of the anammox bacteria; (iii) attachment and growth of biofilm on the small particles; (iv) ongoing growth of the biofilm; and (v) development of new granules from the detached biofilm. In this process, it is clear that the first step of providing a suitably large attachment surface is important to enhanced granulation. The experimental results also showed that small particles core from the disintegrated seeding granules can serve as an effective and attachment material for development of anammox granules.

In order to obtain evidence that the process occurring in the granular reactors fitted to the stoichiometry of the anammox process, nitrogen balances were performed during the reactor operational period. The mean ratio of NO_2^-–N consumed to NH_4^+–N consumed was about 1.21 and the mean ratio of NO_3^-–N produced to NH_4^+–N consumed was about 0.22. These values agree with those obtained by Strous et al. (1999a, b) for the anammox process, in which the ratio of

Fig. 12.3 Size distribution of the ANAMMOX granules by number on 250th day (Reproduced from Ni et al. (2010) with permission from American Society for Microbiology)

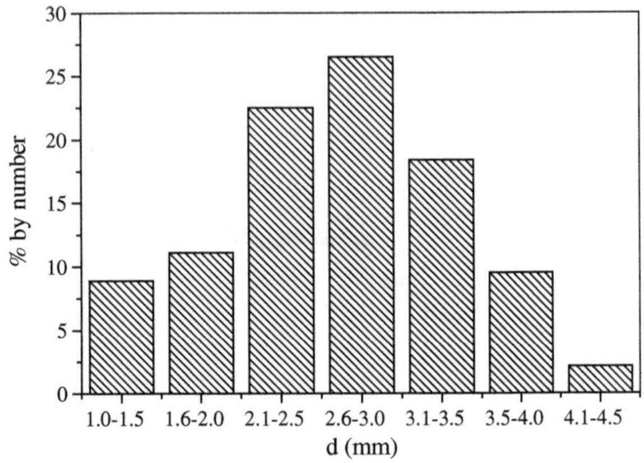

NO_2^-–N consumed to NH_4^+–N consumed and the ratio of NO_3^-–N produced to NH_4^+–N consumed were 1.3 and 0.26, respectively. In addition, the pH of the effluent (7.5–8.0) in the anammox reactor was always higher than the influent (7.0).

12.1.3.3 Characteristics of the Anammox Granules

The mature anammox granules had diameters between 1.0 and 4.5 mm with the majority (approximately 67 %) having a diameter of 2.0–3.5 mm (Fig. 12.3). Their porous inner structure could facilitate the passage of substrates as well as the release of N_2. The scanning electron micrographs showed typical coccoid shaped cells as the dominant microorganisms on the granule surface (Fig. 12.4a), very similar to those reported by others (van de Graaf et al. 1996; Strous et al. 1999a, b; Jetten et al. 2001). Figure 12.4b showed the typical crescent shaped anammox cells obtained via transmission electron micrography of chemically fixed anammox cells (van Niftrik et al. 2008).

This matrix structure showed excellent settleability. The settling velocities of the anammox granules were in a range of 41–79 m h^{-1}, showing their excellent settling ability, compared to conventional sludge flocs. Their settling velocities were similar to those of the methanogenic granules (e.g. 52.9 m h^{-1}), and were at least two times greater than those of the usual flocculated sludge. The significant increase in settling velocities indicated that the anammox granules had a highly dense and compact structure. This property would allow a greater biomass accumulation in the reactor and more effective sludge-effluent separation in the treatment system.

The extracellular polysaccharide of each gram of anammox granule (in VSS) contained 83.2 ± 7.9 mg carbohydrates and 42.7 ± 6.5 mg proteins, with a ratio of proteins/carbohydrates of 0.51. The CLSM analysis (Fig. 12.4c) showed mixed patterns of cells and EPS distributions in the anammox granules. The EPS were

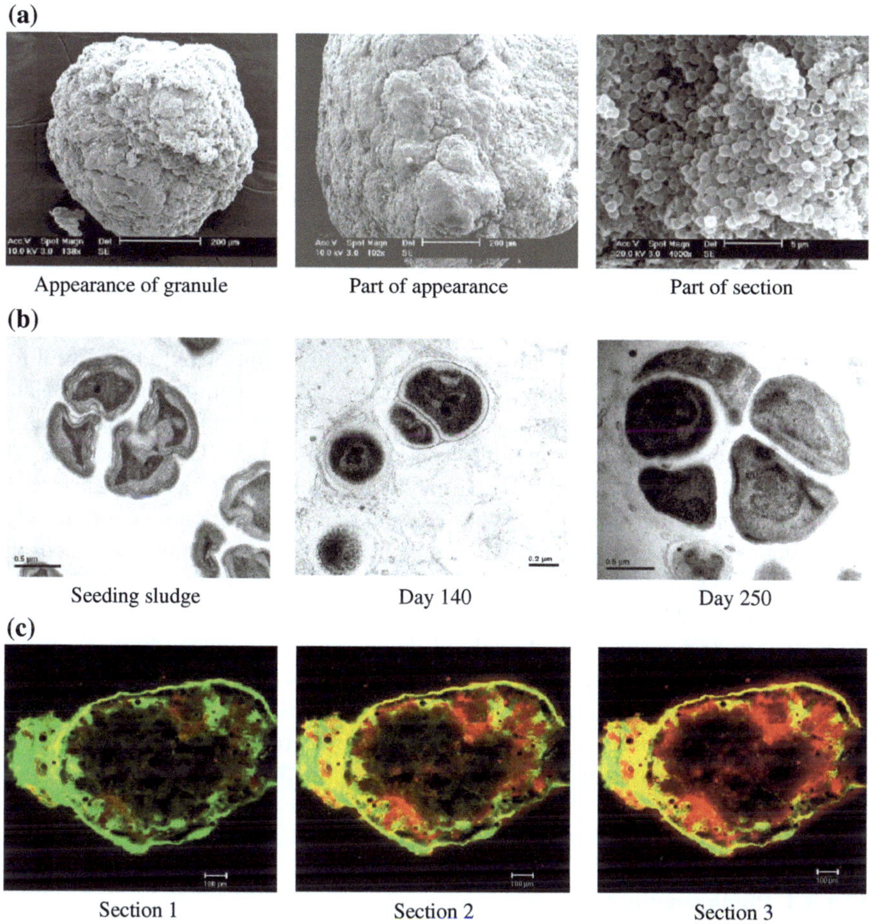

Fig. 12.4 **a** SEM images; **b** TEM images; and **c** CLSM images of the 50-μm cryosections through the anammox granules from surface to center. Cells were stained with SYTO9 (*green*) and polysaccharides were stained with ConA (*red*) (Reproduced from Ni et al. (2010) with permission from American Society for Microbiology)

distributed throughout the granules, while the bacteria were mainly situated in the outer layer of the granule. The total EPS content in methanogenic granules was substrate dependent, ranging 10–91 mg EPS g^{-1} VSS (27). This was substantially lower than that for the anammox granules (The total EPS content was about 125 mg EPS g^{-1} VSS) found in this study. The ratio of proteins/carbohydrates was 1.2–4.0 for the EPS of methanogenic granules (Morgan et al. 1990), much greater than that of the anammox ones. This significant difference suggested that proteins might be the key EPS constituents for the methanogenic granules, but that carbohydrates, rather than proteins, might play a more important role in the formation of the anammox granules. The high amount of glycosylation proteins encoded in the Kuenenia genome is supportive of this suggestion (Morgan et al. 1990).

After granulation, the contact angle became 49 ± 3°, which was higher than the seeding sludge of 44 ± 2°, indicating a high hydrophobicity. Usually, a higher hydrophobicity of the cell surfaces would result in a stronger cell-to-cell interaction, and is considered beneficial for the formation of more dense and stable structures (Thaveesri et al. 1995). From a thermodynamic point of view, an increase in the hydrophobicity of cell surface causes a decrease in the excess Gibbs energy of the surface (Bos et al. 1999), which is in favor of the formation of granules.

This is the first study in which the physical properties of the anammox granules were investigated. The comparisons of the physical characteristics of anammox granules in this study and other types of anaerobic granules (anammox, H_2-producing, and methanogenic granules) reported in the literature would be helpful for better understanding of the physical properties of the anammox granules. The diameter of the anammox granules in this work (2.0–3.5 mm) was much larger than most of other anammox granules reported in literature (Dapena-Mora et al. 2004a, b) and also larger than normal methanogenic granules with mean size of 1.5 mm (Schmidt and Ahring 1996). However, it was comparable to the H_2-producing granules reported by Mu and Yu (2006) (1.0–3.5 mm). The settling velocities of our anammox granules (41–79 m h^{-1}) were higher than methanogenic granules (52.9 m h^{-1}), but similar to H_2-producing granules (32–75 m h^{-1}) (Mu and Yu 2006). The contact angle of the anammox granules (49 ± 3°) was a little lower than that of the H_2-producing granules (54 ± 2°) (Mu and Yu 2006). The contents of carbohydrates and proteins in the EPS of the anammox granules were 83.2 ± 7.9 and 42.7 ± 6.5 mg g^{-1} VSS, respectively, which were both lower than those of Mu and Yu (2006) (115.6 ± 5.2 and 70.9 ± 4.5, respectively).

12.1.3.4 Phylogenetic Analysis

After the anammox activity was observed in the anammox granules, we applied a Planctomycetes-specific full cycle rRNA approach to identify the 16S rRNA gene sequence of the dominant bacteria. Primers Pla46 and 630R were used to amplify near-complete 16S rRNA genes from total genomic DNA extracted from the granules at day 250. Ten clones were randomly selected for sequencing. All of the analyzed 16S rRNA genes had the same sequence and subsequent phylogenetic analysis (Fig. 12.5) showed that the phylotype present in the reactor had 99 % sequence similarity to an anammox clone obtained from a Japanese wastewater plant (AB269934) and was 95 % similar to *Candidatus Brocadia anammoxidans*. (Tsushima et al. 2007).

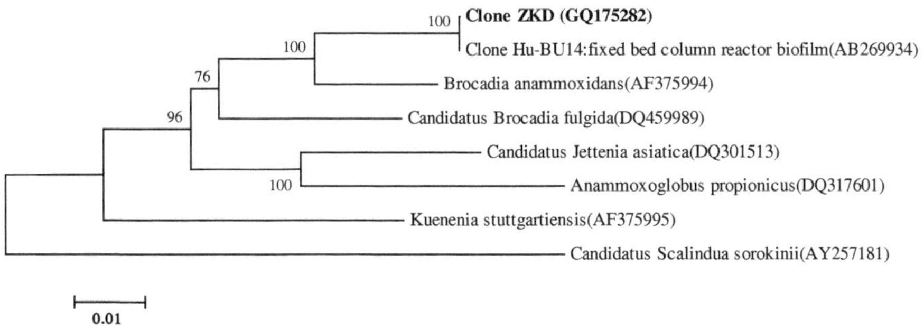

Fig. 12.5 Phylogenetic tree of the clones obtained from the enrichment culture, based on neighbor joining analysis for almost full length of 16S rRNA gene sequences. PCR amplification of 16S rRNA genes was conducted with the primer set of Pla46 and 630R. Bootstrap values (> 50%) are indicated at branch points. The scale bar represents 10 % estimated sequence divergence (Reproduced from Ni et al. (2010) with permission from American Society for Microbiology)

Table 12.1 List of primers used in this study

Target	Primer name	Sequence (5′-3′)	Target site	Reference
Planctomycetes 16S rRNA	Pla46	GGATTAGGCATGCAAGTC	43–63	Neef et. al. (1998)
Planctomycetes 16S rRNA	630R	CAKAAAGGAGGTGATCC	1529–1545	Stephen et. al. (1998)
Anammox 16S rRNA	Amx694F	GGGGAGAGTGGAACTTCGG	694–713	This study
Anammox 16S rRNA	Amx960R	GCTCGCACAAGCGGTGGAGC	960–979	This study

12.1.3.5 Quantification of Anammox Bacteria By Real-Time PCR

Based on the 16S rRNA gene sequences obtained from the anammox granules, a real-time PCR primer set (AMX658F-AMX924R) was designed to quantify the anammox bacteria in the sludge (see Table 12.1). The primers used for quantitative PCR were designed based on the near full-length 16S rRNA gene sequences of anammox bacteria from the NCBI database and nine different reactor cultures. The specificity of the primers was checked using PROBE_MATCH tool of the ARB program. The primers were tested on environmental samples and enrichment cultures available in laboratory. The standard curves for anammox 16S rRNA gene copies were constructed from a series of 10-fold dilutions (from 3.6×10^1 copies to 3.6×10^7 copies) of plasmid DNA carrying the 16S rRNA gene of *Kuenenia stuttgartiensis* for the primer sets of AMX658F-AMX924R. The range test was 2.6×10^{-8} to 2.6×10^{-2} ng of DNA per well for each primer set. Amplification efficiency was calculated from the equation, $\varepsilon c = 10^{(1/s)}-1$. The real-time PCR

Table 12.2 16S rRNA gene copies of the granular sludge

Source	16S rRNA gene copies of anammox granules	
	copies mL^{-1}	Copies g^{-1} VSS
Seeding sludge	1.9×10^5	2.4×10^7
Day 140	2.3×10^7	3.8×10^9
Day 250	3.5×10^8	5.4×10^{10}
Day 300	4.6×10^8	7.1×10^{10}

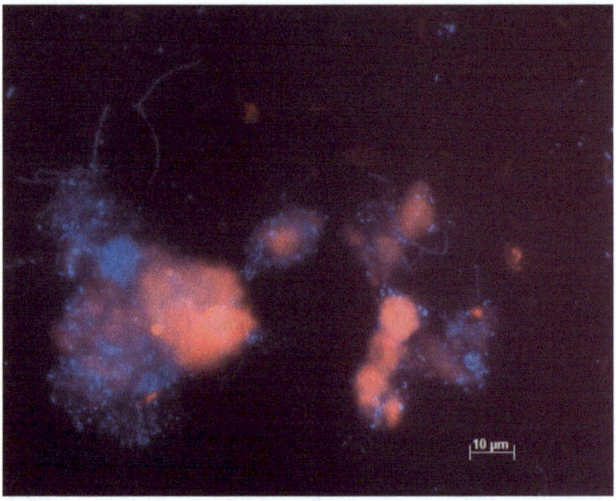

Fig. 12.6 FISH micrographs with Cy3-labeled amx368 (targeting all anammox bacteria), counterstained with DAPI, for the enriched anammox granules: total active bacteria (*blue*) and anammox (*purple*) (Reproduced from Ni et al. (2010) with permission from American Society for Microbiology)

assay with these primers was very consistent as shown by the strong inverse linear relationship between the threshold cycle numbers and the copy numbers of 16S rRNA gene ($R^2 = 0.990$). The amplification efficiencies were between 95 and 105 % with the slopes of −3.30. The copy numbers of 16S rRNA gene of anammox bacteria in the reactor enrichment was quantified as 4.6×10^8 (see Table 12.2) copies mL^{-1}. This was corroborated by FISH analysis using the general anammox probe S-*-AMX-0368-a-A-18 (Fig. 12.6).

12.1.4 Conclusions

In this work, the anammox granules were successfully cultivated in a UASB reactor by seeding with aerobic granular sludge. The average total nitrogen removal efficiency exceeded 94 % after the formation of the granules. The heterogeneous composition of the seed material contributed to a considerable shortening of the start-up period to less than 160 days at a loading rate of 0.064 kg N per kg VSS per day. The mature granules with porous inner structure, which were dominated by cocci microorganisms, had a diameter ranging from 1.0 to 4.5 mm with red color, and had good settling ability and a high settling velocity of

41.4–79.4 m h^{-1}. Cells and extracellular polymeric substances were well mixed and distributed in the granules. The anammox phylotype in the reactor was 99% similar to a clone retrieved from a Japanese waste water treatment plant (AB269934.1) and was 95 % similar to *Candidatus Brocadia anammoxidans*. PCR and FISH analysis shows that 4.6×10^8 cells mL^{-1} were present. The results reported in this work can be used for a better design, shorter start-up, and more stable operation of anammox reactors for the treatment of nitrogen-rich wastewaters.

12.2 Modeling the Granule-Based Anammox Process

12.2.1 Introduction

Biological nitrogen removal from wastewater with a high N content can be costly when using conventional nitrification–denitrification systems, particularly when the wastewater contains only small amounts of biologically degradable carbon compounds. The anammox process is a novel biological nitrogen removal process, in which the ammonium is oxidized to nitrogen gas using nitrite as the electron acceptor (Mulder et al. 1995; van de Graaf et al. 1995; Strous et al. 1997, 1999a; Jetten et al. 2001; Dapena-Mora et al. 2004b; Lopez et al. 2008). This microbiological reaction is very attractive as it presents the opportunities for an N-removal process that improves the overall energy and material balance (Imajo et al. 2004). Since the anammox process is autotrophic, there is no need for carbon source dose to support denitrification (van de Graaf et al. 1996). Furthermore, if the anammox process is combined with a partial nitrification step, only part of the ammonium needs to be nitrified to nitrite, while in the anammox process nitrite and the remaining ammonium will be converted into nitrogen gas. This will reduce oxygen demand in the nitrification reactor and lead to a second reduction in costs. The biomass yield of the anammox process is very low (van de Graaf et al. 1996; Strous et al. 1998), resulting in a small amount of excess sludge production. The low sludge production is another factor that contributes to the substantially lower operation costs compared to the conventional denitrification systems.

For an activated sludge wastewater treatment system, modeling is always a useful tool to understand and optimize substrate utilization in the system. Model simulation and prediction can provide a solid foundation for design and operation of biological treatment systems. Thus, a mathematical model to describe the anammox process is highly desirable. However, few studies have been conducted to model the start-up and dynamic behavior of the anammox process. The behavior of a partial nitrification- anammox system at varying temperatures and dissolved oxygen concentrations was simulated by Hao et al. (2002a, b), but no verification was performed with experimental data and the start-up dynamics were not taken into account in the study. A simulation on a similar reactor was carried out by

Koch et al. (2000), but no long-term dynamic operation of the process was modeled.

The main goal of this study was to develop a mathematic model to describe the anammox process in a UASB reactor. Anammox granular sludge was cultivated in a lab-scale UASB reactor with seeding aerobic granular sludge. The accelerated start-up, microbial community analysis, and characterization of the granules in the anammox reactor were reported elsewhere (Ni et al. 2009). The present paper focused on the modeling work. The established model was then calibrated and validated using the experimental results. Simulation results from both start-up and 1-year dynamic operation of the UASB reactor were compared with the measured values. The kinetic properties of the granular anammox biomass were also quantified in this work. Another feature of our work was the exploration of the substrate diffusion dynamics in the anammox granules by using microelectrode measurement and model evaluation.

12.2.2 Materials and Methods

12.2.2.1 Reactor, Influent, and Seeding Sludge

The reactor, influent, and seeding sludge are described in the previous section.

12.2.2.2 Operating Conditions

After inoculation, the reactor was sparged with nitrogen gas for 30 min at a flow rate of 0.1 L min^{-1}. In the start-up, the UASB reactor was initially fed with the synthetic wastewater with a relatively low nitrogen loading rate (NLR) at an HRT of 30 h for 1 month with the influent flowrate of 0.15 L h^{-1}. Later, the HRT was then shortened to 24 h for 1-month operation. Thereafter, the HRT was kept constant but the NLR was elevated through increasing the influent ammonium and nitrite concentrations. The NLR was increased after the complete consumption of the ammonium and nitrite in the reactor. The reactor temperature was maintained at 30 °C using a ribbon heater and a temperature controller throughout the experiments.

Experiments were carried out in two stages: collecting data for model calibration, followed by collecting data for model verification. The collected data included the influent and effluent NH_4^+–N, NO_2^-–N and NO_3^-–N concentrations of the UASB reactor. The ratio of NH_4^+–N to NO_2^-–N was kept constant throughout the experiments, including the model validation period.

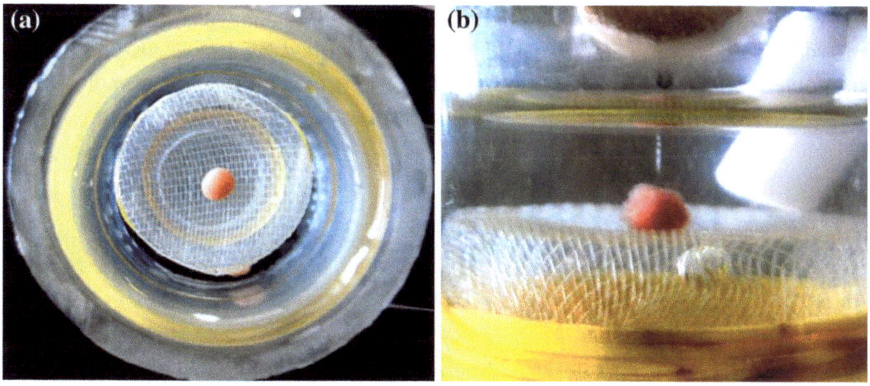

Fig. 12.7 **a** Images of test chamber used for microelectrode measurements; and **b** microelectrode measurement process (From Ni et al. (2009), reprinted with permission from Wiley-Blackwell)

12.2.2.3 Analytical Methods

The anammox granule size was measured as described previously. Approximately 20 mL of biomass granules were randomly removed from the lower-, middle-, and upper part of the sludge-bed zone of UASB. The granules were then washed with distilled water for three times. Their average diameter was measured with a representative sample, in which at least 100 particles were analyzed for granule radius with the image analyzer. The microstructure and predominant bacterial morphologies of the granules were observed using an SEM (Quantn 200, FEI Ltd.). Measurement of total suspended solids (TSS), VSS, NH_4^+–N, NO_2^-–N and NO_3^-–N was performed according to the Standard Methods (APHA 1995). In addition, the volume of granules in the reactor was calculated using the measured granule mean size and VSS concentration in the reactor.

For the measurement of NH_4^+–N, NO_2^-–N, and NO_3^-–N concentration distributions in granules, granule samples of 30 ml were taken from the reactor, and were placed uniformly on the thin nylon thread sieve. The thread sieve was immediately transferred and fixed to the small open glass box that was designed for microelectrode measurement. The movement of the NH_4^+–N, NO_2^-–N, and NO_3^-–N microelectrodes, with a tip diameter of 0.2 mm, was perpendicular to the granule surface when they were used to measure nitrogen-species gradients in the granules. The test chamber used for the microelectrode measurements is shown in Fig. 12.7. In the microelectrode measurement, a micromanipulator was used to adjust the position of the microelectrode tip at a spatial resolution of 5 μm and a microscope was used for precisely locating the granule surface. The movement of the microelectrode tip was perpendicular to the granule surface (Fig. 12.7a, b). No buckling of the needle and no significant changes of the granule were observed in the measurement. The open glass box was supplied with the medium containing 50 mg $L^{-1}$of NH_4^+–N, 30 mg $L^{-1}$of NO_2^-–N, and 35 mg $L^{-1}$of NO_3^-–N. In addition, the assay was conducted in a continuous mode, and the concentrations of ammonia, nitrite and nitrate in the bulk solution did not change along the time. The microelectrodes were prepared and used as described by Liu et al. (2007). The microelectrode profiles were determined over ten times along the granule axis in the tests and the average results are reported here.

12.2.3 Model Development

12.2.3.1 General Assumptions

The granule-based model described here has the following basic assumptions: (1) the granules are spherical in shape and uniform in size; (2) the number of granules and the size of the granules are constant in time; (3) only radial diffusion transport is considered and is described by Fick's law; (4) the diffusion coefficient is constant; and (5) there are no active biomass gradients in the granules at time zero. Also, a model is first developed to describe the hydrodynamics of the UASB reactor. The sludge bed and blanket in the UASB reactor are represented by a non-ideal CSTR with a dead volume and bypass flow. This CSTR is in series with a dispersion plug flow reactor that represents the clarification zone above the sludge blanket.

12.2.3.2 Biological Reactions

The ASM1 (Henze et al. 1987) with a consideration of a two-step denitrification was modified to describe the anammox process. In this model the death–regeneration concept was preferred over the endogenous respiration concept as used in ASM3 (Gujer et al. 1999) to describe the microbial decay. This was because the heterotrophs were found to be always active in the UASB reactor for 1-year operation. Such an observation could be explained only by the death–regeneration concept, as no biodegradable substrate was present the influent. On the other hand, aerobic granules were used as the seeding sludge for the UASB reactor. In order to simulate the anammox reactor start-up, the heterotrophic bacteria had to be included in our model.

The developed model described the relationships among the four solid species: active heterotrophic bacteria (X_H), active anammox bacteria (X_{AN}), released slowly biodegradable substrate (X_S), and residual inert biomass (X_I); five soluble species: ammonia–N (S_{NH_4}), nitrite–N (S_{NO_2}), nitrate–N (S_{NO_3}), dissolved nitrogen (S_{N_2}), and hydrolyzed readily biodegradable substrate (S_S). The units for S_S, X_S, X_H, X_{AN}, and X_I are oxygen demand, which was directly proportional to electron equivalents (8 g O_2 per e^- equivalent). The units for S_{NH_4}, S_{NO_2}, S_{NO_3}, and S_{N_2} are nitrogen.

In this model, both nitrite and nitrate could be used as an electron acceptor by X_H for growth and energy generation, while S_S was used as an electron donor. No biodegradable substrate was present in the influent, but a value of 2 mg L^{-1} of chemical oxygen demand (COD) in the influent was assumed to account for any biodegradable substrate present in the water used for medium preparation, and the biodegradable substrate originating from EDTA in the trace element solution (Dapena-Mora et al. 2004a). The readily degradable substrates in the reactor were also produced through the hydrolysis of X_S. This slowly degradable substrate was produced in the microbial decay, along with inert biomass (X_I).

Monod kinetics was used to describe the dependency of the growth rate of the anammox bacteria (r_{AN}) on NH_4^+–N and NO_2^-–N concentrations as follows:

$$r_{AN} = \mu_{AN} \frac{S_{W4}}{K_{NH_4}^{AN} + S_{NH_4}} \frac{S_{NO_2}}{K_{NO_2} + S_{NO_2}} X_{AN} \tag{12.1}$$

It should be noted that, as a substrate, nitrite could also inhibit the anammox process (Strous et al. 1999a, b; Dapena-Mora et al. 2004). However, the measured nitrite concentrations in the reactor were too low to cause an inhibitory effect. For the decay rate of anammox bacteria, the expression $b_{AN}X_{AN}$ was used.

The stoichiometry and kinetic equations for the heterotrophic and anammox bacteria in the developed model are listed in Table 12.3, while all the parameters used in the model, their symbols, and their units are defined in Tables 12.4 and 12.5.

12.2.3.3 Substrate Diffusion

For the soluble components involved in the biological reactions, the first step is their diffusion into the granules where the reactions take place. Discretization in time of the partial-differential equation describing the reaction-diffusion kinetics in a spherical particle (i.e., granule) was described as the following equation:

$$D_i \cdot \left(\frac{d^2 S_i(r)}{dr^2} + \frac{2}{r} \cdot \frac{dS_i(r)}{dr} \right) + r_i = 0 \tag{12.2}$$

with two boundary conditions:

$$\frac{dS_i}{dr} = 0 \text{ at } r = 0$$
$$S_i = S_{i,sur} \; r = R$$

where S_i is the substrate concentration of component i in the granule, $S_{i,sur}$ is the substrate concentration of component i in the granule surface, r_i is the volumetric substrate conversion rate in the granule, and D_i is the diffusion coefficient of substrate i, r is the distance from the granule center.

Table 12.3 Stoichiometric and kinetic matrix for the anammox model

Component	S_S	S_{NO_2}	S_{NO_2}	S_{NO_3}	S_{N_2}	X_S	X_{AN}	X_H	X_I	Kinetics rate expressions
Process	COD	N	N	N	N	COD	COD	COD	COD	
Hydrolysis	1					-1				$k_H \frac{X_S/X_H}{K_X+X_S/X_H} X_H$
Anammox microorganisms										
Growth of X_{AN}		$-\frac{1}{Y_{AN}} - i_{NBM}$	$-\frac{1}{Y_{AN}} - \frac{1}{1.14}$	$\frac{1}{1.14}$	$\frac{2}{Y_{AN}}$		1			$\mu_{AN} M_{NH_4}^{AN} M_{NO_2} X_{AN}$
Decay of X_{AN}		$i_{NBM} - i_{NXI} f_I$				$1 - f_I$	-1		f_I	$b_{AN} X_{AN}$
Heterotrophic microorganisms										
Growth on S_{NO_2}	$-\frac{1}{Y_H}$	$-i_{NBM}$	$-\frac{1-Y_H}{1.71Y_H}$		$\frac{1-Y_H}{1.71Y_H}$			1		$\mu_H \eta_{NOx} M_S M_{NO_2} M_{NH_4}^{H} X_H$
Growth on S_{NO_3}	$-\frac{1}{Y_H}$	$-i_{NBM}$		$-\frac{1-Y_H}{2.86Y_H}$	$\frac{1-Y_H}{2.86Y_H}$			1		$\mu_H \eta_{NOx} M_S M_{NO_2} M_{NH_4}^{H} X_H$
Decay of X_H		$i_{NBM} - i_{NXI} f_I$				$1 - f_I$		-1	f_I	$b_H X_H$

Notes: $M_{NH_4}^{AN} = \frac{S_{NH_4}}{K_{NH_4}^{AN} + S_{NH_4}}$, $M_{NO_2} = \frac{S_{NO_2}}{K_{NO_2} + S_{NO_2}}$, $M_S = \frac{S_S}{K_S + S_S}$, $M_{NH_4}^{H} = \frac{S_{NH_4}}{K_{NH_4}^{H}} + S_{NH_4}$, $M_{NO_3} = \frac{S_{NO_3}}{K_{NO_3} + S_{NO_3}}$

Table 12.4 Kinetic and stoichiometric parameters used in the anammox model

Parameter	Definition	Values	Unit
Y_H	Anoxic yield coefficient for X_H growth	0.67	g COD g^{-1} COD
Y_{AN}	Yield coefficient for X_{AN} growth	0.164	g COD g^{-1} N
η_{NOx}	Anoxic reduction factor	0.60	–
f_I	Fraction of X_I in respiration	0.08	g COD g^{-1} COD
i_{NBM}	Nitrogen content of biomass	0.07	g N g^{-1} COD
i_{NXI}	Nitrogen content of X_I	0.02	g N g^{-1} COD
μ_H	Maximum growth rate of X_H on S_S	0.5	h^{-1}
K_S	Biomass affinity constant for S_S	20.0	g COD m^{-3}
b_H	Decay rate coefficient of X_H	0.05	h^{-1}
μ_{AN}	Maximum growth rate of X_{AN}	0.003	h^{-1}
b_{AN}	Decay rate coefficient of X_{AN}	0.00016	h^{-1}
$K_{NH_4}^{H}$	Biomass NH_4 affinity constant for X_H	0.01	g N m^{-3}
$K_{NH_4}^{AN}$	Biomass NH_4 affinity constant for X_{AN}	0.73	g N m^{-3}
$K_{NO_2}^{AN}$	Biomass NO_2 affinity constant for X_{AN}	0.55	g N m^{-3}
K_{NO_2}	Biomass NO_2 affinity constant for X_H	0.50	g N m^{-3}
K_{NO_3}	Biomass NO_3 affinity constant for X_H	0.50	g N m^{-3}
k_H	Maximum hydrolysis rate	0.188	h^{-1}
K_X	Hydrolysis affinity constant	1.0	g COD g^{-1} COD

Table 12.5 Parameters for the anammox granules

Parameter	Description	Value	Reference
Granules			
R_{min}	Minimum radius, mm	0.50	This study
R_{max}	Maximum radius, mm	2.25	This study
R_{mean}	Mean radius, mm	1.25	This study
Mass transport			
D^{Ss}	Effective diffusivity of *S*s, dm^2 h^{-1}	0.000576	Beun et al. 2001
D^{NH_4}	Effective diffusivity of NH_4^+, dm^2 h^{-1}	0.000625	Picioreanu et al. 1997
D^{NO_2}	Effective diffusivity of NO_2^-, dm^2 h^{-1}	0.000583	Picioreanu et al. 1997
D^{NO_3}	Effective diffusivity of NO_3^-, dm^2 h^{-1}	0.000583	Picioreanu et al. 1997
D^{N_2}	Effective diffusivity of N_2, dm^2 h^{-1}	0.000917	Picioreanu et al. 1997
Density of solid phase			
ρ_{XH}	Density of X_H, g COD m^{-3}	350000	de Kreuk et al. 2007
ρ_{XI}	Density of X_I, g COD m^{-3}	350000	de Kreuk et al. 2007
ρ_{XAN}	Density of X_{AN}, g COD m^{-3}	350000	de Kreuk et al. 2007

The number of anammox granules and their diameter has an effect on the simulation results, because they influence the overall liquid/granule interfacial area (de Kreuk et al. 2007). The granules grown in this UASB reactor had a diameter range of 1.0–4.5 mm with an average of 2.5 mm. However, the use of a granule size distribution was avoided in this model, because it would significantly increase the complexity of the numerical computations and had no substantial contribution to a better understanding of the system (de Kreuk et al. 2007). Therefore, in the simulations the diameter of the matured anammox granules was chosen to be 2.5 mm, which was the most representative for the granules in this UASB reactor.

The number of anammox granules (*PNo*) in the reactor was calculated according to Eq. 12.3 (Gonzalez-Gil et al. 2001):

$$\mathrm{PN_0} = \frac{S_{\mathrm{TX},R} \times V_{\mathrm{TR}}}{S_{\mathrm{TX},G} \times V_{\mathrm{TG}}} \tag{12.3}$$

where $S_{\mathrm{TX,G}}$ and $S_{\mathrm{TX,R}}$ are the total biomass concentration in the granules and reactor, respectively; V_{TG} is the total volume of the granules; and V_{TR} refers to the total effective reactor volume.

12.2.3.4 Model Simulation

The simulation was performed with the AQUASIM software package (Reichert 1998), as described above.

12.2.4 Results and Discussion

12.2.4.1 Experimental Observations

The UASB reactor had been operated for more than 300 days continuously. The applied NLR increased from 0.042 to 0.5 g L^{-1} per day through an increase in the NH_4^+–N and NO_2^-–N concentrations. The NH_4^+–N removal efficiency gradually increased and reached up to 90 % on day 200th. After that, the NH_4^+–N removal efficiency was kept stably over 90 %. NO_2^-–N was removed together with NH_4^+–N. The average total nitrogen removal efficiency exceeded 94 % and the nitrite was almost completely depleted (~99 %). The typical matured anammox granules are shown in Figs. 12.8a. The anammox granules had a diameter ranging from 1.0 to 4.5 mm. Their color was red as shown in the figure. The SEM images show that a type of cocci with irregular morphology was the dominating microorganisms in the surface of the anammox granules (Fig. 12.8b).

12.2.4.2 Model Calibration

The models developed in this study for simulation tasks incorporated a number of stoichiometry and kinetics parameters relating to both X_{H} and X_{AN}. Some parameters responsible for X_{H} and X_{AN}, such as rates of hydrolysis, saturation and decay coefficients, and stoichiometric parameters listed in Table 12.4, were set to their default values as proposed in the literature, in order to reduce the number of parameters to be calibrated. Model calibration procedure is a process of adjusting

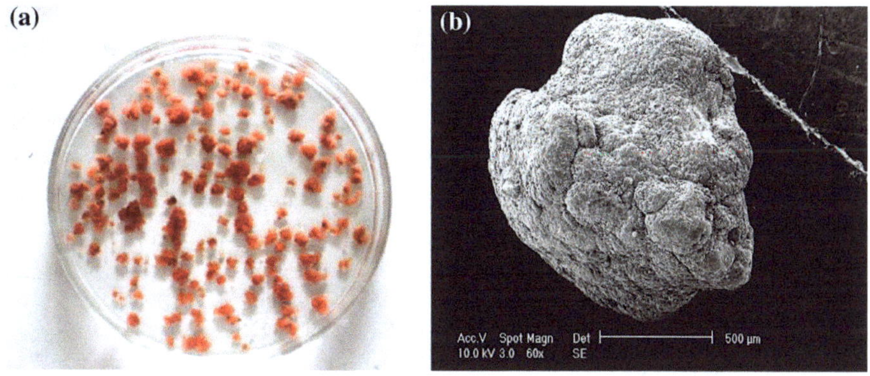

Fig. 12.8 **a** Image of the anammox granules; and **b** SEM image of the ANAMMOX granules (From Ni et al. (2009), reprinted with permission from Wiley-Blackwell)

selected parameters values of the model, so that the results produced by the model with these coefficients closely agree with the measured data. The calibration approach in this work was to fit the model simulation results on the measured components such as effluent nitrite-N, nitrate-N, and ammonia-N concentrations, which were measured within 300 days operation of the UASB reactor, by changing the calibrated parameters. Four parameters (e.g., Y_{AN}, b_{AN}, $K_{NH_4}^{AN}$ and $K_{NO_2}^{AN}$) were changed based on the causality of the parameters on the experimental data.

The profiles of the model predictions and experimental measurements for the effluent nitrite-N, nitrate-N, and ammonia-N concentrations with the influent data in 300-day reactor operation are shown in Fig. 12.9a, b and c, respectively, and the calibrated values of the associated parameters are listed in Table 12.4. As shown in Fig. 12.9, the models and the calibrated parameters were able to adequately simulate the variations of the effluent nitrite-N and ammonia-N concentrations as well as the nitrate-N concentrations for the continuous operation of the anammox reactor.

As shown in Fig. 12.9, the reactor start-up was achieved within 8 months, as evidenced by the formation of the matured anammox granules with a high nitrogen removal. The predicted effluent N levels matched the measured ones during the whole 8-month start-up period. These simulations include the sharp drop off of the effluent NH_4^+–N concentrations, the peaking of the effluent NO_2^-–N, and the leveling off of the effluent NO_3^-–N for modeling of the start-up and dynamic behavior of the anammox system. Moreover, the model was capable of simulating the short-term effects resulting from the variation of the influent N loading, in which high simulated peaks of NH_4^+–N concentrations were observed at high influent NH_4^+–N levels (Fig. 12.9a). A good prediction was also observed with respect to the effluent NO_2^-–N concentration and its disturbance (Fig. 12.9b). Although the difference between the simulated and measured NO_3^-–N concentrations is slightly high, the model was able to predict the NO_3^-–N changing trends reasonably (Fig. 12.9c).

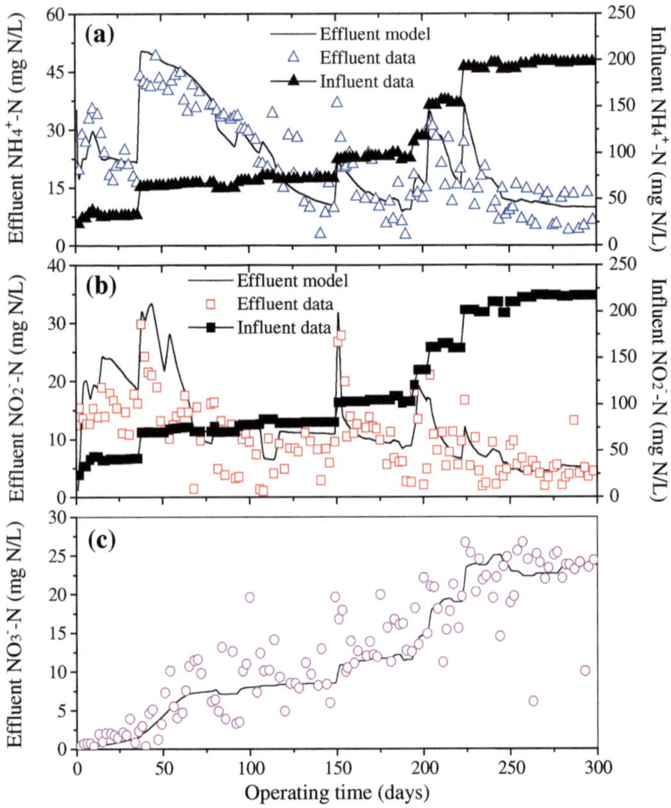

Fig. 12.9 Model calibration using the measured reactor effluent $NH_4{}^+$–N, $NO_2{}^-$–N and NO_3^-–N concentrations data in the initial 300 days of operation with their influent data (dot for measurement and line for simulation) (From Ni et al. (2009), reprinted with permission from Wiley-Blackwell)

The stoichiometric and kinetic parameter values for X_{AN} calibrated in this study were generally comparable to those reported in the literature (van de Graaf et al. 1996; Strous et al. 1999a, b; Hao et al. 2002; Dapena-Mora et al. 2004a). The yield coefficient for X_{AN} growth and the decay rate coefficient of X_{AN} were estimated as 0.164 g COD g^{-1} N and 0.00016 h^{-1}, respectively. However, the $NH_4{}^+$–N affinity constant for X_{AN} calibrated in this study was considerably higher than that reported by Hao et al. (2002), i.e., 0.07 g N m^{-3}. The affinity constants and the specific growing rates obtained in our work were different from the ones obtained in the literature (van de Graaf et al. 1996; Strous et al. 1999a, b; Hao et al. 2002; Dapena-Mora et al. 2004). The model parameters are dependent greatly on the microbial growth conditions. The affinity constant value represents the affinity for the substrate and its significant change indicates substrate-transport limitations. The higher values of the affinity constants and specific growing rates suggest the diffusion limitations in the anammox granules. In fact, the distribution of affinity constants and specific growing rates should vary substantially from the surface layer of the granules to the substratum layer of the granules, attributed to the reduction of biomass activity or the diverse distribution of microbial species. These complex distributions could be lumped into the calibrated parameters values by fitting the experimental data of the anammox granules, which reflect the overall kinetics of the anammox process. The higher values of affinity constants indicated larger mass transport limitations with penetration into the granules. Different model structures might be also responsible for such a difference. Additionally, different wastewater characteristics, reactor operating conditions and microbial communities could also be responsible for the difference.

12.2.4.3 Model Evaluation

The model verification is based on the comparison between the experimental results and model predictions. The experimental data, which were not previously used for model calibration, were employed for model verification. The model verification strategy in this work was based on a check of its capability to predict the effluent components concentrations (S_{NH_4}, $S_{NO_2}$and S_{NO_3}) of the anammox reactor using the dynamic influent data. The data of the continuous reactor operation were used for model verification. The model was solved for the set of parameters shown in Tables 12.4 and 12.5 in order to validate the developed model. The simulation results of the model provide an accurate prediction of S_{NH_4} and S_{NO_2} consumption in the UASB reactor, as well as the S_{NO_3} accumulation, as shown in Figs. 12.10a, b and c. The good agreement between the measured and predicted results suggests the validity of the model developed in this work. Thus, the calibrated model could be used to calculate all other results.

12.2.4.4 Substrate Diffusion within Anammox Granules

The measured NH_4^+–N, NO_2^-–N and NO_3^-–N distributions in an anammox granule are illustrated in Fig. 12.11a. Point 0 on the x-axis indicates the granular surface. In the upper layer of the granule with a thickness of 400 μm, a high NO_3^-–N concentration was observed coupled with a rapid NH_4^+–N and NO_2^-–N reduction, indicating that most of the active anammox bacteria were aggregated on this layer. In the inner part of the granule, the NH_4^+–N and NO_2^-–N levels decreased gradually. NH_4^+–N and NO_2^-–N were converted into N_2 and NO_3^-–N by the anammox bacteria. Since the decreasing rate of NH_4^+–N and NO_2^-–N was low, the abundance of the anammox bacteria in the inner part should be low.

In addition to the internal transport (within the granule, solid–solid), the external transport (in the limit film, liquid–solid) was also considered during the measurement. At the outside of the granular surface, the nitrate concentration dropped rapidly because of its diffusion from the granular surface to the bulk

Fig. 12.10 Model verification using the measured reactor effluent NH_4^+–N, NO_2^-–N and NO_3^-–N concentrations between 320th day and 380th day (From Ni et al. (2009), reprinted with permission from Wiley-Blackwell)

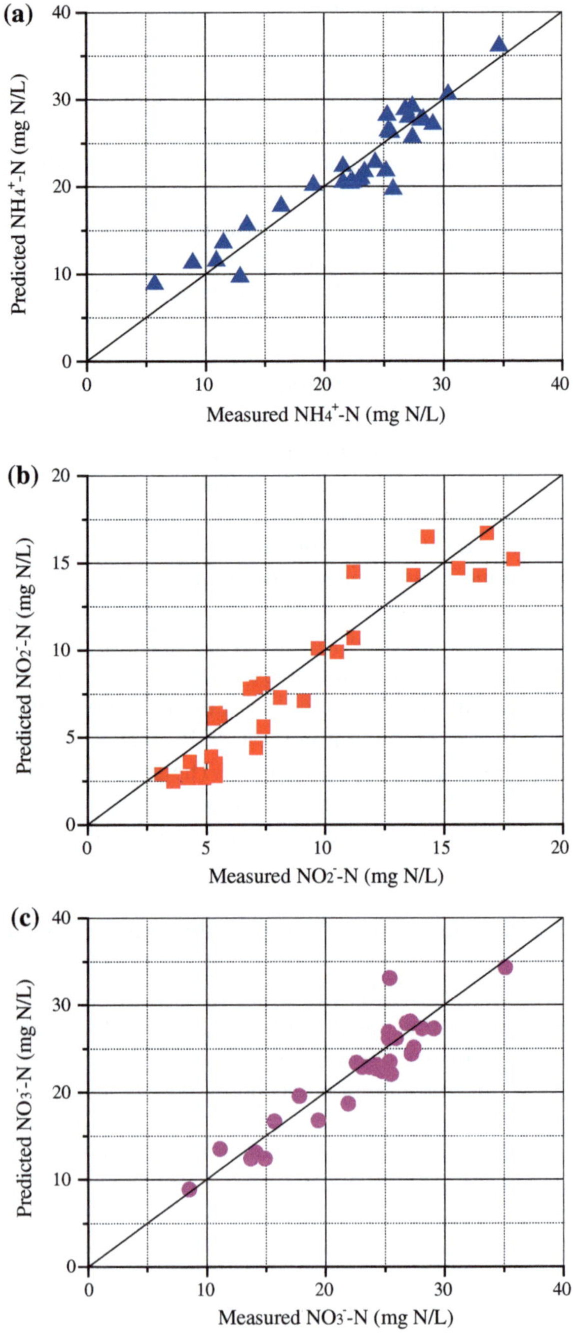

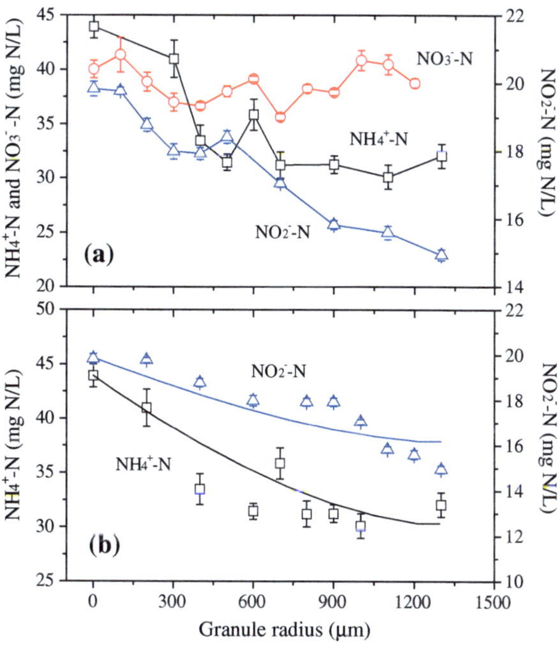

Fig. 12.11 **a** NH_4^+–N, NO_2^-–N and NO_3^-–N microprofiles in an anammox granule; and **b** comparison of the measured and simulated concentration profiles of NH_4^+–N and NO_2^-–N in an ANAMMOX granule. Point 0 on the x-axis indicates the granular surface (From Ni et al. (2009), reprinted with permission from Wiley-Blackwell)

solution. Similar distributions of the AOB and NOB were observed in an autotrophic nitrifying biofilm (Okabe et al. 2004), but the nitrite and nitrate microprofiles were different. The nitrite concentration decreased monotonically from the bulk solution into the granule, and nitrate only changed slightly into the granule. Such a difference might be attributed to the low nitrate concentration and rapid nitrite consumption in the granule.

The model simulation results for the NH_4^+–N and NO_2^-–N microprofile measurements in an anammox granule are shown in Fig. 12.11b. The simulation results match the measured results well. The NH_4^+–N and NO_2^-–N concentrations tended to decrease from the surface because of the diffusion limitation. The anammox process was likely to be mainly restricted to the upper and middle layers of the granules.

12.2.4.5 Effect of Granule Diameter on N Removal

Model simulations for the anammox granules with diameters in the range from 0.25 to 2.75 mm were conducted to evaluate the effect of the granule diameter on N removal. The simulation results in Fig. 12.12 show that the effluent N concentrations decreased and N-removal efficiency (both ammonia and nitrite) increased with the increasing granule diameter. At a granule diameter beyond the optimal value, the N-removal efficiency would not increase further. Therefore, there should be an optimal granule diameter for the maximal N-removal efficiency.

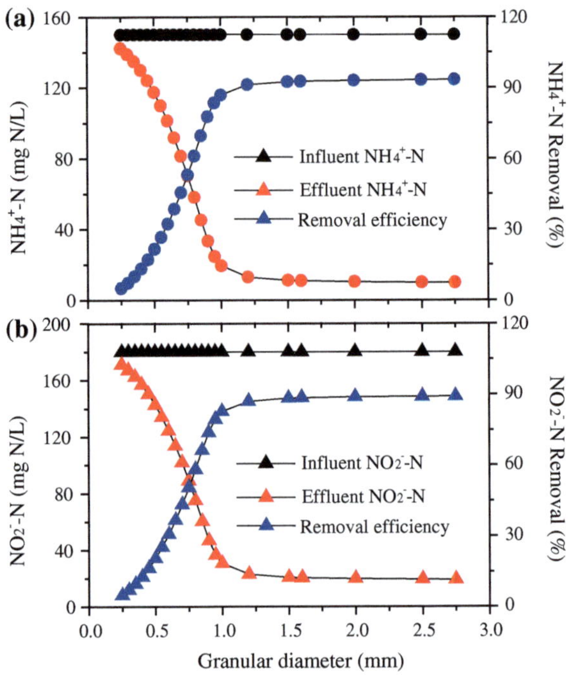

Fig. 12.12 Model simulation for the N-removal by anammox granules at different granule diameters: **a** ammonia removal; and **b** nitrite removal (From Ni et al. (2009), reprinted with permission from Wiley-Blackwell)

For a larger granule diameter, a higher diffusion-resistant level had to be overcome. Because of substrate limitations, cells inside the granules might decay and leak out inert cellular products. Thus, granules with a larger diameter contained a higher inert fraction and thereby a decreased volumetric activity. For this reason, the point on the plateaus of the N-removal efficiency in Figs. 12.12a, b could be recognized as the critical point, to which an optimal granule diameter corresponds. The simulation reveals that the optimum diameter for N removal in our reactor under the simulated operating conditions should be between 1.0 and 1.3 mm.

It should be noted that the optimum diameter between 1.0 and 1.3 mm here was not always suitable to all anammox granular system. In our laboratory scale experiments, it was found that the average granule diameter fluctuated in time between 1.0 and 4.5 mm. This granule diameter is not controllable in current systems, but simulation data strongly suggests that it is correlated with N removal. Thus, the optimum diameter for N removal of the anammox granule was related to the operation conditions of the reactor. In addition, with the proceeding of granulation, the biomass was progressively stratified with the granules settled in the lower part of the reaction zone. When granules were formed increasingly in the reaction zone, a dense sludge bed and a thin sludge blanket were formed with a clear interface between them, which resulted in a decreasing of granule size with the increasing of the height of the reactor.

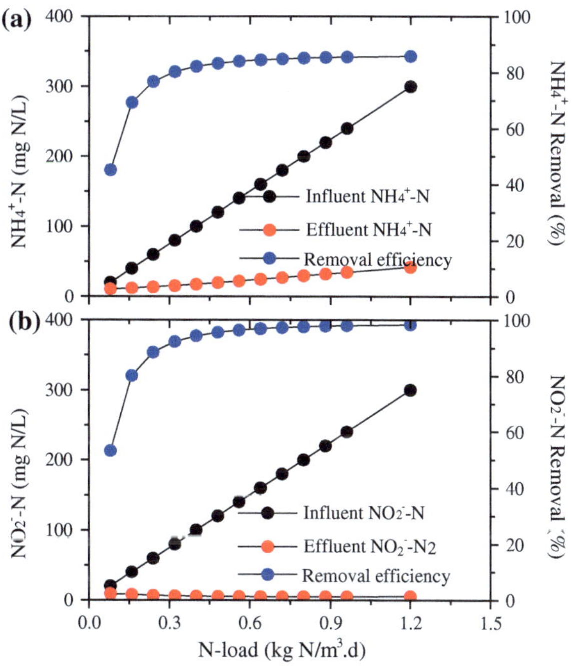

Fig. 12.13 Model simulation for N-removal at different N-loading rates (kg N m^{-3} d^{-1}): **a** ammonia removal; and **b** nitrite removal (From Ni et al. (2009), reprinted with permission from Wiley-Blackwell)

12.2.4.6 Effect of Nitrogen Loading Rate on N Removal

Wastewater treatment plants are confronted with night and day fluctuations, dry weather, and storm water influent, leading to a fluctuation in the sludge loading of the bioreactor (de Kreuk et al. 2007). The model was thus used to evaluate the effect of nitrogen loading on N-removal. In the simulation, sludge loading was varied by changing the influent ammonia and nitrite concentrations, leading to a variation in the sludge loading rate of 0.08–1.3 kg N m^{-3} d^{-1}. The simulation results in Fig. 12.13 indicate that the N-removal efficiency decreases with the increasing loading rate, although the effluent N concentrations increase slightly, with an optimum value of 0.8 kg N m^{-3} d^{-1}. At a higher sludge loading rate, the N-removal efficiency quickly reduced. Lowering the sludge loading rate would largely benefit the N-removal. At a sludge loading rate of 0.8 kg N m^{-3} d^{-1}, the N-removal reached to 91 % (85 % for the ammonium removal and 97% for the nitrite removal). However, a N-removal of 49 % (45 % for the ammonium removal and 53 % for the nitrite removal) was only achieved at a sludge loading rate of 1.2 kg N m^{-3} d^{-1}. At a low sludge loading rates, N-removal efficiency did not fluctuate substantially. At a sludge loading rate higher than 0.8 kg N m^{-3} d^{-1}, the amount of anaerobic ammonium oxidizers was too low to withstand the high N concentration. This would result in a rapid decrease in the N-removal efficiency and the accumulation of N in the reactor.

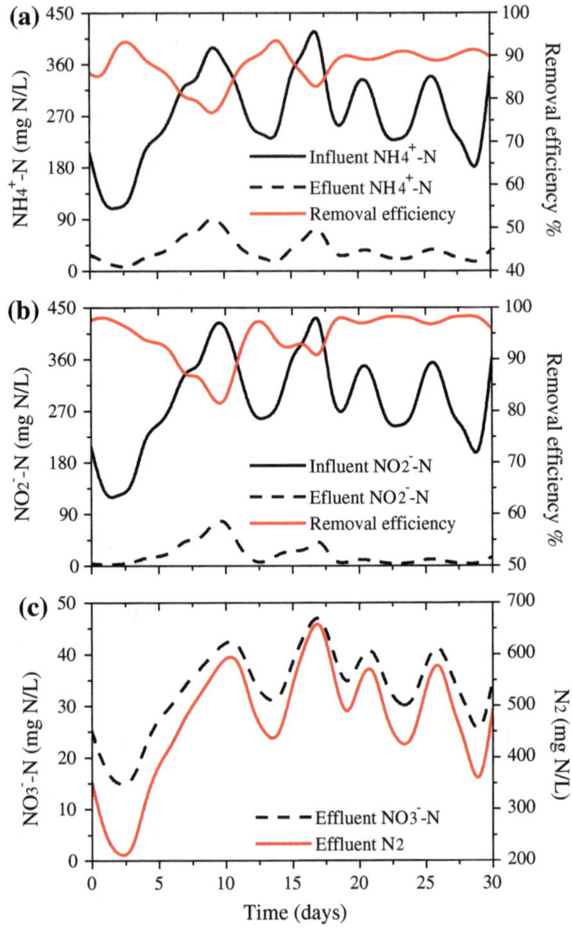

Fig. 12.14 Model prediction for the performance of the anammox reactor: **a** ammonia **b** nitrite; and **c** nitrate and nitrogen (From Ni et al. (2009), reprinted with permission from Wiley-Blackwell)

12.2.4.7 Dynamic Simulation of Long-Term Reactor Operation

The experimental results demonstrate that nitrogen loading rate of the UASB reactor was proven to be the most important parameters for effluent quality in the granular anammox reactor. Thus, it was essential to carry out the simulation of a short-term reactor performance with a variable nitrogen loading rate (NH_4^+–N and NO_2^-–N).

To explore the stability of the granule-based process, a great fluctuation of nitrogen loading rate was applied for model simulations. The simulation results in Fig. 12.14 show that good effluent quality could be obtained even though the nitrogen loading rate input fluctuated in a wide range. In the simulation, the NH_4^+–N removal efficiency was kept stably over 90 % (Fig. 12.14a), and the NO_2^-–N was almost completely. These simulation results demonstrate that the granule-based anammox reactor had a great N-loading resistant capacity. The nitrogen loading rate had a relatively higher effect on the N_2 concentration (Fig. 12.14c), which changed from the minimum of 350 mg N L^{-1} to the maximum of 470 mg N L^{-1} in the simulation.

12.2.5 Conclusions

A mathematical model to describe the anammox process in a UASB reactor was developed in this work. Anammox granules were cultivated in a UASB reactor inoculated with aerobic granular sludge. The matured anammox granules, which were dominated by cocci microorganisms, had a diameter ranging from 1.0 to 4.5 mm with red color. The ASM1 extended with a two-step denitrification consideration was incorporated with the anammox process to formulate this new model. The experimental results of 1-year operation of this UASB reactor were used to calibrate and validate the established model. The model was able to simulate the reactor performance well. The yield coefficient for X_{AN} growth and the decay rate coefficient of X_{AN} were estimated as 0.164 g COD g^{-1} N and 0.00016 h^{-1}, respectively. The measured substrate microprofiles of the anammox granules with the microelectrode matched the predicted results. Simulation results indicate that the optimum granule diameter for maximum N-removal in the case should be between 1.0 and 1.3 mm and that the optimum N loading rate should be 0.8 kg N m^{-3} d^{-1}. The granular anammox process in UASB reactor had a great N-loading resistant capacity.

References

APHA: Standard Methods for the Examination of Water and Wastewater, 19th edn. American Public Health Association, New York (1995)

Beun, J.J., Heijnen, J.J., van Loosdrecht, M.C.M.: N-removal in a granular sludge sequencing batch airlift reactor. Biotechnol. Bioeng. **75**, 82–92 (2001)

Bos, R., van der Mei, H.C., Busscher, H.J.: Physico chemistry of initial microbial adhesive interactions—its mechanisms and methods for study. FEMS Microbiol. Rev. **23**, 179–230 (1999)

Chamchoi, N., Nitisoravut, S.: Anammox enrichment from different conventional sludges. Chemosphere **66**, 2225–2232 (2007)

Dapena-Mora, A., Arrojo, B., Campos, J.L., Mosquera-Corral, A., Mendez, R.: Improvement of the settling properties of Anammox sludge in an SBR. J. Chem. Technol. Biotechnol. **79**, 1417–1420 (2004a)

Dapena-Mora, A., van Hulle, S.W.H., Campos, J.L., Mendez, R., Vanrolleghem, P.A., Jetten, M.: Enrichment of anammox biomass from municipal activated sludge: experimental and odeling results. J. Chem. Technol. Biotechnol. **79**, 1421–1428 (2004b)

De Kreuk, M.K., Picioreanu, C., Hosseini, M., Xavier, J.B., van Loosdrecht, M.C.M.: Kinetic model of a granular sludge SBR—Influences on nutrient removal. Biotechnol. Bioeng. **97**, 801–815 (2007)

Gonzalez-Gil, G., Seghezzo, L., Lettinga, G., Kleerebezem, R.: Kinetics and mass-transfer phenomena in anaerobic granular sludge. Biotechnol. Bioeng. **73**, 125–134 (2001)

Gujer, W., Henze, M., Mino, T., van Loosdrecht, M.C.M.: Activated sludge model no. 3. Water Sci. Technol. **39**, 183–193 (1999)

Gujer, W., Larsen, T.A.: The implementation of biokinetics and conservation principles in ASIM. Water Sci. Technol. **31**, 257–266 (1995)

Hao, X.D., Heijnen, J.J., van Loosdrecht, M.C.M.: Model-based evaluation of temperature and inflow variations on a partial nitrification–Anammox biofilm process. Water Res. **36**, 4839–4849 (2002a)

Hao, X.D., Heijnen, J.J., van Loosdrecht, M.C.M.: Sensitivity analysis of a biofilm model describing a one-stage completely autotrophic nitrogen removal (CANON) process. Biotechnol. Bioeng. **77**, 266–277 (2002b)

Henze, M., Grady, C.P.L. Jr., Gujer, W., Marais, G.V.R., Matsuo, T.: Activated sludge model no. 1. Scientific and Technical Report No. 1. IAWPRC, London (1987)

Imajo, U., Tokutomi, T., Furukawa, K.: Granulation of anammox microorganisms in up-flow reactors. Water Sci. Technol. **49**, 155–163 (2004)

Jetten, M.S.M., Horn, S.J., van Loosdrecht, M.C.M.: Towards a more sustainable municipal wastewater treatment system. Water Sci. Technol. **35**, 171–180 (1997)

Jetten, M.S.M., Schmid, M.C., van de Pas-Schoonen, K., Sinninghe Damste, J., Strous, M.: Anammox organisms: enrichment, cultivation and environmental analysis. Methods Enzymol. **397**, 34–57 (2005)

Jetten, M.S.M., Wagner, M., Fuerst, J., van Loosdrecht, M., Kuenen, G., Strous, M.: Microbiology and application of the anaerobic ammonium oxidation ('anammox') process. Curr. Opin. Biotechnol. **12**, 283–288 (2001)

Kartal, B., Rattray, J., van Niftrik, L.A., van de Vossenberg, J., Schmid, M.C., Webb, R.I., et al.: Candidatus 'Anammoxoglobus propionicus' a new propionate oxidizing species of anaerobic ammonium oxidizing bacteria. Syst. Appl. Microbiol. **30**, 39–49 (2007)

Kartal, B., van Niftrik, L., Rattray, J., van de Vossenberg, J., Schmid, M., Sinninghe Damsté, J.S., Jetten, M.S.M., Strous, M.: Candidatus 'Brocadia fulgida': an autofluorescent anaerobic ammonium oxidizer. FEMS Microbiol. Ecol. **63**, 46–55 (2008)

Koch, G., Egli, K., van der Meer, J.R., Siegriest, H.: Mathematical modeling of autotrophic denitrification in a nitrifying biofilm of a rotating biological contactor. Water Sci. Technol. **41**, 191–198 (2000)

Kowalchuk, G.A., de Bruijn, F.J., Head, I.M., Akkermans, A.D.L., van Elsas, J.D. (eds.): Molecular Microbial Ecology Manual, 2nd edn. Kluwer Academic Publishers, Dordrecht, The Netherlands (2004)

Lettinga, G., Pol, L.W.H.: UASB-Process design for various types of wastewaters. Water Sci. Technol. **24**, 87–107 (1991)

Liu, S.Y., Liu, G., Tian, Y.C., Chen, Y.P., Yu, H.Q., Fang, F.: An innovative microelectrode fabricated using photolithography for measuring dissolved oxygen distributions in aerobic granules. Environ. Sci. Technol. **41**, 5447–5452 (2007)

Lopez, H., Puig, S., Ganigue, R., Ruscalleda, M., Balaguer, M.D., Colprim, J.: Start-up and enrichment of a granular anammox SBR to treat high nitrogen load wastewaters. J. Chem. Technol. Biotechnol. **83**, 233–241 (2008)

Morgan, J.W., Forster, C.F., Evison, L.M.A.: Comparative-study of the nature of biopolymers extracted from anaerobic and activated sludges. Water Res. **24**, 743–750 (1990)

Mu, Y., Yu, H.Q.: Biological hydrogen production in a UASB reactor with granules I: Physicochemical characteristics of hydrogen-producing granules. Biotechnol. Bioeng. **94**, 980–987 (2006)

Mulder, A., van de Graaf, A.A., Robertson, L.A., Kuenen, J.G.: Anaerobic ammonium oxidation discovered in a denitrifying fluidized bed reactor. FEMS Microbiol. Ecol. **16**, 177–184 (1995)

Neef, A., Amann, R., Schlesner, H., Schleifer, K.H: Monitoring a widespread bacterial group in situ detection of planctomycetes with 16S rRNA-targeted probes. Microbiology. **144**, 3257-3266 (1998)

Ni, B.J., Chen, Y.P., Liu, S.Y., Fang, F., Xie, W.M., Yu, H.Q.: Modeling a granule-based anaerobic ammonium oxidizing (ANAMMOX) process. Biotechnol. Bioeng. **103**, 490–499 (2009)

Ni, B.J., Hu, B.L., Fang, F., Xie, W.M., Kartal, B., Liu, X.W., Sheng, G.P., Jetten, M.S.M., Zheng, P., Yu, H.Q.: Microbial and physicochemical characteristics of compact anaerobic ammonium oxidation (anammox) granules in an UASB reactor. Appl. Environ. Microbiol. **76**, 2652–2656 (2010)

Okabe, S., Kindaichi, T., Ito, T., Satoh, H.: Analysis of size distribution and areal cell density of ammonia-oxidizing bacterial microcolonies in relation to substrate microprofiles in biofilms. Biotechnol. Bioeng. **85**, 86–95 (2004)

Picioreanu, C., van Loosdrecht, M.C.M., Heijnen, J.J.: Modelling the effect of oxygen concentration on nitrite accumulation in a biofilm airlift suspension reactor. Water Sci. Technol. **36**, 147–156 (1997)

Pynaert, K., Smets, B.F., Beheydt, D., Verstraete, W.: Start-up of autotrophic nitrogen removal reactors via sequential biocatalyst addition. Environ. Sci. Technol. **38**, 1228–1235 (2004)

Reichert, P. 1998. Aquasim 2.0-User Manual, Computer Program for the Identification and Simulation of Aquatic Systems; EAWAG: Dübendorf, Switzerland (ISBN 3 906484 16 5)

Schmid, M., Walsh, K., Webb, R., Rijpstra, W.I.C., van de Pas-Schoonen, K., Verbruggen, M.J., Hill, T., Moffett, B., Fuerst, J., Schouten, S., Damste, J.S., Harris, J., Shaw, P., Jetten, M., Strous, M.: Candidatus ''Scalindua brodae'', sp nov., Candidatus ''Scalindua wagneri'', sp nov., two new species of anaerobic ammonium oxidizing bacteria. Syst. Appl. Microbiol. **26**, 529–538 (2003)

Schmidt, J.E., Ahring, B.K.: Granular sludge formation in upflow anaerobic sludge blanket (UASB) reactors. Biotechnol. Bioeng. **49**, 229–246 (1996)

Sheng, G.P., Yu, H.Q., Yu, Z.: Extraction of the EPS from a photosynthetic bacterium Rhodopseudomonas acidophila. Appl. Microbiol. Biotechnol. **67**, 125–130 (2005)

Sheng, G.P., Yu, H.Q.: Characterization of extracellular polymeric substances of aerobic and anaerobic sludge using 3-dimentional fluorescence spectroscopy. Water Res. **40**, 1233–1239 (2006)

Stephen, J.R., Kowalchuk, G.A., Bruns, M.A.V., McCaig, A.E., Philips, C.J., Embley, T.M., Prosser, J.I.: Analysis of beta-subgroup proteobacterial ammonia oxidizer populations in soil by denaturing gradient gel electrophoresis analysis and hierarchical phylogenetic probing. Appl. Environ. Microbiol. **64**, 2958–2965 (1998)

Strous, M., Fuerst, J.A., Kramer, E.H., Logemann, S., Muyzer, G., van de Pas-Schoonen, K.T., Webb, R., Kuenen, J.G., Jetten, M.S.M.: Missing lithotroph identified as new planctomycete. Nature **400**, 446–449 (1999a)

Strous, M., Heijnen, J.J., Kuenen, J.G., Jetten, M.S.M.: The sequencing batch reactor as a powerful tool for the study of slowly growing anaerobic ammonium-oxidizing microorganisms. Appl. Microbiol. Biotechnol. **50**, 589–596 (1998)

Strous, M., Kuenen, J.G., Jetten, M.S.M.: Key physiology of anaerobic ammonium oxidation. Appl. Environ. Microbiol. **65**, 3248–3250 (1999b)

Strous, M., van Gerven, E., Kuenen, J.G., Jetten, M.S.M.: Ammonium removal from concentrated waste streams with the Anaerobic Ammonium Oxidation (Anammox) process in different reactor configurations. Water Res. **31**, 1955–1962 (1997)

Strous, M., Kuenen, J.G., Fuerst, J.A., Wagner, M., Jetten, M.S.M.: The anammox case—A new experimental manifesto for microbiological eco-physiology. Antonie van leeuwenhoek Int J. Gen Mol. Microbiol. **81**, 693–702 (2002)

Strous, M., Pelletier, E., Mangenot, S., Rattei, T., Lehner, A., Taylor, M.W., Horn, M., Daims, H., Bartol-Mavel, D., Wincker, P., Barbe, V., Fonknechten, N., Vallenet, D., Segurens, B., Schenowitz-Truong, C., Medigue, C., Collingro, A., Snel, B., Dutilh, B.E., Op den Camp, H.J.M., van der Drift, C., Cirpus, I., van de Pas-Schoonen, K.T., Harhangi, H.R., van Niftrik, L., Schmid, M., Keltjens, J., van de Vossenberg, J., Kartal, B., Meier, H., Frishman, D., Huynen, M.A., Mewes, H.W., Weissenbach, J., Jetten, M.S.M., Wagner, M., Le Paslier, D.:

Deciphering the evolution and metabolism of an anammox bacterium from a community genome. Nature **440**, 790–794 (2006)
Su, K.Z., Yu, H.Q.: Formation and characterization of aerobic granules in a sequencing batch reactor treating soybean-processing wastewater. Environ. Sci. Technol. **39**, 2818–2827 (2005)
Thaveesri, J., Daffonchio, D., Liessens, B., Vandemeren, P., Verstraete, W.G.: Granulation and sludge bed stability in upflow anaerobic sludge bed reactors in relation to surface thermodynamics. Appl. Environ. Microbiol. **61**, 3681–3686 (1995)
Third, K.A., Paxman, J., Schmid, M., Strous, M., Jetten, M.S.M., Cord-Ruwisch, R.: Enrichment of anammox from activated sludge and its application in the CANON process. Microbial Ecol **49**, 236–244 (2005)
Tsushima, I., Kindaichi, T., Ohabe, S.: Quantification of anaerobic ammonium oxidising bacteria in enrichment cultures by real-time PCR. Water Res. **41**, 785–794 (2007)
van de Graaf, A.A., de Bruijn, P., Robertson, L.A., Jetten, M.S.M., Kuenen, J.G.: Autotrophic growth of anaerobic ammonium-oxidizing micro-organisms in a fluidized bed reactor. Microbiology **142**, 2187–2196 (1996)
van de Graaf, A.A., Mulder, A., de Bruijn, P., Jetten, M.S.M., Robertson, L.A., Kuenen, J.G.: Anaerobic ammonium-oxidising bacteria in a fluidizing bed reactor. Microbiology **42**, 2187–2196 (1995)
van der Star, W.R.L., Abma, W.R., Blommers, D., Mulder, J.W., Tokutomi, T., Strous, M., Picioreanu, C., van Loosdrecht, M.C.M.: Startup of reactors for anoxic ammonium oxidation experiences from the first full-scale anammox reactor in Rotterdam. Water Res. **41**, 4149–4163 (2007)
van der Star, W.R.L., Miclea, A.I., van Dongen, U.G.J.M., Muyzer, G., Picioreanu, C., van Loosdrecht, M.C.M.: The membrane bioreactor: a novel tool to grow anammox bacteria as free cells. Biotechnol. Bioeng. **101**, 286–294 (2008)
van Dongen, U.G.J.M., Jetten, M.S.M., van Loosdrecht, M.C.M.: The SHARON((R))-Anammox((R)) process for treatment of ammonium rich wastewater. Water Sci. Technol. **44**, 153–160 (2001)
van Niftrik, L., Geerts, W.J.C., van Donselaar, E.G., Humbel, B.M., Yakushevska, A., Verkleij, A.J., Jetten, M.S.M., Strous, M.: Combined structural and chemical analysis of the anammoxosome: A membrane-bounded intracytoplasmic compartment in anammox bacteria. J. Struc. Biol. **161**, 401–410 (2008)
Xavier, J.B., de Kreuk, M.K., Picioreanu, C., van Loosdrecht, M.C.M.: Multi-scale individual-based model of microbial and bioconversion dynamics in aerobic granular sludge. Environ. Sci. Technol. **41**, 6410–6417 (2007)
Zhang, Y., Ruan, X.H., Op den Camp, H.J.M., Smits, T.J.M., Jetten, M.S.M., Schmid, M.C.: Diversity and abundance of aerobic and anaerobic ammonium-oxidizing bacteria in freshwater sediments of the Xinyi River (China). Environ. Microbiol. **9**, 2375–2382 (2007)

Chapter 13
Summary

Aerobic granular sludge technology will play an important role as an innovative technology alternative to the present activated sludge process in industrial and municipal wastewater treatment in the near future. There are numerous internal interactions among process variables, such as growth, storage, microbial products formation and endogenous respiration, as well as the sludge characteristics, including biomass detachment, oxygen transfer, and diffusion. However, since aerobic granulation is a very complex phenomenon, their mechanisms and mathematical modeling have not been well documented yet. Aimed at the lacks in the studies of aerobic granular sludge, this dissertation is to investigate the formation, the characterization, and the mathematical modeling of aerobic granular sludge. The storage processes of aerobic granules under both aerobic and anoxic conditions were investigated. The production of EPS and SMP in aerobic granular sludge was explored. Furthermore, efforts were made to cultivate aerobic granules on a low-strength municipal wastewater in a pilot-scale SBR. The main conclusions are listed below:

Aerobic granules were successfully cultivated in SBRs fed with both soybean-processing and fatty-acids-rich wastewaters. In the granulation process, the mean diameter of bioparticles gradually increased. Their settling velocity increased to 40 m h^{-1}, while their SVI decreased to 20 mL g^{-1}. The COD removal efficiency increased to 98 %. Three phases in the granule formation process could be clearly distinguished: initial exponential growth phase, linear growth phase afterwards, and final stable phase. The developed model is applicable to describing the aerobic sludge granulation process and substrate diffusion within granules appropriately.

The heterotrophs consumed more oxygen than the autotrophs. More substrate was used for the primary anoxic growth, while a low amount of substrate was diverted to anoxic storage in the denitrifying processes. Biomass content increased with the increasing SRT, but the active biomass ratio decreased. The autotrophs had no significant effect on the total biomass content, despite of their important role in nitrogen removal. The fraction of the heterotrophs considerably increased

B.-J. Ni, *Formation, Characterization and Mathematical Modeling of the Aerobic Granular Sludge*, Springer Theses, DOI: 10.1007/978-3-642-31281-6_13,

as a consequence of increasing influent COD, but the fraction of the autotrophs decreased slowly. The autotrophs were mainly located on the outer layers of granules for oxygen consumption, whereas the heterotrophs occupied the granule center or on the outer layers.

A bioenergetic model for wastewater treatment was developed integrating the thermodynamic methodology with a modified ASM1. The obtained parameters were related to the ASM1 model, which coupled biological reactions with bioenergetics. The model was able to accurately describe the consumption of substrate, the accumulation of MLVSS, and the OUR profiles in the wastewater treatment process and the long-term operation results of the SBR. Model simulation results clearly show the variations of $C_{16}H_{24}O_5N_4$, CH_2O, cells ($C_5H_7O_2N$), H^+, NH_4^+, HCO_3^- concentration, and the release of CO_2.

Aerobic granules subjected to alternative feast and famine environments in an SBR were able to take up the carbon substrate in the soybean wastewater rapidly and to store it as PHA when the substrate was in excess. The model was able to predict the fate of all major model components, i.e., PHA, COD, and OUR. The comparison with ASM3 clearly underlined the merit of incorporating the simultaneous storage and growth process in the model structure. The established model demonstrated its capacity of elucidating the biological processes, such as microbial maintenance, substrate diffusion, simultaneous storage, and growth process.

Hydrolysis, simultaneous anoxic storage and growth, anoxic maintenance, and endogenous decay were found to be the main bioreaction processes governing the anoxic storage in the aerobic granules. Kinetic analysis of NUR indicated that the NUR of granules-based denitrification process included four linear phases of nitrate reduction. Furthermore, the methodology for determining the most important parameter in anoxic storage, i.e., anoxic reduction factor, was established based on an analysis of anoxic storage mechanisms. The performance of storage process in a granule-based denitrification system was accurately simulated.

The amount of produced EPS and their molecular weight distribution depend on the utilization of external substrate. The quantity of produced EPS increased in the substrate utilization process. Two components of EPS were identified by the EEM analysis: proteins at Ex/Em 280/340 nm, and the fulvic acid-like component at 320/400 nm. The model predictions were in qualitative agreement with the experimental observations. Results showed that EPS increased rapidly with the substrate consumption, but decreased slightly after the external substrate was completely consumed. Electrons from the external substrate were distributed in the following order: new biomass synthesis of 61 %, oxygen for respiration of 21 %, and EPS of 18 %.

A novel and convenient approach to evaluate the EPS production kinetics was developed. The weighted least-squares analysis was employed to calculate approximate differences in EPS concentration between model predictions and experimental results. Parameters estimation results indicated that the kinetic coefficients of EPS production by activated sludge and their practical identifiability information could be obtained accurately and conveniently with this approach.

The UAP, produced in the substrate utilization process, were found to be carbonaceous compounds with an MW lower than 290 kDa and were quantified separately from BAP. The BAP were mainly cellular macromolecules with an MW in a range of 290–5,000 kDa, and could be further classified into the GBAP with an MW of 1,000 kDa, which were produced in the microbial growth phase, and the EBAP with an MW of 4,500 kDa, which were generated in the endogenous phase. The GBAP and EBAP had different formation rates from the hydrolysis of EPS and distinct biodegradation kinetics. The major contributions of organic compound in the secondary effluent of activated sludge plants might come from EBAP.

An integrated substrate utilization equation was formulated and used to construct the objective function for determining UAP. On the basis of total SMP measurements, BAP formation was determined with a BAP formation equation. With this approach the fraction of substrate electrons diverted to UAP formation with a BAP formation rate could be calculated. The validity of this approach was confirmed by both independent SMP production tests and results reported in literature. Furthermore, this approach which required only few data sets or even one set could be applied for other biological processes with similar difficulty for quantitative product determination.

The formation of EPS, SMP, and internal storage products (X_{STO}) in aerobic granular sludge was investigated using experimental and modeling approaches. An expanded unified model describing the production and the consumption of EPS, SMP, and X_{STO} was formulated after integrating the electron flows from the external substrate to EPS, SMP, and X_{STO}. The agreement between model outputs and experimental EPS, SMP, and X_{STO} data from distinctly different experiments supported that the expanded unified model properly captured the relationships among the forms of microbial products.

Simulation results underlined the importance of the initial substrate and biomass concentrations for the overall formation and consumption of EPS, SMP, and X_{STO} in the aerobic granules. A higher substrate concentration resulted in a greater concentration of EPS, SMP, and X_{STO}. An accumulation of biomass in a bioreactor led to an increased production rate of EPS, SMP, and X_{STO}. However, there was no direct correlation between the biomass concentration with the total content of the EPS, SMP, and X_{STO}. The model could be used for process understanding and thus for optimization of the aerobic-granule-based reactors.

The growth, maintenance, death, and microbial product formation processes of the autotrophs, with the NOB as an example, in activated sludge were experimentally investigated with an NOB-enriched culture, and they were described well with a model established. The agreement between the model outputs and the experimental measurements of OUR, NO_2^--N, NO_3^--N, EPS, and SMP suggested that the established model was able to capture the connection among the two types of microbial products and nitrite consumption in the NOB-enriched culture.

The released SMP in the nitrite oxidation granules could be utilized as the carbon source for the heterotrophic growth. Microelectrode tests showed that the oxygen diffusion limitation resulted in anoxic microenvironments in the nitrite-oxidizing granules and allowed a sequential utilization of nitrate as electron acceptors for denitrification. Model calibration and validation results indicated that the model developed in this work was capable of simulating the nitrate loss in biological nitrite oxidation process by aerobic granular sludge.

For the first time, aerobic granules with an excellent settling ability were cultivated in a pilot-scale SBR for the treatment of low-strength municipal wastewater. Mathematical modeling of this pilot-scale granule-based reactor was successfully achieved. The granules had a diameter ranging from 0.2 to 0.8 mm and had good settling ability with a settling velocity of 18–40 m h^{-1}. Three bacteria with rod, coccus, and filament morphologies coexisted in the granules. The volume exchange ratio and settling time were found to be two key factors in the granulation of activated sludge grown on such a low-concentration wastewater in an SBR.

The anammox granules were successfully cultivated in a UASB reactor by seeding with aerobic granular sludge. The average total nitrogen removal efficiency exceeded 94 % after the formation of the granules. The heterogeneous composition of the seed material contributed to a considerable shortening of the startup period to less than 160 d at a loading rate of 0.064 kg N per kg VSS per day. The mature granules with porous inner structure, which were dominated by cocci microorganisms, had a diameter ranging from 1.0 to 4.5 mm with red color, and had good settling ability and a high settling velocity of 41.4–79.4 m h^{-1}. The anammox phylotype in the reactor was 99 % similar to a clone retrieved from a Japanese wastewater treatment plant (AB269934.1) and was 95 % similar to *Candidatus Brocadia anammoxidans*. PCR and FISH analysis shows that 4.6×10^8 cells mL^{-1} were present.

The ASM1 extended with a two-step denitrification consideration was incorporated with the anammox process to formulate a new model. The model was able to simulate the 1-year reactor performance well. The yield coefficient for X_{AN} growth and the decay rate coefficient of X_{AN} were estimated as 0.164 g COD g^{-1} N and 0.00016 h^{-1}, respectively. The measured substrate micro-profiles of the anammox granules with the microelectrode matched the predicted results. Simulation results indicated that the optimum granule diameter for maximum N-removal in the case should be between 1.0 and 1.3 mm and that the optimum N loading rate should be 0.8 kg N m^{-3} d^{-1}.

Index

B.-J. Ni, *Formation, Characterization and Mathematical Modeling of the Aerobic Granular Sludge*, Springer Theses, DOI: 10.1007/978-3-642-31281-6,

MIX
Papier aus verantwortungsvollen Quellen
Paper from responsible sources
FSC® C105338

If you have any concerns about our products,
you can contact us on
ProductSafety@springernature.com

In case Publisher is established outside the EU,
the EU authorized representative is:
Springer Nature Customer Service Center GmbH
Europaplatz 3, 69115 Heidelberg, Germany

Printed by Libri Plureos GmbH
in Hamburg, Germany